D0849112

IDENTIFICATION METHODS FOR MICROBIOLOGISTS

THE SOCIETY FOR APPLIED BACTERIOLOGY
TECHNICAL SERIES NO. 14

IDENTIFICATION METHODS FOR MICROBIOLOGISTS

SECOND EDITION

Edited by

F. A. SKINNER

Rothamsted Experimental Station, Harpenden, Herts, UK

AND

D. W. LOVELOCK

H. J. Heinz Co. Ltd., Hayes Park, Hayes, Middlesex, UK

1979

ACADEMIC PRESS

A Subsidiary of Harcourt Brace Jovanovich, Publishers

LONDON · NEW YORK · TORONTO · SYDNEY · SAN FRANCISCO

ACADEMIC PRESS INC. (LONDON) LTD
24/28 OVAL ROAD
LONDON NW1

U.S. Edition Published by
ACADEMIC PRESS INC.
111 FIFTH AVENUE
NEW YORK, NEW YORK 10003

British Library Cataloguing in Publication Data

Identification methods for microbiologists.—2nd ed.
—(Society for Applied Bacteriology. Technical
series; no. 14 ISSN 0300–9610).
1. Micro-organisms—Identification
I. Skinner, Frederick Arthur II. Lovelock,
Dennis William III. Series
576 QR65 79–41203

ISBN 0–12–647750–7

Printed in Great Britain by
Latimer Trend & Company Ltd, Plymouth

Contributors

A. C. BAIRD-PARKER, *Unilever Research, Colworth/Welwyn Laboratory, Unilever Ltd., Colworth House, Sharnbrook, Bedford MK44 1LQ, UK*

ELLA M. BARNES, *ARC Food Research Institute, Colney Lane, Norwich NR4 7UA, UK*

IRENE BATTY, *Bacteriology Department, Wellcome Research Laboratories, Langley Court, Beckenham, Kent, UK*

C. D. BRACEWELL, *Ministry of Agriculture, Fisheries and Food, The Central Veterinary Laboratory, New Haw, Weybridge, Surrey KT15 3NB, UK*

J. G. CARR, *Long Ashton Research Station, Long Ashton, Bristol BS18 9AF, UK*

M. J. CORBEL, *Ministry of Agriculture, Fisheries and Food, The Central Veterinary Laboratory, New Haw, Weybridge, Surrey KT15 3NB, UK*

A. L. FURNISS, *Public Health Laboratory Service, Public Health Laboratory, Preston Hall Hospital, Maidstone, Kent M20 7NH, UK*

K. P. W. GILL, *Ministry of Agriculture, Fisheries and Food, The Central Veterinary Laboratory, New Haw, Weybridge, Surrey KT15 3NB, UK*

M. GOODFELLOW, *Department of Microbiology, The Medical School, The University, Newcastle-upon-Tyne NE1 7RU, UK*

R. J. GROSS, *Computer Identification Laboratory, National Collection of Type Cultures, Central Public Health Laboratory, Colindale Avenue, London NW9 5HT, UK*

P. R. HAYES, *Department of Microbiology, Agricultural Science Building, University of Leeds, Leeds LS2 9JT, UK*

A. C. HAYWARD, *Department of Microbiology, University of Queensland, St. Lucia, Brisbane, 4067, Australia*

MARGARET S. HENDRIE, *Ministry of Agriculture, Fisheries and Food, Torry Research Station, PO Box 31, 135 Abbey Road, Aberdeen AB9 8DG, UK*

L. R. HILL, *National Collection of Type Cultures, Central Public Health Laboratory, Colindale Avenue, London NW9 5HT, UK*

B. HOLMES, *Computer Identification Laboratory, National Collection of Type Cultures, Central Public Health Laboratory, Colindale Avenue, London NW9 5HT, UK*

S. P. LAPAGE, *Computer Identification Laboratory, National Collection of*

Type Cultures, Central Public Health Laboratory, Colindale Avenue, London NW9 5HT, UK

J. V. LEE, *Public Health Laboratory, Preston Hall Hospital, Maidstone, Kent, ME20 7NH, UK*

T. A. MCMEEKIN, *Faculty of Agricultural Science, University of Tasmania, Hobart 7001, Tasmania, Australia*

P. S. NUTMAN, *Soil Microbiology Department, Rothamsted Experimental Station, Harpenden, Hertfordshire, AL5 2JQ, UK*

R. J. OWEN, *National Collection of Type Cultures, Central Public Health Laboratory, Colindale Avenue, London NW9 5HT, UK*

SUSAN M. PASSMORE, *Long Ashton Research Station, Long Ashton, Bristol BS18 9AF, UK*

P. RIDGWAY WATT, *Beecham Pharmaceuticals, Chemotherapeutic Research Centre, Brockham Park, Betchworth, Surrey RH3 7AJ, UK*

B. ROWE, *Computer Identification Laboratory, National Collection of Type Cultures, Central Public Health Laboratory, Colindale Avenue, London NW9 5HT, UK*

K. P. SCHAAL, *Institute of Hygiene, University of Cologne, Cologne, Federal Republic of Germany*

M. ELISABETH SHARPE, *National Institute for Research in Dairying, Shinfield, Reading RG2 9AT, UK*

J. M. SHEWAN, *Ministry of Agriculture, Fisheries and Food, Torry Research Station, PO Box 31, 135 Abbey Road, Aberdeen AB9 8DG, UK*

F. A. SKINNER, *Soil Microbiology Department, Rothamsted Experimental Station, Harpenden, Hertfordshire AL5 2JQ, UK*

P. H. A. SNEATH, *School of Medicine and School of Biological Sciences, University of Leicester, Department of Microbiology, University Road, Leicester LE1 7RH, UK*

E. L. THOMAS, *Ministry of Agriculture, Fisheries and Food, The Central Veterinary Laboratory, New Haw, Weybridge, Surrey KT15 3NB, UK*

P. D. WALKER, *Bacteriology Department, Wellcome Research Laboratories, Langley Court, Beckenham, Kent, UK*

J. M. VINCENT, *Department of Microbiology, University of Sydney, Sydney, N.S.W. 2006, Australia*

Preface

The proven popularity of "Identification Methods for Microbiologists" Parts A and B, which were numbers 1 and 2 in the Technical Series, led the Society for Applied Bacteriology to promote publication of a second edition of this work. This present volume combines chapters from both parts of the first edition which have been fully revised and brought up to date. There have been many significant advances in identification methods for certain genera as well as some changes in the taxonomic structure of genera following the publication of the new Bergey's Manual in 1974. This has necessitated substantial rewriting of some chapters and we would like to thank all contributors for the way in which they have accomplished this often onerous task.

Some changes in authorship have occurred and some of the original chapters have been omitted, usually because the authors were no longer active in those subject areas. This work is therefore not a comprehensive review of modern microbiology, but it does include studies on those bacterial genera that occupy the attention of most of the world's bacteriologists.

The editors hope that this second edition will serve the worker at the laboratory bench at least as well as its forerunners have done.

September 1979

F. A. Skinner
D. W. Lovelock

Contents

LIST OF CONTRIBUTORS v
PREFACE vii

The Identification of Pseudomonads 1
MARGARET S. HENDRIE AND J. M. SHEWAN
Methods 2
Discussion 6
References 12

Isolation and Characterization of *Xanthomonas* . . 15
A. C. HAYWARD
Techniques of Isolation from Plant Material or Soil . . 17
Characteristics of *Xanthomonas* 20
References 28

Methods for Identifying Acetic Acid Bacteria . . . 33
J. G. CARR AND SUSAN M. PASSMORE
Isolation 34
Identification at Generic Level. 36
Identification at Species Level 38
Conclusions 44
References 45

The Identification and Classification of *Rhizobium* . . 49
J. M. VINCENT, P. S. NUTMAN AND F. A. SKINNER
The Genus *Rhizobium* 49
Identification of Rhizobia 52
The Species of *Rhizobium* 55
Strain Differences within Species 59
Modern Views on Classification 62
References 65

Techniques in the Identification and Classification of *Brucella* Species . . . 71
M. J. CORBEL, C. D. BRACEWELL, E. L. THOMAS AND K. P. W. GILL .
Definitions of Genus and Species . . . 72
Basic Materials and Methods . . . 77
Methods for Identification at the Genus Level . . . 90
Methods for Identification at the Species and Biotype Levels 102
References . . . 117

Biochemical Identification of Enterobacteriaceae . . . 123
S. P. LAPAGE, B. ROWE, B. HOLMES AND R. J. GROSS
Methods . . . 123
Biochemical Reactions of the Genera . . . 126
Some Characters of the Genus *Salmonella* . . . 126
References . . . 139

Identification of Human Vibrios . . . 143
A. L. FURNISS
Methods . . . 144
Identification . . . 145
References . . . 149

Identification of *Aeromonas, Vibrio* and Related Organisms 151
J. V. LEE, MARGARET S. HENDRIE AND J. M. SHEWAN .
Methods . . . 155
References . . . 165

Identification Methods Applied to *Chromobacterium* . . . 167
P. H. A. SNEATH
Methods Useful in Recognizing Members of the Genus . 168
Tests for Distinguishing between *Chromobacterium violaceum*, *C. lividum*, *C. fluviatile* and the Marine Group . . . 171
References . . . 174

The Identification of Gram Negative, Yellow Pigmented Rods . . . 177
P. R. HAYES, T. A. MCMEEKIN AND J. M. SHEWAN . .
Test Methods . . . 178
Identification . . . 182
References . . . 185

Methods for the Characterization of the Bacteroidaceae . 189
ELLA M. BARNES
Methods 190
Identification of Organisms 192
General Conclusions 198
References 198

Methods for Identifying Staphylococci and Micrococci . 201
A. C. BAIRD-PARKER
Preliminary Screening and Maintenance of Isolates . . 201
Separation of Members of the Genus *Staphylococcus* from Members of the Genus *Micrococcus* 202
Identification of Species 203
Identification Methods 206
References 209

Colonial Morphology and Fluorescent Labelled Antibody Staining in the Identification of Species of the Genus *Clostridium* 211
P. D. WALKER AND IRENE BATTY
Materials and Methods 212
Results 214
Discussion 230
References 231

Identification of the Lactic Acid Bacteria 233
M. ELISABETH SHARPE
Characteristics Used to Recognize Genera of Lactic Acid Bacteria 233
Streptococcus 235
Pediococcus and *Aerococcus* 241
Leuconostoc 243
Lactobacillus 246
References 255

Identification Methods for *Nocardia, Actinomadura* and *Rhodococcus* 261
M. GOODFELLOW AND K. P. SCHAAL
Identification at Generic Level. 261
Identification at Specific Level 266
References 274

The Estimation of Base Compositions, Base Pairing and Genome Sizes of Bacterial Deoxyribonucleic Acids . 277
R. J. OWEN AND L. R. HILL
DNA Extraction and Purification 277
Base Composition Estimation 281
DNA-DNA Base Pairing 286
Estimation of Genome Size 289
References 292

An Improved Automatic Multipoint Inoculator . . . 297
P. RIDGWAY WATT
Construction 298
Operation 299
Applications 301
References 303

SUBJECT INDEX 305

Erratum The legend for Fig. 5, page 103 should read "On each plate, reading clockwise . . . "

The Identification of Pseudomonads

Margaret S. Hendrie and J. M. Shewan

Torry Research Station, Aberdeen, UK

Since the publication of the previous paper on the identification of *Pseudomonas* species (Hendrie & Shewan 1966) there have been major studies on the taxonomy of the aerobic pseudomonads (Stanier *et al.* 1966; Ballard *et al.* 1968; Ballard *et al.* 1970; Palleroni *et al.* 1970; Palleroni *et al.* 1972). With the publication of the eighth edition of Bergey's Manual (Buchanan & Gibbons 1974) and subsequent taxonomic work by the Berkeley group (Palleroni 1975; Stanier 1976) and others (e.g. de Vos & De Ley 1978) the problems of classification and identification of members of the Pseudomonadaceae have been considerably clarified. The recognition of four sections in Bergey's Manual (Buchanan & Gibbons 1974) with the establishment of diagnostic tables in place of the dichotomous keys used in earlier editions of the Manual are undoubtedly advances and have much to recommend them. There are some obvious gaps in the Manual keys, particularly the omission of the marine pseudomonads and some important spoilage types. Earlier studies (Shewan *et al.* 1960) showed that some marine types were well separated from the terrestrial species, and the finding that some had DNA base ratios markedly different from the terrestrial species indicated that they could not belong to the genus *Pseudomonas* (Mandel *et al.* 1965; Mandel 1966; Herbert *et al.* 1971; Shewan 1971). The subsequent proposal of Baumann *et al.* (1972) that the low mol % guanine + cytosine marine group be assigned to a new genus—*Alteromonas*—seems to us a reasonable one. However, the difference in the DNA base ratio was the only criterion cited by Baumann *et al.* (1972) separating *Alteromonas* from *Pseudomonas*, a somewhat unsatisfactory situation. During this time work was also in progress at Torry Reseaıch Station seeking other criteria which would separate the high mol % G + C pseudomonads from those with a low mol % G + C value. This resulted in the recognition that certain diagnostic tests are useful in separating the pseudomonads into five phenons (Lee 1977) without having to estimate the guanine + cytosine content of the DNA. Lee *et al.* (1977) also extended the circumscription of the genus *Alteromonas* to include non-marine types.

Here we do not propose merely to duplicate the keys given in the eighth edition of Bergey's Manual (Buchanan & Gibbons 1974) but rather to indicate which easily performed tests are of use in the identification of pseudomonads and extend the identification key of Gibson *et al.* (1977). Furthermore, we shall confine our tables to the groups with which we are most familiar—the saprophytes from food, water, soil, etc., i.e. those most frequently encountered by the general microbiologist. We shall not deal in detail with animal and plant pathogens or with the facultative chemolithotrophs. Keys for the identification of clinical isolates can be found in publications such as Cowan and Steel's Manual for the Identification of Medical Bacteria (Cowan 1974) or the contribution of Hugh & Gilardi (1974) to the American Society of Microbiology Manual of Clinical Microbiology or the more recent paper by King & Phillips (1978). The identification scheme of Lelliott *et al.* (1966) is used by plant pathologists.

Methods

Media

For terrestrial isolates (including food, soil and freshwater sources) normal nutrient media are used, but for marine isolates media should be prepared with sea water (natural or artificial) (see Collins *et al.* 1973 for suitable formulae) or be the normal test media with 2·5% NaCl added. However, if it is found that an isolate from a marine source grows better on ordinary media than on seawater media then tests should be carried out in the normal media. In the methods described below media prepared with sea water (75% filtered, aged sea water : 25% distilled water) or supplemented with 2·5% NaCl, should be used where appropriate.

Incubation conditions

Cultures should be tested for ability to grow over a range of temperatures in order to determine both the range and the optimum temperature for growth. Generally speaking 20°C is suitable for the majority of pseudomonads—mesophiles and psychrotrophs—but some psychrophiles may require temperatures lower than 20°C. When determining the temperature range isolates are streaked on nutrient agar plates (8 or 10 isolates per plate) and replicate plates are incubated at 37°C (3 days), 30°C (3–5 days), 25°C (5 days), 15°C and 10°C (up to 2 weeks), 5°C and 1°C (up to 3 weeks). Testing for growth at higher temperatures is done in nutrient broth incubated in a water bath. In determining growth at 41°C (for *Ps. aeruginosa*) growth should be obtained in three successive subcultures (Haynes & Rhodes 1962).

Microscopic examination

(a) Phase contrast (× 20 objective) examination of a 24 h broth culture for motility and morphology of the living cell.

(b) Gram stained smear from a 2 or 3 day old agar culture. Smears are made in either a drop of sterile distilled water or sterile 2% NaCl solution for organisms growing on seawater agar, air dried and heat fixed. The smears made in NaCl solution are washed in tap water for 5 min to remove the salt, and drained before staining in the usual manner by Jensen's modification of the Gram stain (Cruickshank 1968).

(c) Electron microscopy for flagellar attachment is performed only when it is not possible to identify an isolate without confirming the type of flagellation. Usually a 24 h agar slope culture at 20°C is used for electron microscope preparations but in some cases it may be necessary to examine cultures at another temperature or from a broth culture (e.g. *A. putrefaciens* strains may have cells with lateral flagella when grown on a solid medium, but only polar flagella when grown in a liquid medium (W. Hodgkiss, pers. comm.).

Pigment production

Whilst pigmentation may be observed on nutrient media, it is often necessary to use special media to demonstrate or confirm the production of pigments. For diffusible pigments the media of King *et al.* (1954) and Paton (1959) are used and the cultures are examined under u.v. illumination ($\lambda = 365$ nm) to demonstrate fluorescence. Using the media of King *et al.* (1954) as a liquid it is easy to confirm pyocyanin production in medium A by chloroform extraction and acidification of the extract to give a red colour.

Production of non-diffusible pigments may be enhanced or stimulated by various methods e.g. the addition of certain carbohydrates (glucose, starch, glycerol) or skim milk to nutrient agar; or exposure of plates after incubation to either daylight or a low temperature (refrigerator) for a few days.

O-F test

The medium of Hugh & Leifson (1953) with glucose as the carbohydrate is used or the marine oxidation-fermentation (MOF) medium (Leifson 1963) at pH 7·6 and with 0·0025% bromocresol purple indicator in place of the phenol red in the original formulation. Tubes are retained for up to 2 weeks before being discarded. It is helpful in determining that indicator changes are due to the dissimilation of the carbohydrate if tubes of the basal medium without carbohydrate are set up in parallel with the test medium.

Oxidase reaction

Kovács' (1956) test is performed on the young growth (usually 24 h) from the lawn on antibiotic sensitivity plates. A blue colour in 10 s is positive, in 10–15 s, a weak positive.

Antibiotic sensitivity tests

A nutrient agar plate is surface seeded from a young (2–3 day) broth culture, air dried and antibiotic discs applied. Plates are read for zones of inhibition after 24 h incubation, but slow growing strains may require longer incubation. We routinely use Mastrings (Diamed Diagnostics Ltd, Mast House, Derby Road, Bootle, Merseyside L20 1EA) made up to our formulation of antimicrobials from their standard concentrations viz. penicillin G, 4 units; streptomycin, 25 μg; chloramphenicol, 50 μg; tetracycline, 25μg; novobiocin, 5 μg and polymyxin B, 250 units. When screening unknown isolates a Whatman AA disc (6·0 mm) impregnated with 35 μg of 2:4-diamino-6:7-di*iso*propyl pteridine (0/129) phosphate (BDH Chemicals, Poole, Dorset) is placed in the centre of each plate. 0/129 sensitivity indicates that the isolate belongs to the *Vibrio* group.

Extracellular hydrolases

In testing for these enzymes up to six isolates may be streaked on one plate. Where plates have to be treated with a reagent to demonstrate zones of activity they are usually incubated for 4 days, but slow growing strains may be incubated for up to 1 week. Clear zones in either an opaque medium or after adding reagent, are positive.

Gelatin: Nutrient agar + 1% gelatin. After incubation plates are flooded with mercuric chloride (15%) in hydrochloric acid (10%) (Frazier 1926).

Starch: Nutrient agar + 0·8% soluble starch (BDH Chemicals). Incubated plates are treated with Gram's iodine to stain the starch.
Casein: Nutrient agar + 10% dried skim milk. Agar and milk (in 30% of water from total volume) are sterilized separately and mixed before pouring the plates. Incubated plates are examined for clear zones after 2, 4 and 7 days.
DNase: DNase Test Agar (Difco Laboratories, P.O. Box 148, Central Avenue, West Molesey, Surrey) is incubated at 30°C as recommended or for 2 days at 20°C followed by 1 day at 30°C. After incubation the plates are treated with 1% HCl to precipitate the DNA. It was found that for organisms with an optimum growth temperature of 20°C, incubation at 20°C for growth followed by 30°C for enzyme activity gave better zones in positive strains than either 3 days at 20°C or 2 days at 30°C.

Amino acid decarboxylases

Møller's (1955) method is used to test for lysine and ornithine decarboxylases and the arginine dihydrolase system. Difco Decarboxylase Base Moeller is used to prepare the media, and a tube of basal medium without amino acid is always inoculated in parallel with the test media. Cultures are incubated for up to one week, or occasionally longer for slow growing strains. Arginine tubes showing an alkaline reaction are tested with Nessler's reagent for the presence of ammonia before being discarded. Thornley's (1960) method may also be used for the arginine dihydrolase test.

Levan production

Levan production is determined by the growth of mucoid colonies on nutrient agar + 5% sucrose (Lelliott *et al.* 1966).

Hydrogen sulphide production

Blackening of triple sugar iron (TSI) agar (Anon. 1958) or Kligler's Iron agar gives a clear difference between strong H_2S producers and weak H_2S producers which are generally only detected by the more sensitive lead acetate paper method.

β-galactosidase

β-galactosidase activity is detected by the ONPG test (Lowe 1962).

Lipase activity

Lipase activity is detected by the hydrolysis of Tweens, particularly Tween 80 (Sierra 1957) or the production of a pearly layer on and/or detection of free fatty acids in egg yolk agar by staining with saturated $CuSO_4$ solution.

Trimethylamine oxide (TMAO) reduction

Trimethylamine oxide (TMAO) reduction to trimethylamine (TMA) is tested for by the method of Wood & Baird (1943). Lee *et al.* (1977) used the modification of Laycock & Reiger (1971) to perform rapid tests and used 20% KOH in place of the saturated K_2CO_3. In inexperienced hands the Laycock & Reiger rapid method of testing the cultures by placing a filter paper with indicator over several tubes does lead to difficulties as traces of alkali can get on to the rim of the tube when the reagent is added and give a false result.

Na^+ requirement for growth

Na^+ requirement for growth is tested in the medium of Lee *et al.* (1977). Great care must be taken to ensure that the equipment and tubes used in the preparation and dispensing of the medium are chemically clean.

Growth in 7·5% NaCl

Growth in 7·5% NaCl is tested in 1% Trypticase (BBL) or 1% peptone + 7·5% NaCl (Analar quality).

Single carbon source utilization

Although single carbon source utilization results are not included in our tables, they may be required for confirmation of identity by using tables in Bergey's Manual (Buchanan & Gibbons 1974) or the many papers published since the mid-1960s. The simple basal medium of Palleroni & Doudoroff (1972) is recommended rather than the more complex medium of Stanier *et al.* (1966). For marine isolates, a suitable medium was formulated by Lee *et al.* (1977).

Discussion

For the identification of pseudomonads, Tables 1 to 5 have been prepared. These tables are not intended to be a complete definitive identification scheme, but more to give a general overall picture of part of a

TABLE 1. Differentiation of pseudomonads from other heterotrophic, Gram negative bacteria

Motility	Flagella	Oxidase (Kovács)	O-F test (glucose)	Pigments			
				Diffusible		Non-diffusible	
				Fluorescent	Non-fluorescent		
+	Polar	+[a]	Oxidative or no acid	Green or none	Blue-green, red, brown, orange or none	None or yellow, red, brown, blue, pink, black	Pseudomonads
+	Polar	+	Fermentative	—	None or brown	None or yellow, blue, red	Vibrionaceae
+	Peritrichous	+	No acid or oxidative	—	—	—	*Alcaligenes* / *Agrobacterium*
+	Peritrichous	+	No acid or oxidative	—	—	Yellow	*Flavobacterium*
+	Peritrichous	—	Fermentative	—	—	None or yellow, red	Enterobacteriaceae
—	None	—	Fermentative	—	—	—	Enterobacteriaceae
—	None	+ or —	Oxidative or no acid	—	—	—	*Moraxella*-like / *Acinetobacter*
—	None	+ or —	No acid or oxidative or weakly fermentative	—	—	Yellow, orange or red	*Flavobacterium* / *Cytophaga* / *Flexibacter*

[a] Most species are oxidase positive, but some well-defined species are oxidase negative.

TABLE 2. First stage in the identification of pseudomonads

Pigments: diffusible	Green or none; red-brown, brown-black, or orange more rare	None	None
non-diffusible	Usually none; yellow, blue or purple strains occur occasionally	Usually none; yellow strains occur occasionally	Usually none or pinky-brown, but yellow, red, purple, or black strains occur (rare)
O-F test (glucose)	Oxidative	No acid	No acid or oxidative
DNase	—	(—)	(+)
Arginine dihydrolase	(+)	(—)	—
Ornithine decarboxylase	—	—	(—)
H_2S production	—	—	(—)
Sensitivity to chloramphenicol	—	+	+
Na^+ required for growth	—	(—)	(+)
DNA base ratio (range)	58–70	55–70	42–54
	Pseudomonas (see Table 3)	*Pseudomonas* (see Table 4)	*Alteromonas* (see Table 5)

+: all species positive; —: all species negative; (+): most species positive; (—): mos species negative.

large and diverse group of organisms and to indicate the source of more detailed information. It should also be realized that there are exceptions to every rule and that a test result for any one group or species is seldom 100% positive or negative.

The pseudomonads are differentiated from other Gram negative heterotrophic bacteria in Table 1, and Table 2 further differentiates the group into the genera *Pseudomonas* and *Alteromonas*. Tables 3 and 4 differentiate some of the more frequently encountered *Pseudomonas* spp., whilst Table 5 differentiates the majority of *Alteromonas* spp. so far named in the literature.

In our experience, working with the named strains from culture collections, the species allocated to *Alteromonas* by Baumann *et al.* (1972) are not homogeneous. There are at least two distinct groups corresponding to the two groups in the dendrograms in the original paper, viz. A-1, A-2 (*Alteromonas communis* and *A. vaga*) and E-1, E-2 (*A. macleodii* and *A. marinopraesens*, later renamed *A. haloplanktis* (Reichelt & Baumann

TABLE 3. Differentiation of some oxidative *Pseudomonas* spp.

	Ps. aeruginosa	*Ps. fluorescens* group	*Ps. putida*	*Ps. fragi*	*Ps. cepacia*	*Ps. cichorii*	*Ps. syringae*	*Ps. marina*
Pigments								
diffusible:								
fluorescent	Green, occasionally none	Green	Green	None	None	Green	Green	None
non-fluorescent	Blue-green, red, brown-black or none	None or orange	None	None	Pink or purple in certain media	None	None	None
non-diffusible	None	Usually none, orange or blue (rare) occasionally occur	Usually none, yellow very occasionally occurs	None	Yellow or none or purple on certain media	None	None	None
Oxidase (Kovács)	+	+	+	+	+	+	—	—
Growth at 41°C	+	—	—	—	d	—	—	—
Gelatin hydrolysis	+	+	—	—	d	—	d	—
Casein hydrolysis	+	+	—	—	+			—
Arginine dihydrolase	+	+	+	+	—			—
Lysine decarboxylase	—	—	—	—	+			—
Levan from sucrose[a]	—	d	—	—	—	—	d	
ONPG[b]	—	—	—	—	+			
Sensitivity to polymyxin B	+	+	+	+	—			+

+: over 90% strains positive; —: over 90% strains negative; d: between 10% and 90% strains positive.
[a] Data from Bergey's Manual (Buchanan & Gibbons 1974).
[b] Data from Hugh & Gilardi (1974).

TABLE 4. Differentiation of some *Pseudomonas* spp. which do not produce acid in glucose O-F medium

	Ps. acidovorans	*Ps. testosteroni*	*Ps. alcaligenes*	*Ps. pseudoalcaligenes*	*Ps. stutzeri*	*Ps. mendocina*	*Ps. maltophilia*	*Ps. paucimobilis*[a]	*Ps. indigofera*	*Ps. doudoroffi*	*Ps. nautica*
Non-diffusible pigments	None	None	None	None	None	None	None or yellow	Yellow	Blue or none	None	None
Oxidase (Kovács)	+	+	+	+	+	+	—	+	+	+	+
DNase	—	—	—	—	—		+	+	—	—	—
Growth at 41°C	—	—	+	+	d	+		—		+	—
Gelatin hydrolysis	—	—	+	d	+	—	+	—	d	—	—
Starch hydrolysis	—	—	—	—	+	—	—	d	—	—	—
Tween 80 hydrolysis	+	+				+	+	+		—	+
Arginine dihydrolase	—	—	+[b]	—	—	+[b]	—	—	—	—	—
Lysine decarboxylase	—	—	—	—	—	—	+[b]	—	—	—	—
Indol production	—	—	—	—	—	—	—	—	+	—	—
Sensitivity to penicillin	—	—	—	—	—	—	—	—	+	+	—
novobiocin	—	—	—	—	—	—	—	+	+	—	—
polymyxin B	+	+	+	+	+		+	—	+	+	+

+: over 90% strains positive; —: over 90% strains negative; d: between 10% and 90% strains positive.
[a] All data from Holmes *et al.* (1977).
[b] Data from Hugh & Gilardi (1974).

TABLE 5. Differentiation of *Alteromonas* spp.

	A. communis	*A. vaga*	*A. macleodii*	*A. haloplanktis*	*A. luteoviolaceus*[a]	*A. rubra*[a]	*A. citrea*[a]	*A. putrefaciens*	*A. piscicida*	*Alteromonas* sp. (Lee *et al.* 1977 phenon D)
Non-diffusible pigment	None	None	None	None	Purple	Red	Yellow	Pinky-brown	Yellow or orange	None
Oxidase (Kovács)	+	—	+	+	+	+	+	+	+	+
Catalase	+	+	+	+	—	+	+ (weak)	+	+	+
O-F test (glucose)	Oxidative[b]	Oxidative[b]	No acid[b]	No acid	Oxidative or no acid	No acid		No acid	Oxidative (weak)	No acid
DNase	—	—	+	+	+	+	+	+	+	+
Gelatin hydrolysis	—	—	+	+	+	+	+	+	+	+
Starch hydrolysis	—	—	+	—	+	+	+	—	+	d
Lipase	—	—	+	+	+	+	+	d	+	+
Ornithine decarboxylase	—	—		—	—	—	—	+	—	—
Lysine decarboxylase	—	—	—	—	—	d	—	—	—	—
Trimethylamine oxide reduced to trimethylamine	—	+		—	—		—	+		+
H_2S production	—	—	—	—	—	—	—	+	—	+
Sensitivity to penicillin	—	+	—	—	d	—	d	—	—	+
Na^+ required for growth	+	+	+	+	+	+	+	—	+	+
Growth in 7·5 % NaCl	+	+	+	+	—	+	+	—	+	+

+: over 90% strains positive; —: over 90 % strains negative; d: between 10% and 90% strains positive.
[a] Data from Gauthier (1976*a*, *b*, 1977).
[b] Result on designated type strain only in MOF medium with bromocresol purple indicator.

1973)). Kersters (pers. comm.) has also reported that the *A. communis/A. vaga* group does not appear to be closely related to the *A. macleodii/A. haloplanktis* group. The species described by Gauthier (1976*a*, *b*, 1977), *A. luteoviolaceus*, *A. rubra* and *A. citrea*; Lee *et al.* (1977), *A. putrefaciens* and *Ps. piscicida* (*A. piscicida*—paper in preparation) are closer to the *A. macleodii/A. haloplanktis* group than to the *A. communis/A. vaga* group. Chan *et al.* (1978) have described two more species, *A. undina* and *A. espejiana*. The position of these two species has not been evaluated by us, but the published description of the species and our tests on one strain assigned to *A. espejiana* indicate that they are also close to the *A. macleodii/A. haloplanktis* group.

References

ANON. 1958 Report of the Enterobacteriaceae Subcommittee of the Nomenclature Committee of the International Association of Microbiological Societies. *International Bulletin of Bacteriological Nomenclature and Taxonomy* **8,** 25–70.

BALLARD, R. W., DOUDOROFF, M., STANIER, R. Y. & MANDEL, M. 1968 Taxonomy of the aerobic pseudomonads: *Pseudomonas diminuta* and *P. vesiculare. Journal of General Microbiology* **53,** 349–361.

BALLARD, R. W., PALLERONI, N. J., DOUDOROFF, M., STANIER, R. Y. & MANDEL, M. 1970 Taxonomy of the aerobic pseudomonads: *Pseudomonas cepacia, P. marginata, P. alliicola* and *P. caryophylli. Journal of General Microbiology* **60,** 199–214.

BAUMANN, L., BAUMANN, P., MANDEL, M. & ALLEN, R. D. 1972 Taxonomy of aerobic marine eubacteria. *Journal of Bacteriology* **110,** 402–429.

BUCHANAN, R. E. & GIBBONS, N. E. (eds) 1974 *Bergey's Manual of Determinative Bacteriology*, 8th edn. Baltimore: The Williams & Wilkins Co.

CHAN, K. Y., BAUMANN, L., GARZA, M. M. & BAUMANN, P. 1978 Two new species of *Alteromonas*: *Alteromonas espejiana* and *Alteromonas undina. International Journal of Systematic Bacteriology* **28,** 217–222.

COLLINS, V. G., JONES, J. G., HENDRIE, M. S., SHEWAN, J. M., WYNN-WILLIAMS, D. D. & RHODES, M. E. 1973 Sampling and estimation of bacterial populations in the aquatic environment. In *Sampling—Microbiological Monitoring of Environments*, eds Board, R. G. & Lovelock, D. W. London & New York: Academic Press.

COWAN S. T. 1974 *Cowan and Steel's Manual for the Identification of Medical Bacteria*, 2nd edn. Cambridge: Cambridge University Press.

CRUICKSHANK, R. (ed.) 1968 *Medical Microbiology*, 11th edn (revised reprint) p. 650, Edinburgh & London: E. & S. Livingstone.

DE VOS, P. & DE LEY, J. 1978 Heterogeneity of *Pseudomonas* rRNA cistrons and its significance for taxonomy. *Journal of Applied Bacteriology* **45,** xii–xiii.

FRAZIER, W. C. 1926 A method for the detection of changes in gelatin due to bacteria. *Journal of Infectious Diseases* **39,** 302–309.

GAUTHIER, M. J. 1976*a* Morphological, physiological, and biochemical characteristics of some violet-pigmented bacteria isolated from seawater. *Canadian Journal of Microbiology* **22,** 138–149.

GAUTHIER, M. J. 1976*b* *Alteromonas rubra* sp. nov., a new marine antibiotic producing bacterium. *International Journal of Systematic Bacteriology* **26,** 459–466.

Gauthier, M. J. 1977 *Alteromonas citrea,* a new Gram-negative, yellow-pigmented species isolated from seawater. *International Journal of Systematic Bacteriology* **27,** 349–354.

Gibson, D. M., Hendrie, M. S., Houston, N. C. & Hobbs, G. 1977 The identification of some Gram negative, heterotrophic aquatic bacteria. In *Aquatic Microbiology*, eds Skinner, F. A. & Shewan, J. M. London, New York & San Francisco: Academic Press.

Haynes, W. C. & Rhodes, L. J. 1962 Comparative taxonomy of crystallogenic strains of *Pseudomonas aeruginosa* and *Pseudomonas chlororaphis. Journal of Bacteriology* **84,** 1080–1084.

Hendrie, M. S. & Shewan, J. M. 1966 The identification of certain *Pseudomonas* species. In *Identification Methods for Microbiologists*, Part A, eds Gibbs, B. M. & Skinner, F. A. London & New York: Academic Press.

Herbert, R. A., Hendrie, M. S., Gibson, D. M. & Shewan, J. M. 1971 Bacteria active in the spoilage of certain sea foods. *Journal of Applied Bacteriology* **34,** 41–50.

Holmes, B., Owen, R. J., Evans, A., Malnick, H. & Willcox, W. R. 1977 *Pseudomonas paucimobilis,* a new species isolated from human clinical specimens, the hospital environment and other sources. *International Journal of Systematic Bacteriology* **27,** 133–146.

Hugh, R. & Gilardi, G. L. 1974 *Pseudomonas.* In *Manual of Clinical Microbiology*, 2nd edn, eds Lennette, E. H., Spaulding, E. H. & Truant, J. P. Washington D.C.: American Society for Microbiology.

Hugh, R. & Leifson, E. 1953 The taxonomic significance of fermentative versus oxidative metabolism of carbohydrates by various Gram-negative bacteria. *Journal of Bacteriology* **66,** 24–26.

King, A. & Phillips, I. 1978 The identification of pseudomonads and related bacteria in a clinical laboratory. *Journal of Medical Microbiology* **11,** 165–176.

King, E. O., Ward, M. K. & Raney, D. E. 1954 Two simple media for the demonstration of pyocyanin and fluorescin. *Journal of Laboratory and Clinical Medicine* **44,** 301–307.

Kovács, N. 1956 Identification of *Pseudomonas pyocyanea* by the oxidase reaction. *Nature, London* **178,** 703.

Laycock, R. H. & Reiger, L. W. 1971 Trimethylamine-producing bacteria on haddock (*Melanogrammus aeglefinus*) fillets during refrigerated storage. *Journal of the Fisheries Research Board of Canada* **28,** 305–309.

Lee, J. V., Gibson, D. M. & Shewan, J. M. 1977 A numerical taxonomic study of some *Pseudomonas*-like marine bacteria. *Journal of General Microbiology* **98,** 439–450.

Leifson, E. 1963 Determination of carbohydrate metabolism of marine bacteria. *Journal of Bacteriology* **85,** 1183–1184.

Lelliott, R. A., Billing, E. & Hayward, A. C. 1966 A determinative scheme for the fluorescent plant pathogenic pseudomonads. *Journal of Applied Bacteriology* **29,** 470–489.

Lowe, G. H. 1962 The rapid detection of lactose fermentation in paracolon organisms by the demonstration of β-D-galactosidase. *Journal of Medical Laboratory Technology* **19,** 21–25.

MANDEL, M. 1966 Deoxyribonucleic acid base composition in the genus *Pseudomonas*. *Journal of General Microbiology* **43,** 273–292.

MANDEL, M., WEEKS, O. B. & COLWELL, R. R. 1965 Deoxyribonucleic acid base composition of *Pseudomonas piscicida*. *Journal of Bacteriology* **90,** 1492–1493.

MØLLER, V. 1955 Simplified tests for some amino acid decarboxylases and for the arginine dihydrolase system. *Acta pathologica et microbiologica scandinavica* **36,** 158–172.

PALLERONI, N. J. 1975 General properties and taxonomy of the genus *Pseudomonas*. In *Genetics and Biochemistry of Pseudomonas*, eds Clarke, P. H. & Richmond, M. H. London & New York: John Wiley & Sons.

PALLERONI, N. J., BALLARD, R. W., RALSTON, E. & DOUDOROFF, M. 1972 Deoxyribonucleic acid homologies among some *Pseudomonas* species. *Journal of Bacteriology* **110,** 1–11.

PALLERONI, N. J. & DOUDOROFF, M. 1972 Some properties and taxonomic subdivisions of the genus *Pseudomonas*. *Annual Review of Phytopathology* **10,** 73–100.

PALLERONI, N. J., DOUDOROFF, M., STANIER, R. Y., SOLÁNES, R. E. & MANDEL, M. 1970 Taxonomy of the aerobic pseudomonads: the properties of the *Pseudomonas stutzeri* group. *Journal of General Microbiology* **60,** 215–231.

PATON, A. M. 1959 Enhancement of pigment production by *Pseudomonas*. *Nature, London,* **184,** 1254.

REICHELT, J. L. & BAUMANN, P. 1973 Change of the name *Alteromonas marinopraesens* (ZoBell and Upham) Baumann *et al.* to *Alteromonas haloplanktis* (ZoBell and Upham) comb. nov. and assignment of strain ATCC 23821 (*Pseudomonas enalia*) and strain c A 1 DeVoe and Oginsky to this species. *International Journal of Systematic Bacteriology* **23,** 438–441.

SHEWAN, J. M. 1971 The microbiology of fish and fishery products—a progress report. *Journal of Applied Bacteriology* **34,** 299–315.

SHEWAN, J. M., HOBBS, G. & HODGKISS, W. 1960 A determinative scheme for the identification of certain genera of Gram-negative bacteria, with special reference to the Pseudomonadaceae. *Journal of Applied Bacteriology* **23,** 379–390.

SIERRA, G. 1957 A simple method for the detection of lipolytic activity of microorganisms and some observations on the influence of the contact between cells and fatty substrates. *Antonie van Leeuwenhoek* **23,** 15–22.

STANIER, R. Y. 1976 Réflexions sur la taxonomie des *Pseudomonas*. *Bulletin de l'Institut Pasteur* **74,** 255–270.

STANIER, R. Y., PALLERONI, N. J. & DOUDOROFF, M. 1966 The aerobic pseudomonads: a taxonomic study. *Journal of General Microbiology* **43,** 159–271.

THORNLEY, M. J. 1960 The differentiation of *Pseudomonas* from other Gram-negative bacteria on the basis of arginine metabolism. *Journal of Applied Bacteriology* **23,** 37–52.

WOOD, A. J. & BAIRD, E. A. 1943 Reduction of trimethylamine oxide by bacteria. I. The Enterobacteriaceae. *Journal of the Fisheries Research Board of Canada* **6,** 194–201.

Isolation and Characterization of *Xanthomonas*

A. C. Hayward

Department of Microbiology, University of Queensland, St. Lucia 4067, Queensland, Australia

Most workers with extensive experience of the handling and recognition of representatives of the genus *Xanthomonas* have found them only in association with plants or plant materials as plant pathogens (Dye & Lelliott 1974); it is doubtful whether the reports of *Xanthomonas* from other sources such as clinical materials (Von Graevenitz 1973) or water are correct. Gavini & Leclerc (1975) made a systematic study of 128 Gram negative rods producing non-diffusible pigments, isolated from diverse surface waters, and concluded that none could be correctly identified as xanthomonads. By contrast, Calomiris *et al.* (1977) obtained isolates of several petroleum-degrading bacteria from estuarine sediments which they identified as *Xanthomonas* spp. However, these isolates have several characteristics not found in *Xanthomonas* including capacity to grow at 5°C and failure to oxidize some of the carbohydrates which are typically metabolized by members of the genus.

Identification of aerobic, yellow-pigmented Gram negative bacteria with polar flagella poses many problems because of the paucity of biochemical reactions which serve to differentiate xanthomonads from allied genera. Separation largely depends on knowledge of plant source, pigmentation and the nature of the growth upon nutrient media containing a utilizable carbohydrate. In the past the genera *Cytophaga*, *Flavobacterium*, *Pseudomonas* and *Xanthomonas* have become the dumping ground for an assortment of poorly characterized yellow-pigmented bacteria (Goodfellow *et al.* 1976).* The salient features differentiating *Xanthomonas*, *Flavobacterium* and *Cytophaga* are shown in Table 1. Relatively fast-growing yellow- or orange-pigmented bacteria which occur on moribund plant materials have frequently been incorrectly

* Yellow, Gram negative bacteria with a single polar flagellum from clinical sources have been placed in a new taxon *Pseudomonas paucimobilis* (Holmes *et al.* 1977. *International Journal of Systematic Bacteriology* **27,** 133–146).

referred to the genus *Xanthomonas* (Dye 1963*a*, *b*, *c*). On further investigation these isolates have usually been found to be the ubiquitous saprophyte *Erwinia herbicola*, or a closely related organism. This confusion has usually arisen because of failure to recognize the gross difference in growth rate between *E. herbicola* and xanthomonads, as well as the fermentative metabolism and peritrichous flagellation of the former species (Pon *et al.* 1954; Hayward & Hodgkiss 1961; Dye 1963*a*). There are many examples of diseases of plants attributed to *E. herbicola*—like organisms, which have subsequently been found to be caused by xanthomonads. For example, bacterial blight of vine (*Vitis vinifera*) was for many years thought to be caused by *E. vitivora* (Baccarini) Du Plessis, an organism resembling *E. herbicola*. The true cause of the disease is now

TABLE 1. The properties of the genera *Cytophaga, Flavobacterium* and *Xanthomonas*[a]

Property	*Cytophaga*	*Flavobacterium*	*Xanthomonas*
Morphology	Gram negative rod; flexible, short or elongate	Gram negative rods	Gram negative rods
Pigment	Carotenoid; yellow-orange or pink for certain species	Carotenoid; yellow-orange	Brominated aryl octanes; yellow-orange; rarely non-pigmented
Xanthomonas pigment absorption spectrum	—	—	+
Respiration type	Strict aerobe or facultative anaerobe	Strict aerobe; few facultative anaerobes	Strict aerobe
Gliding motility	+	—	—
Motility due to flagella	—	+ (peritrichous or non-motile)	+ (one polar)
Mode of attack on glucose	Oxidative	Feeble or absent	Oxidative
Plant pathogenicity	—	—	(+)
G + C content of the DNA (Mol %)	33–42	30–42 63–70	63–72

[a] Based on Park & Holding (1966), Buchanan & Gibbons (1974) and Gavini & Leclerc (1975).
+: all strains positive; —: no strains positive; (+): most strains positive.

known to be a slow-growing xanthomonad of unusual character which is often obscured on isolation plates by more vigorous saprophytic bacteria such as *E. herbicola* (Panagopoulos 1969; Erasmus *et al.* 1974).

Techniques of Isolation from Plant Material or Soil

General techniques of isolation

The longevity of xanthomonads in seed or pressed, dried leaves may be prolonged (Schnathorst 1969), but their numbers in soil decline rapidly as has been clearly shown in studies on *X. campestris* (Danso *et al.* 1973; Schaad & White 1974*b*). Isolations are most commonly made from young leaf spot or stripe lesions. Portions of diseased material 3–4 mm in size are cut from the boundary between diseased and healthy tissue, and mounted in a little water under a coverslip on a microscope slide. The cut edge of the lesion should be examined microscopically, preferably under phase contrast at a magnification of at least × 400, for the presence of the bacterial ooze which is characteristic of most bacterial diseases of plants. Gram staining of heat fixed smears of the ooze may be necessary to differentiate bacteria from plastids, plant debris or latex. The nature of the ooze varies somewhat with different plant pathogens; in general bacterial masses from tissues invaded by a xanthomonad are tightly coherent, even in freshly collected material. When old, dried specimens are examined ooze appears gradually after mounting in water, whereas in fresh specimens the ooze flows more freely. Frequently the ooze is of such a density that the mounting liquid is of a milky cloudiness when held to the light.

Portions of leaf or stem lesions containing *Xanthomonas* spp. may be suspended in sterile distilled water, one-quarter strength Ringer solution, dilute phosphate buffer or a nutrient medium, allowed to diffuse for 1–2 h and loopfuls of the suspension then streaked out on a suitable nutrient medium such as sucrose peptone (SP) agar (*vide infra*). Isolation plates should be incubated at 25–30°C and examined daily for a period of 3–4 days for the *X. campestris* group, or for up to 14 days for the slow-growing xanthomonads such as *X. albilineans*. There are marked differences in growth rate among the bacteria present in plants, which can serve as a guide in isolation. Almost invariably the common saprophytes such as *E. herbicola* are of faster growth rate than the plant pathogens. Bacteria which produce macroscopic colonies after 24–36 h are likely to be saprophytes. If the material from which the suspension was prepared was fresh and in good condition colonies of the pathogen may predominate on isolation plates. Old, diseased plant material is commonly invaded by a succession of saprophytic fungi and bacteria including *E. herbicola*. In such instances it is advisable to use a selective medium or to prepare several dilutions from the original suspension so that cells of the pathogen and saprophytes are less numerous and therefore more widely separated by streaking on isolation plates. Dense suspensions should be diluted to 1:100 or 1:1000 or by transferring a 3 mm diam. loopful of the original suspension to fresh suspending medium. The purpose of suspending diseased material in a suitable liquid prior to streaking out is to allow the bacteria to separate from the dead plant cells among which they are embedded. This process may be enhanced by teasing apart the tissue in a little sterile suspending medium in a petri dish with sterile seekers, or by grinding the tissue into small fragments with a sterile pestle and mortar (Bradbury 1970). Alternatively, small portions of diseased plant material may be added directly to dried agar surfaces and allowed to diffuse into a few drops of suspending medium for 15–30 min prior to streaking.

The needle-puncture method of isolation of bean pathogens, including *X. phaseoli*, is applicable to many plant diseases in which xanthomonads are involved (Goth 1965). A sterile needle is pushed through the margin of young leaf lesions on to a suitable isolation medium, the needle being forced through the lesion, then downward to make a stab into the agar. For isolations from thick pods and mature stems, the needle is used to pierce the lesion, withdrawn and then pricked into the agar. Drops of sterile liquid medium may then be added to the stab marks prior to streaking out.

A technique has been developed for the detection of low numbers of *X. citri* in the bark of citrus trees affected by citrus canker (Goto 1972*a*). Pieces of bark are surface sterilized, macerated in test tubes containing a nutrient medium and enriched by incubation at 20°C for 20 h. The supernatant liquids are centrifuged at 6000 **g** for 20 min to concentrate bacterial cells. The sediments are resuspended in a nutrient medium and infiltrated with a syringe into the young leaves of a susceptible citrus seedling. When lesions have developed on infiltrated leaves in the glasshouse, the causative bacterium can be reisolated and identified by phage typing.

Isolation of Xanthomonas albilineans, *the cause of leaf scald of sugar cane*

Xanthomonas albilineans is often difficult to isolate from diseased material because of its slow growth rate on artificial medium and the ease with which isolation plates become overgrown by saprophytic bacteria or fungi. The following method is based upon that described by Dean (1974) which is similar to that described by Persley (1972). Leaves bearing the characteristic white streaks or 'pencil lines' are used as source of the pathogen. A leaf section, about 3 × 10 mm, including a young 'pencil line' is placed in drop of water on a microscope slide, covered with a coverslip, and the presence of ooze from the cut ends confirmed microscopically. The purpose of this examination is to find an infected vein that is at the right stage to produce ooze. An additional leaf section, 2–3 cm long, is then cut from an area immediately adjacent to the section from which the bacteria were just observed. Both ends of this section are sealed by touching them to a drop of rubber cement. The section is then surface-sterilized in *ca.* 2·5% sodium hypochlorite solution for 2–3 min, rinsed in sterile water, and transferred to a dry area in the bottom of a sterile petri dish near a small drop (*ca.* 0·3 ml) of sterile water. The ends bearing the rubber cement are cut off with a sterile blade; the remaining central portion of the section is moved into the sterile water and cut into several short sections. After about 5 min, a loop of the water is streaked over the surface of a plate of SP agar (Martin & Robinson 1961), which has the following composition (g l^{-1} in distilled water): peptone, 5; K_2HPO_4, 0·5; $MgSO_4.7H_2O$, 0·25; sucrose, 10 or 20; agar, 15; pH adjusted to 6·8. On SP agar translucent, confluent, pale yellow bacterial growth typical of *X. albilineans* should be visible at 6–10 days (Persley 1972), with single colonies appearing on the second or third set of streaks a few days after confluent growth is seen at the top of the plate.

Selective media may also be used to overcome the problem of fast-growing fungi and saprophytic bacteria on isolation plates. Persley (1972) used SP agar containing 100 mg l^{-1} of the fungal inhibitor, cycloheximide, and found that selectivity was further improved by the addition of penicillin G at a concentration of 200 iu ml^{-1}.

Isolation may also be carried out from sugar cane stem samples. For this purpose portions of the stalk which are several cm in length are partially decontaminated by immersing in 95% ethyl alcohol followed by flaming off the excess. The stalk is then split open and *ca.* 5 mm^2 portions of tissue removed from the node where red discoloured vascular bundles are present. The tissue is then macerated in 5 ml of sterile distilled water in a petri dish and allowed to stand at room temperature for 2–4 h before streaking out from the supernatant liquid on to SP agar. Alternatively the juice may be expressed from the stalk by squeezing between pliers into a sterile container. The juice is then immediately streaked out. Since the concentration of saprophytes as well as of the pathogen is often high in the expressed juice it is advisable to use the selective medium when using this method (Persley 1972).

Selective media for the isolation of Xanthomonas campestris *from soil or from plant material*

Information on the dynamics of survival of xanthomonads in soil has been generally lacking due to the absence of suitable selective media for isolation. A medium known as SX agar, based mainly upon the ability of *X. campestris*, the cause of black rot of crucifers, to produce wide zones of starch hydrolysis, has been devised for this purpose (Schaad & White 1974*a*) and has the following composition (g l^{-1} in distilled water): K_2HPO_4, 2; NH_4Cl, 5; beef extract (Difco), 1; soluble potato starch, 10; methyl violet B, 0·01 (1 ml of a 1% solution in 20% ethanol); methyl green, 0·02 (2 ml of a 1% solution); cycloheximide, 0·25; agar, 15. All ingredients are added prior to autoclaving, and the pH of 6·8 left unadjusted; plates are stored and incubated in the dark. About 10% of the population of *X. campestris* added to soil could be detected on this medium. Other soil micro-organisms were suppressed sufficiently to allow successful assay of the pathogen on SX agar using 1:10 dilutions of soil (Schaad & White 1974*a*). This medium, although moderately inhibitory to certain nomenspecies of the *X. campestris* group, may be useful for studies on the survival in soil of other starch-hydrolysing *Xanthomonas* spp.

A less complex selective medium has been used successfully for the detection of low numbers of *X. campestris* in crucifer seed. The assay procedure makes use of water-enriched cabbage seeds and a nutrient-starch-cycloheximide agar medium of the following composition (Schaad & Kendrick 1975). This medium contains (g l^{-1} in distilled water): Nutrient Agar (Difco), 10; soluble starch, 10; cycloheximide, 0·25. All ingredients are added prior to autoclaving.

Characteristics of *Xanthomonas*

Classification in the eighth edition of Bergey's Manual

For some years there has been a trend towards reduction in the number of species of the plant pathogenic bacteria in general, particularly among the plant pathogenic representatives of *Pseudomonas* and in *Xanthomonas*. This trend is shown in the eighth edition of Bergey's Manual (Dye and Lelliott 1974) in which the number of species of *Xanthomonas* has been reduced to five: *X. campestris* with which 102 nomenspecies are listed as synonyms, *X. ampelina*, *X. fragariae*, *X. albilineans* and *X. axonopodis*. Where previously such species as *X. citri* and *X. malvacearum* had separate status they are now nomenspecies of the *X. campestris* group.*

* Nomenspecies of the *X. campestris* group have been formally designated as *pathovars* of *X. campestris* (Young *et al.* 1978 *New Zealand Journal of Agricultural Research* **21**, 153–177).

There is a marked difference between the five *Xanthomonas* spp. in rate of growth; whereas the nomenspecies of the *X. campestris* group are relatively fast-growing and produce colonies which are 1–3 mm in diam. after 3–5 days incubation on a nutrient medium containing glucose or sucrose, the remaining four taxospecies are relatively slow-growing on first isolation from infected plant material, and 4–14 days incubation may be required before colonies appear which are 1 mm in diam.

Description of the genus

The following description of the genus is based upon the work of Dye (1962, 1966*a*, *b*) and Bergey's Manual (Dye & Lelliott 1974). Xanthomonads are Gram negative, strictly aerobic, polarly flagellate, monotrichous rods, which occur as single cells, measuring 0·2–0·8 μm × 0·6–2·0 μm. They are chemo-organotrophs with a respiratory but never fermentative metabolism, molecular oxygen being the terminal electron acceptor. They are catalase positive; the oxidase reaction is negative or delayed (15–60 s). Nitrates are not reduced to nitrite; acetoin and indole are not produced. Most species hydrolyse gelatin, starch and Tween 80 rapidly; hydrogen sulphide is produced from cysteine and by most species from thiosulphate and peptone. Sodium hippurate is not hydrolysed. The nutritional requirements are complex and usually include methionine, glutamic acid and nicotinic acids in various proportions. Asparagine is not utilized as sole source of carbon and of nitrogen. The optimum temperature for growth is 25–27°C; none grow at 5°C, most grow at 7°C but some are unable to grow at 9°C; all grow at 30°C and almost all fail to grow at 40°C. Growth on agar medium is usually yellow; fluorescent pigments are not produced. The G + C content of the DNA of 29 nomenspecies of the *X. campestris* group has been found by De Ley (1968) to range from 63·5–69·2 mol % and by Murata & Starr (1973) for 13 nomenspecies to range from 63·9–71·6 mol % using measurements of buoyant density and *Tm* values to determine the base composition. The slow-growing xanthomonads possess DNA with the following % G + C: *X. albilineans* (63·1–63·5), *X. ampelina* (66·9), *X. axonopodis* (62·6–64·4) and *X. fragariae* (62·6) according to Murata & Starr (1973).

According to Leifson (1960) non-motile variants are common in *Xanthomonas*; however, Dye (1962) found all of 290 isolates representing 57 nomenspecies to be actively motile. Fimbriae have not been observed except in the case of *X. manihotis* (Lozano & Sequeira 1974). Poly-β-hydroxybutyrate inclusions are not known to be formed except in *X. manihotis* (Lozano & Sequeira 1974) and *X. albilineans* (Hayward 1960).

Xanthomonas ampelina is distinct from the other four species of *Xanthomonas* in failure to grow at 35°C, in giving negative reactions for aesculin hydrolysis, production of acid from glucose, and for utilization of propionate, and in giving a positive reaction for hydrolysis of urea and for utilization of tartrate. All of these characteristics are contrary to an earlier definition of *Xanthomonas* (Dye 1966*a*). Recognition of *X. ampelina* has therefore resulted in a broadened concept of the genus.

Pigmentation

Petroleum ether extracts of the yellow pigment in representatives of the *X. campestris* group possess an absorption spectrum with maxima at 418, 437 and 463 nm (Starr & Stephens 1964). The pigments are unique aryloctane methyl esters distinguished by carrying one or two bromine atoms and not carotenoid in nature as was previously believed (Andrewes *et al.* 1973). The assumption has been made that brominated aryloctanes constitute the group-specific pigments in the genus *Xanthomonas* in spite of the fact that detailed studies have been limited to a single isolate of *X. juglandis*. Pigments of this type are not known to occur in other bacteria. Not all of the nomenspecies of the *X. campestris* group are pigmented; there are 'albino' strains which lack pigment. Some isolates of *X. ricini* are pigmented and others non-pigmented; a similar variation has been found in *X. uppalii* (Durgapal 1969), and *X. azadirachtae* (Chakravarti & Gupta 1975), whereas all isolates of *X. manihotis* (Lozano & Sequeira 1974) and of *X. pedalii* (Patel & Jindal 1972) are non-pigmented. Bacterial plant pathogens producing a slimy growth on sucrose-containing medium but which lack the yellow, water-insoluble pigment characteristic of the *X. campestris* group may be confused with *Pseudomonas*. Their identification as xanthomonads depends upon recognition of the physiological and biochemical properties typical of the genus, including starch hydrolysis, proteolytic and lipolytic activity, failure to reduce nitrate, range of carbohydrates from which acid is produced under aerobic conditions, as well as possession of a single polar flagellum, and susceptibility to *Xanthomonas* bacteriophages (Goto & Starr 1972*a*). Resort may also be made to nucleic acid hybridization studies.

The nature of the yellow, water-insoluble pigment of the slow-growing xanthomonads, *X. albilineans*, *X. axonopodis*, *X. fragariae* and *X. ampelina* has not been examined.*

A diffusible brown pigment is produced by a few members of the *X. campestris* group (Basu & Wallen 1967). A high glucose concentration in culture media causes inhibition of pigment development, apparently be-

* Xanthomonadin pigments have subsequently been identified in representatives of all species except *X. ampelina* (Starr *et al.* 1977 *Archives of Microbiology* **113,** 109).

cause of the repressive effect of glucose on the synthesis or secretion of tyrosinase or related enzymes (Basu 1974).

Colony form and extracellular polysaccharide slime production

Xanthomonas fragariae and the *X. campestris* group produce a colourless polysaccharide slime on media containing peptone and a utilizable carbohydrate such as glucose or sucrose. The colonies are domed, shining and of a mucoid or viscid consistency. Colony variation occurs on artificial medium and includes the occurrence of non-mucoid colony forms or those of intermediate mucoidness (Goto 1972*b*). The mucoid nature of the colonies is a function of extracellular slime production. In a few cases the slime has been fully studied and characterized. That of *X. campestris* is an anionic heteropolysaccharide known as xanthan gum. The covalent structure of the polysaccharide has been determined (Jansson *et al.* 1975; Melton *et al.* 1976; Holzwarth 1976) and consists of a main chain of β-1,4-linked D-glucose units, as in cellulose, but with a three-sugar side chain attached to alternate glucose residues. The gum contains D-glucose, D-mannose, D-glucuronic acid, acetal-linked pyruvic acid and O-acetyl. It is known that the extracellular slime from small colony variants of the mucoid wild type contains the same molar ratio of sugars, but differs from that of the parent strain in its solution properties and in having a lower content of pyruvic acid and O-acetyl substituents (Cadmus *et al.* 1976). Slimes of similar composition have been examined in several other nomenspecies of the *X. campestris* group (Lesley & Hochster 1959; Orentas *et al.* 1963); however, in *X. vesicatoria* the mannose moiety was replaced by galactose. There is insufficient systematic information available at present to indicate whether differences in structure of the heteropolysaccharide are of taxonomic importance.

Xanthan gum has unusual properties in solution including high viscosity at low concentration, which is stable to changes in temperature and pH. There are many industrial applications (Rocks 1971). The functional role of the slime for xanthomonads as plant pathogens is less well understood. The slime may confer upon xanthomonads an unusual degree of desiccation resistance to account for their remarkable longevity in dried leaf specimens at room temperature (Schnathorst 1969). In those xanthomonads which invade the vascular system, such as *X. campestris*, the slime restricts water flow in the xylem vessels (Sutton & Williams 1970). The ordered form of the polymer can bind co-operatively to certain plant cell wall polysaccharides in a way which suggests a role in the colonization of the plant host by the bacterial pathogen (Morris *et al.* 1977).

Variation between species in nutritional and biochemical properties

Unlike the plant pathogenic pseudomonads, *Xanthomonas* spp. commonly exhibit growth factor requirements which may be of importance in identification (Starr 1946). Most species grow to some extent in a simple medium containing ammonium chloride, glucose and salts. Methionine, glutamic acid and nicotinic acid in various mixtures are either stimulatory or essential; for example, *X. albilineans* has an obligate requirement for methionine, and *X. ampelina* for glutamic acid (Panagopoulos 1969). *Xanthomonas pruni* may show an obligate requirement for nicotinic acid in a minimal defined medium containing either glutamic acid or methionine, but this requirement is not reproducible in all strains. Occasionally growth occurs in a medium which is free from nicotinic acid. It is possible that spontaneous mutation to nicotinic acid independence occurs at high frequency in minimal medium (Brown & Wagner 1970). Growth of *X. pruni* may be obtained on tryptophan in the absence of nicotinic acid (Wilson & Henderson 1963).

There are several xanthomonads in which the basic growth factor requirements are not fully understood. *Xanthomonas axonopodis* requires a mixture of amino acids; in *X. oryzae*, glutamic acid is the most favourable nitrogen source followed by cystine (Watanabe 1963).

Nomenspecies of the *X. campestris* group are relatively homogeneous in the physiological and biochemical properties used in classification. There have been few detailed systematic studies of collections of isolates of a single nomenspecies from many different localities. Where this has been done the variation shown between isolates of a nomenspecies has been as great as that between species. Reddy & Ou (1976) made a detailed study of 40 isolates of *X. oryzae*, the cause of bacterial leaf blight of rice, from widely separated geographical regions. No distinct biochemical groups or types were found to exist, which showed that there was much greater homogeneity in biochemical reactions than indicated by the literature. They considered that their data would serve to identify *X. oryzae* when separated from the host and also to differentiate it from other nomenspecies of *Xanthomonas*. Another nomenspecies of the *X. campestris* group, *X. translucens* f.sp. *oryzicola*, the cause of bacterial leaf streak of rice, may be distinguished from *X. oryzae* on the basis of certain biochemical characteristics. The leaf streak pathogen gives positive reactions for β-glucosidase (using arbutin as substrate), phenylalanine deaminase and tyrosinase, whereas *X. oryzae* gives negative reactions in all three tests (Reddy & Ou 1974).

Xanthomonas malvacearum, the cause of bacterial blight of cotton, has been shown to be divisible into two biochemical groups differing in ability to produce acid from lactose, proteolytic activity, colony form and phage susceptibility (Hayward 1964; Singh and Verma 1974, 1975).

Xanthomonas spp. are variable in their ability to break down complex polysaccharides. Starch is rapidly hydrolysed by many species. In *X. vesicatoria* (Dye *et al.* 1964) and *X. vasculorum* (Hayward 1962) different isolates exhibit strongly positive, weakly positive or negative reactions on starch agar. Other nomenspecies of the *X. campestris* group such as *X. holcicola*, *X. vitians* and *X. oryzae* (Reddy & Ou 1976) give consistently negative reactions for starch hydrolysis. Many xanthomonads liquefy pectate gel and will produce a soft rot in uncooked potato slices (Dye 1960). The nature of the pectolytic enzymes has been investigated (Prunier & Kaiser 1964; Knösel & Garber 1967, 1968; Reddy *et al.* 1969; Hildebrand 1971). Starr & Nasuno (1967) examined the pectolytic activity of 19 nomenspecies; none produced hydrolytic polygalacturonase or pectin *trans*-eliminase; 17 showed some pectinesterase activity and 10 liquefied a nutrient pectate gel. Of these 10, seven excreted polygalacturonic acid *trans*-eliminase in pectin-containing liquid medium, but the mode of action of the other three strains could not be ascertained. Various xanthomonads have been shown to hydrolyse carboxymethyl cellulose (Goto & Okabe 1958; Knösel & Garber 1967; Lange & Knösel 1970; Porwal & Chakravarti 1971; Reddy & Ou 1976). Xylanase activity has been detected in *X. alfalfae* (Reddy *et al.* 1974) and shown to be absent in *X. incanae* and *X. phaseoli* (Maino *et al.* 1974).

Bacteriophage relationships

Bacteriophages have been frequently used in attempts to confirm the identity of *Xanthomonas* spp. Phages with lytic activity on xanthomonads vary in their specificity depending on their source; phages isolated from soil or sewage produce plaques of pin-head size and are frequently of wide host range, whereas those obtained from plant material usually produce large clear plaques on the nomenspecies affecting the plant from which the phage was isolated, and are usually highly specific. Sutton *et al.* (1958) obtained a phage of the former type which was capable of lysing many nomenspecies of the *X. campestris* group; phages of similar character were obtained by Stolp & Starr (1964). Goto & Starr (1972*a*) have shown that phages obtained from plant material infected with a nomenspecies of the *X. campestris* group are usually highly specific with activity on the homologous nomenspecies but little or no activity on other nomenspecies. Bacteriophages of both types have a role to play in identification, typing schemes and in establishing interrelationships between different nomenspecies (Stolp & Starr 1964; Sutton & Wallen

1967; Goto & Starr 1972*a*; Civerolo 1976). Several nomenspecies have been shown to carry temperate phages (Goto & Starr 1972*b*).

Nucleic acid hybridization

Nucleic acid hybridization experiments are a powerful tool by which the relationship between organisms may be assessed (De Ley 1968; Mandel 1969). However, the experimental conditions under which hybridization is carried out between single stranded DNA or RNA is important in determining the extent of reassociation. It has been found that the thermal stability of the hybrid nucleic acid molecules formed varies with the incubation temperature at which the reassociation takes place, and that a thermal stability much lower than that of the parent DNAs indicates that many sequences of the bases are imperfectly in register. Temperatures of reassociation which are greater than *Tm* 25°C, where *Tm* is the temperature at the mid-point of the thermal denaturation curve of the parent DNA, are considered to be permissive, and to allow reassociation where there is incomplete base sequence homology, whereas *Tm* 15°C is considered to be a restrictive temperature allowing reassociation where there is almost exact base sequence homology. There are differing opinions on the taxonomic value of reassociation experiments carried out under permissive or restrictive conditions (Johnson & Ordal 1968; De Ley *et al.* 1973). The early DNA hybridization tests with nomenspecies of the *X. campestris* group and various *Pseudomonas* spp. (De Ley & Friedman 1965; Friedman & De Ley 1965; De Ley *et al.* 1966) probably gave a distorted picture of the relationship between these organisms because of the abundance of non-specific binding which occurred under the conditions in which these tests were carried out. There was a very high homology between many *Xanthomonas* and *Pseudomonas* spp. which has not been confirmed by later work. The conclusion that the classification of phytopathogenic bacteria on host specificity has little phylogenetic basis seemed to be supported by these early studies on DNA hybridization, and the results influenced the discussion of the taxonomy of *Xanthomonas* leading up to the treatment of the genus in the eighth edition of Bergey's Manual (Murata 1975).

Murata & Starr (1973) used new techniques and principles in their studies of the segmental homology of the DNA from various *Pseudomonas* and *Xanthomonas* spp. Twenty nomenspecies of the *X. campestris* group together with *X. fragariae*, *X. albilineans*, *X. axonopodis* and *X. ampelina* and 13 *Pseudomonas* spp. comprising saprophytes as well as plant pathogens were compared for their genetic relatedness. Their results showed that all species of *Xanthomonas* possess DNA segments unique to the genus which reveal themselves in partial homology among *Xanthomonas* spp. Only *Ps. maltophilia* showed partial homology with *Xanthomonas* spp., a conclusion which was reached independently through use of hybridization of bulk DNA with ribosomal RNA (rRNA) by Palleroni *et al.* (1973). Murata & Starr (1973) found that the segments common to all xanthomonads did not exceed 50% of the entire genome. Each species possessed large amounts of nomenspecies-specific DNA segments. The segments specific to nomenspecies of the *X. campestris* group were as large as those in the four slow-growing taxospecies. It has been concluded that most *Xanthomonas* spp. for which no criterion but host specificity has existed for identification can now be identified in terms of DNA homology (Murata 1975).

Serology and electrophoresis

Serology and electrophoresis of soluble proteins and enzymes are two approaches to identification at the level of species and subspecies which hold promise. Charudattan *et al.* (1973) designated two serotypes among 72 isolates of *X. vesicatoria* on the basis of presence or absence of certain precipitin bands in gel diffusion plates. However, there was no correlation between serotype and host specificity, since isolates from tomato and pepper included both serotypes. A similar conclusion regarding the lack of correlation between pathotype and serotype in *X. vesicatoria* was reached by Schaad (1976) on the basis of immunological comparison and characterization of the ribosomes obtained from 25 isolates. Using Ouchterlony double-diffusion tests, two lines of precipitation were detected. One precipitin, designated R-1, was always present in homologous systems, was specific at the subspecies level, and was identified as a trypsin and RNase-insensitive immunogen of the 50S subunit core particle. Based on the R-1 immunogen, the 25 isolates of *X. vesicatoria* were grouped into three serotypes. Ribosomes of this nomenspecies failed to react with ribosomes from 15 other closely and distantly related bacteria including six species of *Xanthomonas*. The only ribosomes from other bacteria for which evidence of relatedness was found was with the closely related species, *X. campestris*. These results suggest a method of differentiation of nomenspecies of the *X. campestris* group which may be more fruitful than conventional agglutination or other tests.

El-Sharkawy & Huisingh (1971) examined 69 isolates of *Xanthomonas* representing 21 species by polyacrylamide gel electrophoresis of soluble proteins. The patterns of esterases, phosphatases, malate, glucose-6-phosphate and α-glycerophosphate dehydrogenases, and other soluble proteins were determined. Protein patterns were generally identical for all isolates of the same species but dissimilar for different species. Most species possessed multiple forms of enzymes. Whereas the pattern of α-glycerophosphate dehydrogenase was identical in all species, suggesting genus-specificity, esterases and phosphatases varied widely among species. In most cases, however, taxon-specific patterns, with slight quantitative variations, were evident. No lactate dehydrogenase activity was detected. This technique like that of serology of ribosomes appears useful in differentiating among nomenspecies of *Xanthomonas* which until now have been identifiable only by their specific pathogenicity.

References

ANDREWES, A. G., HERTZBERG, S., LIAAEN-JENSEN, S. & STARR, M. P. 1973 *Xanthomonas* pigments. 2. The *Xanthomonas* 'carotenoids'—non-carotenoid brominated arylpolyene esters. *Acta chemica scandinavica* **27,** 2383–2395.

BASU, P. K. 1974 Glucose inhibition of the characteristic melanoid pigment of *Xanthomonas phaseoli* var. *fuscans*. *Canadian Journal of Botany* **52,** 2203–2206.

BASU, P. K. & WALLEN, V. R. 1967 Factors affecting virulence and pigment production of *Xanthomonas phaseoli* var. *fuscans*. *Canadian Journal of Botany* **45,** 2367–2374.

BRADBURY, J. F. 1970 Isolation and preliminary study of bacteria from plants. *Review of Plant Pathology* **49,** 213–218.

BROWN, A. T. & WAGNER, C. 1970 Regulation of enzymes involved in the conversion of tryptophan to NAD in a colourless strain of *Xanthomonas pruni*. *Journal of Bacteriology* **101,** 456–463.

BUCHANAN, R. E. & GIBBONS, N. E. (eds) 1974 *Bergey's Manual of Determinative Bacteriology* 8th edn. Baltimore: Williams & Wilkins.

CADMUS, M. C., ROGOVIN, S. P., BURTON, K. A., PITTSLEY, J. E., KNUTSON, C. A. & JEANES, A. 1976 Colonial variation in *Xanthomonas campestris* NRRL B-1459 and characterization of the polysaccharide from a variant strain. *Canadian Journal of Microbiology* **22,** 942–948.

CALOMIRIS, J. J., AUSTIN, B., WALKER, J. D. & COLWELL, R. R. 1977 Enrichment for estuarine petroleum-degrading bacteria using liquid and solid media. *Journal of Applied Bacteriology* **42,** 135–144.

CHAKRAVARTI, B. P. & GUPTA, D. K. 1975 A non-pigmented strain of *Xanthomonas azadirachtii* Moniz & Raj causing leaf spot of neem (*Azadirachta indica* A. Juss). *Current Science* **44,** 240–241.

CHARUDATTAN, R., STALL, R. E. & BATCHELOR, D. L. 1973 Serotypes of *Xanthomonas vesicatoria* unrelated to its pathotypes. *Phytopathology* **63,** 1260–1265.

CIVEROLO, E. L. 1976 Phage sensitivity and virulence relationships in *Xanthomonas pruni*. *Physiological Plant Pathology* **9,** 67–75.

DANSO, S. K. A., HABTE, M. & ALEXANDER, M. 1973 Estimating the density of individual bacterial populations introduced into natural ecosystems. *Canadian Journal of Microbiology* **19,** 1450–1451.

DEAN, J. L. 1974 A method for isolating *Xanthomonas albilineans* from sugar cane leaves. *Plant Disease Reporter* **58,** 439–441.

De Ley, J. 1968 DNA base composition and hybridization in the taxonomy of phytopathogenic bacteria. *Annual Review of Phytopathology* **6,** 63–90.

De Ley, J. & Friedman, S. 1965 Similarity of *Xanthomonas* and *Pseudomonas* deoxyribonucleic acid. *Journal of Bacteriology* **89,** 1306–1309.

De Ley, J., Park, I. W., Tijtgat, R. & Van Ermengen, J. 1966 DNA homology and taxonomy of *Pseudomonas* and *Xanthomonas*. *Journal of General Microbiology* **42,** 43–56.

De Ley, J., Tijtgat, R., De Smedt, J. & Michiels, M. 1973 Thermal stability of DNA:DNA hybrids within the genus *Agrobacterium*. *Journal of General Microbiology* **78,** 241–252.

Durgapal, J. C. 1969 A note on 'albino' *Xanthomonas*. *Current Science* **38,** 391–392.

Dye, D. W. 1960 Pectolytic activity in *Xanthomonas*. *New Zealand Journal of Science* **3,** 61–69.

Dye, D. W. 1962. The inadequacy of the usual determinative tests for the identification of *Xanthomonas* spp. *New Zealand Journal of Science* **5,** 393–416.

Dye, D. W. 1963*a* The taxonomic position of *Xanthomonas uredovorus* (Pon *et al.* 1954.) *New Zealand Journal of Science* **6,** 146–149.

Dye, D. W. 1963*b* Comparative study of the biochemical reactions of additional *Xanthomonas* spp. *New Zealand Journal of Science* **6,** 483–486.

Dye, D. W. 1963*c* The taxonomic position of *Xanthomonas stewartii* (E. F. Smith 1914) Dowson 1939. *New Zealand Journal of Science* **6,** 495–506.

Dye, D. W. 1966*a* A comparative study of some atypical 'xanthomonads' *New Zealand Journal of Science* **9,** 843–854.

Dye, D. W. 1966*b* Cultural and biochemical reactions of additional *Xanthomonas* spp. *New Zealand Journal of Science* **9,** 913–919.

Dye, D. W. & Lelliott, R. A. 1974 Genus *Xanthomonas*. In *Bergey's Manual of Determinative Bacteriology* 8th edn, eds Buchanan, R. E. & Gibbons, N. E. Baltimore: Williams & Wilkins.

Dye, D. W., Starr, M. P. & Stolp, H. 1964 Taxonomic clarification of *Xanthomonas vesicatoria* based upon host specificity, bacteriophage sensitivity, and cultural characteristics. *Phytopathologische Zeitschrift* **51,** 394–407.

El-Sharkawy, T. A. & Huisingh, D. 1971 Differentiation among *Xanthomonas* species by polyacrylamide gel electrophoresis of soluble proteins. *Journal of General Microbiology* **68,** 155–165.

Erasmus, H. D., Matthee, F. N. & Louw, H. A. 1974 A comparison between plant pathogenic species of *Pseudomonas*, *Xanthomonas* and *Erwinia* with special reference to the bacterium responsible for bacterial blight of vines. *Phytophylactica* **6,** 11–18.

Friedman, S. & De Ley, J. 1965 'Genetic species' concept in *Xanthomonas*. *Journal of Bacteriology* **89,** 95–100.

Gavini, F. & Leclerc, H. 1975 Étude de bacilles gram negatif pigmentés en jaune, isolés de l'eau. *Revue internationale d'océanographie médicale* **37/38,** 17–68.

Goodfellow, M., Austin, B. & Dickinson, C. H. 1976 Numerical Taxonomy of some yellow-pigmented bacteria isolated from plants. *Journal of General Microbiology* **97,** 219–233.

Goth, R. W. 1965 Puncture method for isolating bacterial blights of bean. *Phytopathology* **55,** 930–931.

GOTO, M. 1972*a* Survival of *Xanthomonas citri* in the bark tissues of citrus trees. *Canadian Journal of Botany* **50,** 2629–2635.

GOTO, M. 1972*b* Interrelationships between colony type, phage susceptibility and virulence in *Xanthomonas oryzae*. *Journal of Applied Bacteriology* **35,** 505–515.

GOTO, M. & OKABE, N. 1958 Cellulolytic activity of phytopathogenic bacteria. *Nature, London* **182,** 1516.

GOTO, M. & STARR, M. P. 1972*a* Phage-host relationships of *Xanthomonas citri* compared with those of other xanthomonads. *Annals of the Phytopathological Society of Japan* **38,** 226–248.

GOTO, M. & STARR, M. P. 1972*b* Lysogenization of *Xanthomonas phaseoli* and *X. begoniae* by temperate *X. citri* bacteriophages. *Annals of the Phytopathological Society of Japan* **38,** 267–274.

HAYWARD, A. C. 1960 A method for characterizing *Pseudomonas solanacearum*. *Nature, London* **186,** 405–406.

HAYWARD, A. C. 1962 Studies on bacterial pathogens of sugar cane. I Differentiation of isolates of *Xanthomonas vasculorum*, with notes on an undescribed *Xanthomonas* sp. from sugar cane in Natal and Trinidad. *Occasional Paper, Sugar Industry Research Institute, Mauritius* **13,** 1–12.

HAYWARD, A. C. 1964 Bacteriophage sensitivity and biochemical group in *Xanthomonas malvacearum*. *Journal of General Microbiology* **35,** 287–298.

HAYWARD, A. C. & HODGKISS, W. 1961 Taxonomic relationships of *Xanthomonas uredovorus*. *Journal of General Microbiology* **26,** 133–140.

HILDEBRAND, D. C. 1971 Pectate and pectin gels for differentiation of *Pseudomonas* sp. and other bacterial plant pathogens. *Phytopathology* **61,** 1430–1436.

HOLZWARTH, G. 1976 Conformation of the extracellular polysaccharide of *Xanthomonas campestris*. *Biochemistry* **15,** 4333–4339.

JANSSON, P.-E., KENNE, L. & LINDBERG, B. 1975 Structure of the extracellular polysaccharide from *Xanthomonas campestris*. *Carbohydrate Research* **45,** 275–282.

JOHNSON, J. L. & ORDAL, E. J. 1968 Deoxyribonucleic acid homology in bacterial taxonomy: effect of incubation temperature on reaction specificity. *Journal of Bacteriology* **95,** 893–900.

KNÖSEL, D. & GARBER, E. D. 1967 Pektolytische und cellulolytische Enzyme bei *Xanthomonas campestris* (Pammel) Dowson. *Phytopathologische Zeitschrift* **59,** 194–202.

KNÖSEL, D. & GARBER, E. D. 1968 Separation of pectolytic and cellulolytic enzymes in culture filtrates of phytopathogenic bacterial species by starch gel zone electrophoresis. *Phytopathologische Zeitschrift* **61,** 292–298.

LANGE, E. & KNÖSEL, D. 1970 Zur Bedeutung pektolytischer, cellulolytischer und proteolytischer Enzyme für die virulenze phytopathogener Bakterien. *Phytopathologische Zeitschrift* **69,** 315–329.

LEIFSON, E. 1960 *Atlas of Bacterial Flagellation*. London: Academic Press.

LESLEY, S. M. & HOCHSTER, R. M. 1959 The extracellular polysaccharide of *Xanthomonas phaseoli*. *Canadian Journal of Biochemistry and Physiology* **37,** 513–529.

LOZANO, J. C. & SEQUEIRA, L. 1974 Bacterial blight of cassava in Colombia: etiology. *Phytopathology* **64,** 74–82.

MAINO, A. L., SCHROTH, M. N. & PALLERONI, N. J. 1974 Degradation of xylan by bacterial plant pathogens. *Phytopathology* **64,** 881–885.

MANDEL, M. 1969 New approaches to bacterial taxonomy: perspectives and prospects. *Annual Review of Microbiology* **23,** 239–274.

MARTIN, J. P. & ROBINSON, P. E. 1961 Leaf Scald. In *Sugar Cane Diseases of the World* Vol I, eds Martin, J. P., Abbott, E. V. & Hughes, C. G. Amsterdam: Elsevier.

MELTON, L. D., MINDT, L., REES, D. A. & SANDERSON, G. R. 1976 Covalent structure of the extracellular polysaccharide from *Xanthomonas campestris*: evidence from partial hydrolysis studies. *Carbohydrate Research* **46,** 245–257.

MORRIS, E. R., REES, D. A., YOUNG, G., WALKINSHAW, M. D. & DARKE, A. 1977 Order-disorder transition for a bacterial polysaccharide in solution. A role for polysaccharide conformation in recognition between *Xanthomonas* pathogen and its plant host. *Journal of Molecular Biology* **110,** 1–16.

MURATA, N. 1975 Improved DNA-DNA hybridization as a taxonomic tool for phytopathogenic bacteria. *Japan Agricultural Research Quarterly* **9,** 131–136.

MURATA, N. & STARR, M. P. 1973 A concept of the genus *Xanthomonas* and its species in the light of segmental homology of deoxyribonucleic acids. *Phytopathologische Zeitschrift* **77,** 285–323.

ORENTAS, D. G., SLONEKER, J. H. & JEANES, A. 1963 Pyruvic acid content and constituent sugars of exocellular polysaccharides from different species of the genus *Xanthomonas*. *Canadian Journal of Microbiology* **9,** 427–430.

PALLERONI, N. J., KUNISAWA, R., CONTOPOULOU, R. & DOUDOROFF, M. 1973 Nucleic acid homologies in the genus *Pseudomonas*. *International Journal of Systematic Bacteriology* **23,** 333–339.

PANAGOPOULOS, C. G. 1969 The disease 'tsilik marasi' of grapevine: its description and identification of the causal agent (*Xanthomonas ampelina* sp. nov.) *Annales de l'Institut Phytopathologique Benaki* **9,** 59–81.

PARK, R. W. A. & HOLDING, A. J. 1966 Identification of some common gram-negative bacteria. *Laboratory Practice* **15,** 1124–1127.

PATEL, P. N. & JINDAL, J. K. 1972 Bacterial leaf spot of *Pedalium murex* L. caused by a new albino species of *Xanthomonas*. *Indian Phytopathology* **25,** 318–320.

PERSLEY, G. J. 1972 Isolation methods for the causal agent of leaf scald disease. *Sugar Cane Pathologists' Newsletter* **8,** 24.

PON, D. S., TOWNSEND, C. E., WESSMAN, G. E., SCHMITT, C. G. & KINGSOLVER, C. H. 1954 A *Xanthomonas* parasitic on uredia of cereal rusts. *Phytopathology* **44,** 707–710.

PORWAL, S. & CHAKRAVARTI, B. P. 1971 *In vitro* and *in vivo* production of cellulolytic enzymes by *Xanthomonas cyamopsidis*. *Biochemie und Physiologie der Pflanzen* **A 162,** 368–370.

PRUNIER, J. P. & KAISER, P. 1964 Study of the pectinolytic activity in phytopathogenic and saprophytic bacteria of plants. I. Research on pectinolytic enzymes. *Annales des Épiphyties* **15,** 205–219.

REDDY, M. N., STUTEVILLE, D. L. & SORENSEN, E. L. 1969 Pectolytic activity by *Xanthomonas alfalfae*. *Phytopathology* **59,** 887–888.

REDDY, M. N., STUTEVILLE, D. L. & SORENSEN, E. L. 1974 Xylanase activity of *Xanthomonas alfalfae* in culture and during pathogenesis of bacterial leaf spot of alfalfa. *Phytopathologische Zeitschrift* **80,** 267–278.

REDDY, O. R. & OU, S. H. 1974 Differentiation of *Xanthomonas translucens* f. sp. *oryzicola* (Fang *et al.*) Bradbury, the leaf streak pathogen, from *Xanthomonas oryzae* (Uyeda & Ishiyama) Dowson, the blight pathogen of rice, by enzymatic tests. *International Journal of Systematic Bacteriology* **24,** 450–452.

Reddy, O. R. & Ou, S. H. 1976 Characterization of *Xanthomonas oryzae* (Uyeda & Ishiyama) Dowson, the bacterial blight pathogen of rice. *Annals of the Phytopathological Society of Japan* **42,** 124–130.

Rocks, J. K. 1971 Xanthan gum. *Food Technology* **25,** 476–483.

Schaad, N. W. 1976 Immunological comparison and characterization of ribosomes of *Xanthomonas vesicatoria*. *Phytopathology* **66,** 770–776.

Schaad, N. W. & Kendrick, R. 1975 A qualitative method for detecting *Xanthomonas campestris* in crucifer seed. *Phytopathology* **65,** 1034–1036.

Schaad, N. W. & White, W. C. 1974*a* A selective medium for soil isolation and enumeration of *Xanthomonas campestris*. *Phytopathology* **64,** 876–880.

Schaad, N. W. & White, W. C. 1974*b* Survival of *Xanthomonas campestris* in soil. *Phytopathology* **64,** 1518–1519.

Schnathorst, W. C. 1969 Extension of the period of survival of *Xanthomonas malvacearum* in dried cotton tissue. *Phytopathology* **59,** 707.

Singh, R. P. & Verma, J. P. 1974 The distribution and the lysis patterns of the bacteriophages of *Xanthomonas malvacearum*, the incitant of bacterial blight of cotton. *Proceedings of the Indian National Science Academy* **B 40,** 363–368.

Singh, R. P. & Verma, J. P. (1975) Biochemical variation in Indian isolates of *Xanthomonas malvacearum*. *Indian Phytopathology* **28,** 179–183.

Starr, M. P. 1946 The nutrition of phytopathogenic bacteria. I. Minimal nutritive requirements of the genus *Xanthomonas*. *Journal of Bacteriology* **51,** 131–143.

Starr, M. P. & Nasuno, S. 1967 Pectolytic activity of phytopathogenic xanthomonads. *Journal of General Microbiology* **46,** 425–433.

Starr, M. P. & Stephens, W. I. 1964 Pigmentation and taxonomy of the genus *Xanthomonas*. *Journal of Bacteriology* **87,** 293–302.

Stolp, H. & Starr, M. P. 1964 Bacteriophage reactions and speciation of phytopathogenic xanthomonads. *Phytopathologische Zeitschrift* **51,** 442–478.

Sutton, J. C. & Williams, P. H. 1970 Relation of xylem plugging to black rot lesion development in cabbage. *Canadian Journal of Botany* **48,** 391–401.

Sutton, M. D. & Wallen, V. R. 1967 Phage types of *Xanthomonas phaseoli* isolated from beans. *Canadian Journal of Botany* **45,** 267–280.

Sutton, M. D., Katznelson, H. & Quadling, C. 1958 A bacteriophage that attacks numerous phytopathogenic *Xanthomonas* species. *Canadian Journal of Microbiology* **4,** 493–497.

Von Graevenitz, A. 1973 Clinical microbiology of unusual *Pseudomonas* species. *Progress in Clinical Pathology* **5,** 185–218.

Watanabe, M. 1963 Studies on the nutritional physiology of *Xanthomonas oryzae* (Uyeda et Ishiyama) Dowson. *Annals of the Phytopathological Society of Japan* **28,** 201–208.

Wilson, R. G. & Henderson, L. M. 1963 Tryptophan-niacin relationship in *Xanthomonas pruni*. *Journal of Bacteriology* **85,** 221–229.

Methods for Identifying Acetic Acid Bacteria

J. G. CARR AND SUSAN M. PASSMORE

University of Bristol, Research Station, Long Ashton, Bristol, UK

Acetic acid bacteria are found in association with fruits and flowers but most frequently in fermented products where their presence may be a nuisance since they can cause disorders and off-flavours. Alternatively, they may be found in vinegar generators where various forms of alcoholic wash (beer, cider, wine or alcohol plus nutrients) are converted to vinegar. Their most distinctive activity is the oxidation of ethyl alcohol to acetic acid. However, they are capable of many other chemical activities and these are utilized in their identification.

There are two types of acetic acid bacteria which are currently known as *Acetobacter* and *Gluconobacter* and are listed as such in *Bergey's Manual* (De Ley & Frateur 1974*a*, *b*). The previous name of the latter genus was *Acetomonas*, a name coined by Leifson in 1954 (Leifson 1954) when he discovered that what had hitherto been considered a homogeneous group described under the name of *Acetobacter*, was in fact composed of two similar but distinct kinds of organism. We do not regard the use of the name *Gluconobacter* as very satisfactory. However, as this name is now used in the current edition of *Bergey's Manual* (Buchanan & Gibbons 1974) it will be used in this paper to avoid confusion.

There are a number of techniques now available which have proved to be very useful in establishing relationships between apparently unrelated groups of bacteria. Some of these methods, such as % G + C ratios, DNA homology or protein electrophoretic patterns, deal with genetic constitution or protein composition. Such methods are specialized and complex and cannot be advocated for the routine identification of bacteria. Furthermore, in an Adansonian system as advocated by Sneath (1957), where all characteristics carry the same weight, % G + C ratios and similar properties are no more significant than the ability of an organism to produce a particular product from a specified substrate.

The identification of acetic acid bacteria has different objectives from taxonomy and it is necessary to bias our views. For identification, the first question of practical significance is to ask if the organisms produce acetic acid from ethanol. If the answer is 'yes', then it is proper to ask what kind of acetic acid bacteria they are. It is the aim of this paper to deal with these two questions.

Isolation

Before an identification can be made it is necessary to isolate the organisms. There are several media available for the isolation of acetic acid bacteria and one useful characteristic that can be used for this purpose is their tolerance of high acidity. Most of these organisms will grow readily on media of pH 4·0 and less, but such a low pH sometimes causes problems since the agar is partially hydrolysed during autoclaving and does not form a firm gel. To overcome this, an increase in agar concentration to 3% and a reduction in sterilization time are recommended.

The low pH makes the medium selective so that it only supports the growth of yeasts, acetic and lactic acid bacteria and some acid tolerant bacilli if these are present. Selective agents may be added (Beech & Carr 1955), but, unless the load of organisms other than acetic acid bacteria is heavy, the use of such substances is not necessary.

The type of medium used will depend, to some extent, upon the habitat being investigated. For malt vinegar, a medium composed of malt extract fortified with 1% of yeast extract and adjusted to pH 4·0 is suitable in liquid or solid form. Alternatively, malt wort, removed after fermentation but before acetification, fortified with 1% of yeast extract and adjusted to pH 4·0 is equally suitable. For ciders, the use of apple juice-yeast extract medium (AJYE) (Carr 1953), which grows acetic acid bacteria well, is recommended. The use of a medium containing ethanol is not essential since most of these bacteria can utilize carbohydrates as well as alcohols for energy gain. However, a medium containing ethanol is useful since the increasing aroma of acetic acid indicates the presence of acetic acid bacteria.

Organisms grow fairly well on the media described previously at their optimum growth temperature of 25–30°C. Their colony shapes are variable and even within a pure culture it is quite common to find two or three colony forms. Most organisms are motile in the early stages of growth but this sometimes ceases after a few hours' incubation. All acetic acid bacteria are Gram negative ranging from almost coccoid in shape to fairly long rods. Some cultures, particularly those from natural habitats, contain long swollen involution forms (Fig. 1) and throw off a variety of colonies. In liquid media most of these bacteria grow best at

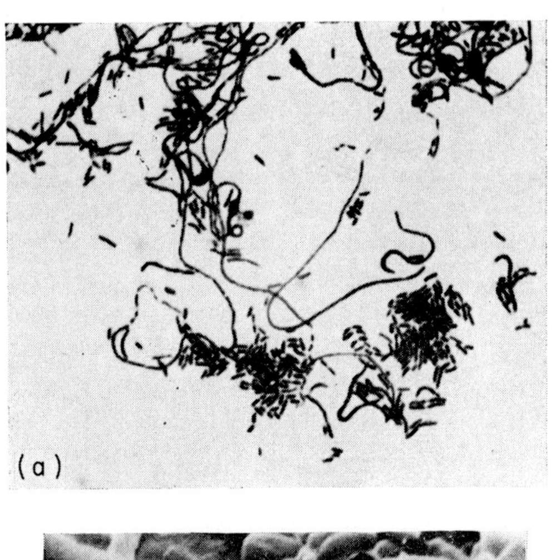

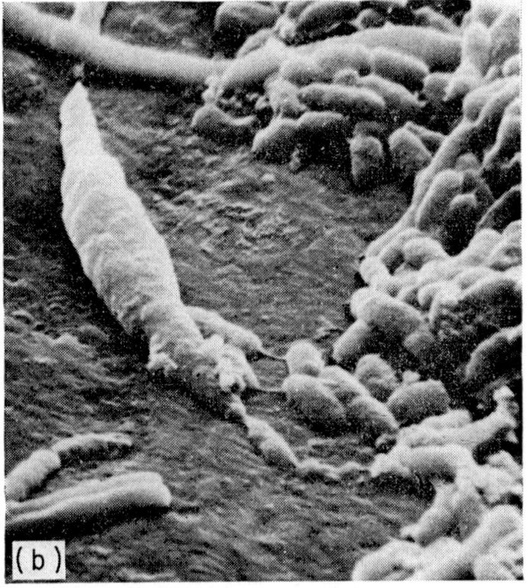

FIG. 1. Involution forms of acetic acid bacteria. (a) Normal cells, thickened filaments and small sacs in a 10 day culture of *Gluconobacter* sp. (LM, × 3000). (b) Normal cells, thickened filament and immature sac in a 4 day culture of *A. rancens* (SEM, × 6100).

the air/liquid interface indicating that they are strongly aerobic, but this does not, however, prevent them from surviving in conditions of severe anaerobiosis. It is often possible to isolate acetic acid bacteria from fermenting apple juice even when the yeasts are at their most active.

Identification at Generic Level

The production of acetic acid from ethanol can be monitored by the use of the following medium: Difco yeast extract, 3%; agar, 2%; bromocresol green (added at the rate of 1 ml of a 2·2% solution in ethanol per litre). This is conveniently dispensed as 6·5 ml amounts in McCartney bottles, and then sterilized. When molten, 1 ml of sterile-filtered ethanol (15% v/v) is added to give a final concentration of 2% alcohol. These are made into slopes and when solidified may be inoculated with the test organisms and incubated at 28°C. Best growth is obtained if the bottle caps are loosened.

When first prepared, the slopes are bluish green but, as the acetic acid bacteria grow, they change to yellow due to the production of acid. If the organism is a *Gluconobacter* sp. the slope will remain yellow. If, however, the organism is an *Acetobacter* sp. the yellow colour will return to the original bluish green. This is because *Acetobacter* spp. can carry the oxidation a stage further and convert acetic acid into carbon dioxide and water, thus raising the pH and changing the indicator back to the characteristic blue of the upper end of its range. This process has been called 'over-oxidation'. Frequent observations must be made since the process of acetification followed by over-oxidation can be missed if the strain oxidizes particularly vigorously. Some strains are very slow over-oxidizers, especially certain strains of *A. xylinum*.

A confirmatory test may be performed by growing the organisms on the following medium (Frateur 1950): Difco yeast extract, 3%; very finely powdered calcium carbonate, 2%; agar, 2%. This is dispensed as 13 ml amounts into McCartney bottles. When required, 2 ml of sterile-filtered 15% ethanol is added and the medium thoroughly mixed before pouring plates. A number of test organisms can be inoculated as smears of 1 cm diam. As acetic acid is produced the chalk is cleared (Fig. 2*a*), but over-oxidizing *Acetobacter* spp. give 'irisation' (Frateur 1950) as chalk is gradually redeposited (Fig. 2*b*). Organisms other than *Acetobacter* spp. may show the same effect and the test using the bromocresol green-ethanol slopes should be carried out first.

These tests not only answer the first question—does the organism produce acetic acid from ethanol?—but they also separate the genera *Acetobacter* and *Gluconobacter* by the way they produce and dissimilate

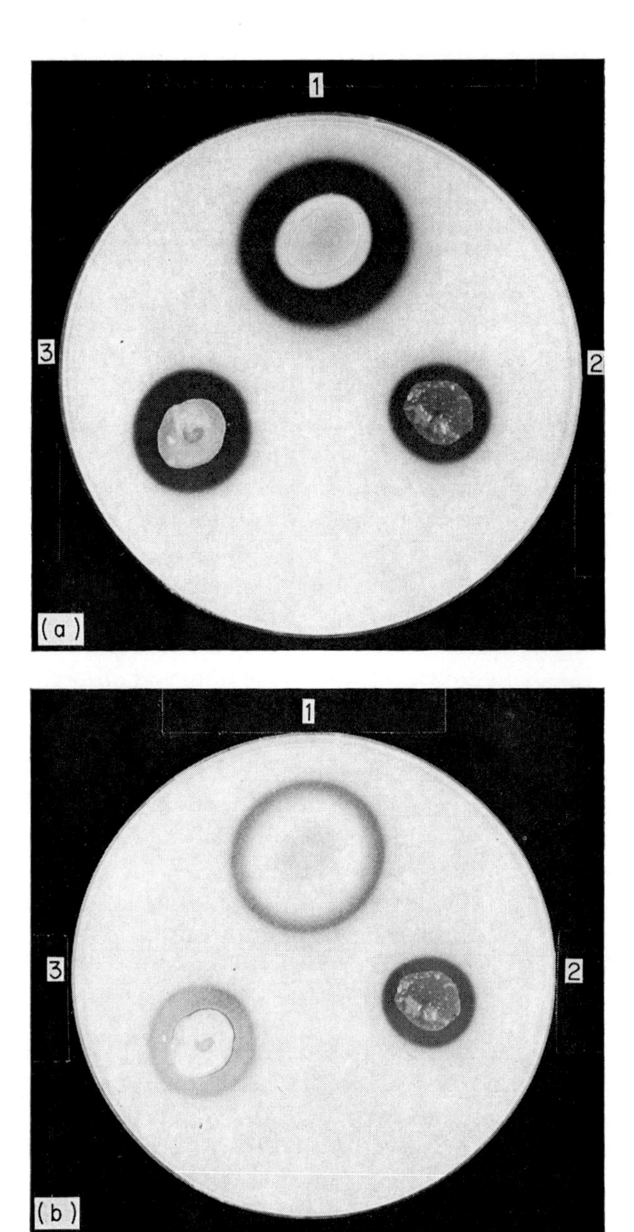

FIG. 2. Chalk-ethanol test for ethanol oxidation. (1) and (3) 'over-oxidizing' *Acetobacter mesoxydans* strains. (2) *Acetomonas* sp.—acid production only. (a) Two days' incubation. (b) Four days' incubation—note irisation around (1) and (3).

acetic acid. Should further proof of acetic acid production be required, the organism can be grown in a liquid form of the bromocresol green-ethanol medium and the end products submitted to gas liquid chromatographic analysis (James & Martin 1952; Cochrane 1975).

Identification at Species Level

Acetobacter

At first glance the Table of *Acetobacter* spp. published in the 8th edition of Bergey's Manual (Buchanan & Gibbons 1974; De Ley & Frateur 1974*b*) looks more complicated than that published by Carr (1968), but there are a great number of similarities. For those only interested in identifying *Acetobacter* spp., some of the tests presented by De Ley & Frateur (1974*b*) are unnecessary. However, it may be useful to review these tests and to discuss their value as aids to identification.

Catalase

This is a useful test and is easily performed. Various media may be used providing they are not too acid. Catalase is detected by applying hydrogen peroxide to growth on a solid medium and observing evolution of gas bubbles. One suitable medium, described later, is that used to detect acid produced from glucose. It is surprising to see in De Ley & Frateur's (1974*b*) table that they present this property as 'usually' positive. In the authors' experience of acetic acid bacteria from fermented beverages or vinegar generators, all freshly isolated strains are invariably catalase positive with the exception of *A. paradoxus* and *A. peroxydans* which are consistently negative. It is possible that strains may become catalase negative after prolonged laboratory cultivation as these organisms are notoriously 'mutable'.

Ketogenesis in glycerol or erythritol

De Ley & Frateur (1974*b*) do not make clear whether glycerol and erythritol are alternative substrates or whether some organisms metabolize one and not the other. Carr (1968) described the use of glycerol to identify the ketogenic properties of these organisms and called the test 'Dihydroxyacetone from glycerol'. It is intended to retain this title for the test which is performed as follows. The medium consists of Difco yeast extract, 3%; glycerol, 3% (v/v); agar, 2%. Several smears of different organisms may be placed on a plate. After incubation at 28°C for 48 h the plates are flooded with Fehling's solution and left for a short time. Positives are rapidly surrounded by a halo of orange cuprous oxide while negatives remain unaltered. Overcrowding of smears or prolonged incubation allows powerful dihydroxyacetone producers to fill the plate, thus obscuring any negative organisms that may be present.

Formation of 5-ketogluconate, 2-ketogluconate or gluconate

Although not mentioned by De Ley & Frateur (1974*b*) it is the formation of these substances from glucose which is important for identification. While it may be of interest to know the type of oxo-acids formed from glucose this characteristic does not aid identification. In practice all that it is necessary to know is whether these organisms produce acid from glucose. It is intended to retain the characteristic, 'acid from glucose', which may be determined with the aid of the following medium: Difco yeast extract, 3%; glucose, 10%; $CaCO_3$, 3%; agar, 2%. Several organisms may be spread in smears of about 1 cm diam. A positive reaction is indicated by the development of a cleared area around the bacterial growth. Organisms which produce no acid from glucose will often grow but produce no clearing. A temperature of 28°C is recommended for incubation. Organisms vary in the speed at which they produce acid but most just begin to show clearing after 48 h.

Growth on ethanol

What is really meant by this statement is growth on ethanol as a sole carbon source. This was called growth in Hoyer's medium by Carr (1968) and it is the medium devised by this worker that is recommended for this test. Its composition is as follows: $(NH_4)_2 SO_4$, 0·1%; $KHPO_4$, 0·01%; KH_2PO_4, 0·09%; $MgSO_4.7H_2O$, 0·025%; $FeCl_3.6H_2O$, 0·002%; ethanol, 3%. The most convenient way of preparing this medium is to dissolve the salts in 80% of the final volume of distilled water and dispense as 4 ml amounts into tubes, and autoclave. Ethanol may then be added as a 15% (v/v) sterile-filtered solution at the rate of 1 ml per tube. Care must be taken when inoculating this medium to avoid transfer of excess organic material and a straight wire inoculation is recommended. Such a small initial inoculum causes growth to be slow and it may take up to 14 days at 28°C to appear.

Production of cellulose

This test is mainly useful for identifying *A. xylinum* although *A. estunensis* also reacts positively in the first of the two test procedures. The typical appearance of a cellulosic pellicle is as a yellowish-cream cartilaginous pellicle and a clear subnatant fluid. A confirmatory test may be performed by staining the pellicle with Lugol's iodine and then applying 60% H_2SO_4. Threads of cellulose stain bright blue and these can be observed under the microscope. Alternatively optical brightener may be incorporated in a suitable medium and cellulosic colonies grown under these conditions will fluoresce under u.v. light (Passmore 1973; Passmore & Carr 1975).

Production of γ pyrones

These compounds are only produced by a single species, namely *A. aceti* subsp. *liquefaciens*. However, development of these compounds requires 28 days at 28°C, a length of time considered to be too long for routine purposes.

Production of a brown pigment

This test is of limited use amongst the *Acetobacter* spp. since according to De Ley & Frateur (1974*b*) only a single subspecies, namely *liquefaciens* produces any pigment. There are, however, *Gluconobacter* strains which also have this characteristic. A suitable medium for showing pigment is as follows: Difco yeast extract, 2%; glucose, 2%; $CaCO_3$, 2%; agar, 2%. This is best dispensed as slopes. Since the medium tends to brown, it is often helpful to compare organisms of unknown reaction with known negatives and positives. The medium already mentioned tests only for the production of brown pigment from glucose, but this can be modified by adding other sugars. Pigment production is very common amongst *Gluconobacter* (*Acetomonas*) strains and can also be reproduced using fructose as substrate (Carr *et al.* 1963). Other sugars, such as sucrose and maltose, will also act as substrates for brown pigment production in some strains (Shimwell & Carr 1959).

Flagella patterns

This test can be used to decide whether an organism belongs to the genus *Acetobacter* or *Gluconobacter* if it is motile. Young cultures tend to contain more motile cells than older ones. Flagella are easily detached so that considerable care is required in handling the organisms. A suitable stain has been devised by Shimwell (1959). Electron microscopy is also useful for determining the arrangement of flagella, but care must be taken to observe enough cells to obtain reliable information.

Naming the species of Acetobacter

De Ley & Frateur (1974*b*) have chosen to reduce the number of *Acetobacter* spp. to *A. aceti*, *A. pasteurianus* and *A. peroxydans*. The first of these contains the following subspecies: *aceti*, *orleanensis*, *xylinum* and *liquefaciens*. The second contains subspecies *pasteurianus*, *lovaniensis*, *estunensis*, *ascendens* and *paradoxus*, whereas the third, *A. peroxydans*, contains no subspecies at all. It is interesting to note that De Ley & Frateur (1974*b*) have revived the specific name ***pasteurianus***, but make no mention of the fact that this species produces starch. Shimwell (1957) stated that only freshly isolated strains show this property and that there is a marked tendency for these organisms to lose their ability to produce starch when cultivated in a laboratory.

Table 1 shows a comparison between characteristics of the strains of *Acetobacter* as published by Carr (1968) and De Ley & Frateur (1974*b*). The names of each test are given in the terminology of each of these authors as they are sometimes the same and sometimes different. The organisms are grouped in pairs, the first of each pair being the name used by Carr (1968) and the results of the tests as published. The second in each group is the subspecific name allocated by De Ley & Frateur (1974*b*) accompanied by their published test results. The most noticeable feature of these sets of results is how closely they correspond. The only real differences are that De Ley & Frateur (1974*b*) reported some strains of *A. xylinum* and *A. rancens* as able to grow on ethanol as a sole carbon source. The present authors have also encountered this phenomenon with *A. xylinum* but have attributed the growth to a carry over of nutrients with the inoculum rather than to an innate ability to grow on ethanol as a sole carbon source.

There are two additions to the list of organisms that do not appear in Carr's (1968) publication: this is an organism referred to by the present authors as *A. aceti* (pigmented) and by De Ley & Frateur (1974*b*) as *A. aceti* subsp. *liquefaciens*. Their name comes as something of a surprise because two strains (AC-8 and U4) were kindly donated by Professor T. Asai. When examined by Carr & Shimwell (1960) they were found to be like *A. aceti* with the additional property of producing a brown pigment from glucose, a character until that time observed only in the then-named genus *Acetomonas*. These two strains donated by Asai were called by him *Gluconobacter melanogenus* which means that since publication in the 8th edition of Bergey's Manual (Buchanan & Gibbons 1974) the generic name has been changed to *Acetobacter*, the specific name to *aceti* and the subspecific name *liquefaciens* has also been added. The other organism not included in Carr's (1968) list is *A. estunensis*, an *Acetobacter* sp. very similar to *A. lovaniensis*, but giving, in addition, a positive cellulose reaction (Carr 1958*a*). There is one other organism that might also perhaps have been included and that is a strain of *A. aceti* giving a positive cellulose reaction (Carr 1958*b*).

TABLE 1. A comparison of the identification schemes of Carr (1968) and De Ley & Frateur (1974*b*) for *Acetobacter* spp.

Names of tests: Carr (1968)	Names of tests: De Ley & Frateur (1974*b*)	C: *A. aceti*	D: subsp. *aceti*	C: *A. mesoxydans*	D: subsp. *orleanensis*	C: *A. xylinum*	D: subsp. *xylinum*	C: *A. aceti* (pigmented)	D: subsp. *liquefaciens*	C: *A. rancens*	D: subsp. *pasteurianus*	C: *A. lovaniensis*	D: subsp. *lovaniensis*	C: *A. estunensis*	D: subsp. *estunensis*	C: *A. ascendens*	D: subsp. *ascendens*	C: *A. paradoxus*	D: subsp. *paradoxus*	C: *A. peroxydans*	D: subsp. *peroxydans*
Catalase	Catalase	+	+	+	+	+	+	+	+	+	+	+	+	+	+	+	+	—	—	—	—
Growth in Hoyer's medium	Growth on ethanol	+	+	—	—	—	(—)	+	+	—	(—)	+	+	+	+	—	—	—	—	+	+
Acid production	Formation of gluconate	+	+	+	+	+	+	+	+	+	+	+	+	+	+	—	—	—	—	—	—
Dihydroxyacetone from glycerol	Ketogenesis in glycerol	+	+	+	+	+	+	+	+	—	—	—	—	—	—	—	—	—	—	—	—
Cellulose	Cellulose	—	—	—	—	+	+	—	—	—	—	—	—	+	—	—	—	—	—	—	—
Brown pigment	Brown pigment	—	—	—	—	—	—	+	+	—	—	—	—	—	—	—	—	—	—	—	—

(—): usually negative.
C: Carr (1968).
D: De Ley & Frateur (1974*b*).

Other Acetobacter *species*

Another organism that should be mentioned is *A. acidophilum*. Wiame *et al.* (1959) isolated it from a trickling vinegar generator. The most unusual characteristic is its inability to grow above pH 4·3. As these authors pointed out, the organism is difficult to handle and the type of tests described in this paper cannot be applied unless modified. From their description, the organism is of low oxidative capacity, probably resembling a strain of *A. peroxydans* with a low pH optimum. There is reason to suppose that by adapting the tests recorded in Table 1 to a lower pH range it would be possible to establish this organism's relationship with other acetic acid bacteria.

Gluconobacter

This name was one of several coined by Asai (1934, 1935) and applied to what has been shown subsequently to be a somewhat mixed collection of Gram negative organisms. The original papers of Asai (1934, 1935) were written in Japanese and, therefore, remained unknown to the world of bacteriology until they appeared as shortened English translations in 1958. This was after Leifson (1954) had shown for the first time that acetic acid bacteria were not homogeneous but consisted of two groups—the true *Acetobacter* spp. (overoxidizers of acetic acid, with peritrichous flagella when motile) and a group he chose to call *Acetomonas* (did not metabolize acetic acid, propelled by polar flagella when motile). Leifson (1954) never formally proposed the generic name *Acetomonas*: this was left to Shimwell & Carr (1959) who defined this genus as containing the single species *Acetomonas oxydans*. De Ley (1961), describing various biochemical and taxonomic aspects of the acetic acid bacteria, chose to call the non-overoxidizers *Gluconobacter*, presumably on historical grounds. This was followed by further support for this name by Asai *et al.* (1964). That *Gluconobacter* included such organisms as '*Gluconobacter melanogenus*', which proved to be an *Acetobacter* sp. as mentioned earlier, does not appear to have been taken into account. Similarly, in the 8th edition of Bergey's Manual the section on the modified genus *Gluconobacter* (De Ley & Frateur 1974*a*) gives the impression that this is an homogeneous genus, the inconsistencies having been ignored.

The present genus *Gluconobacter*, as recorded in the 8th edition of Bergey's Manual, bears little relationship to the old genus, but is virtually identical with the genus *Acetomonas* as defined by Shimwell & Carr (1959). Even the single species contained therein bears the same specific name as that proposed by these authors, namely *oxydans*. There is also a good measure of agreement between the comparison of *Gluconobacter* and *Pseudomonas* made by De Ley & Frateur (1974*a*) and that made earlier by Shimwell *et al.* (1960) between *Acetomonas* and *Pseudomonas.*

Identification at specific level

Since this genus contains only a single species, once it has been established that the organism produces acetic acid from ethanol but is not an *Acetobacter* sp. no further identification is necessary. However, if these organisms are subjected to the same tests as the *Acetobacter* spp. they will give the results listed in Table 2.

TABLE 2. Some further tests for strains of *Gluconobacter* spp.

Tests	*Gluconobacter*
Catalase	+
Growth in Hoyer's medium	—
Acid from glucose	+
Dihydroxyacetone from glycerol	+
Production of cellulose	—
Brown pigment	±

It should be noted, however, that only some of the strains produce a brown pigment from glucose but may do so particularly from fructose and to a lesser extent from sucrose and maltose (Shimwell & Carr 1959). A few rarer strains produce non-soluble pink pigments. Such organisms were at one time called *Acetobacter roseus.*

In addition to those tests already mentioned there are other tests which differentiate the two groups of acetic acid bacteria and these are listed in Table 3. These are added for completeness since it is considered that most are too time consuming for routine identification of these bacteria.

Conclusions

The foregoing presents a scheme for identifying acetic acid bacteria which, it is hoped, will be of use to the working bacteriologist who is not generally interested in the minutiae of taxonomy, but rather in the sort of organisms he is isolating from a particular source. It is not claimed to be the best scheme, but it has the virtues of simplicity in use and interpretation. It is also a reminder that living organisms are not stable, though some are more stable than others.

TABLE 3. Properties of *Acetobacter* and *Gluconobacter* (*Acetomonas*)

	Acetobacter	*Gluconobacter*	*Reference*
Ethanol oxidized to $CO_2 + H_2O$	+	—	Frateur (1950)
Lactate oxidized to carbonate	+	—	Frateur (1950)
Nutrition type	Lactophilic	Glycophilic	Brown & Rainbow (1956)
Oxidize various amino acids	+	—	Joubert, Bayens & De Ley (1961)
Citric acid cycle	+	—	Cheldelin (1960) [a]
Flagella (if motile)	Peritrichous	Polar	Leifson (1954)
Serology	Heterogeneous	Homogeneous	McIntosh (1962)
Infrared spectra	Differ from strain to strain but all distinguishable from *Acetomonas*	Differ from strain to strain but all distinguishable from *Acetobacter*	Scopes (1962)

[a] Cheldelin uses the name *Acetobacter suboxydans* in this publication. This and *A. melanogenum* were the names most commonly applied to the majority of organisms that now constitute the genus *Gluconobacter*.

References

ASAI T. 1934, 1935 Taxonomic studies of acetic acid bacteria and allied oxidative bacteria in fruits and a new classification of oxidative bacteria. *Journal of the Agricultural Chemical Society of Japan* **10, 621,** and *ibid.*, **11, 50** (in Japanese). English summaries in Jubilee Publication in Commemoration of the 60th Birthday of Prof. Kin-ichiro Sakaguchi, 1958.

ASAI, T., IIZUKA, H. & KAMAGATA, K. 1964 The flagellation and taxonomy of genera *Gluconobacter* and *Acetobacter* with reference to the existence of intermediate strains. *Journal of General and Applied Microbiology* **10,** 95–126.

BEECH, F. W. & CARR, J. G. 1955 A survey of inhibitory compounds for the separation of yeasts and bacteria in apples and ciders. *Journal of General Microbiology* **12,** 85–94.

BROWN, G. D. & RAINBOW, C. 1956 Nutritional patterns in acetic acid bacteria. *Journal of General Microbiology* **15,** 61–69.

BUCHANAN, R. E. & GIBBONS, N. E. (eds) 1974 *Bergey's Manual of Determinative Bacteriology*, 8th edn. Baltimore: The Williams and Wilkins Co.

CARR, J. G. 1953 The lactic acid bacteria of cider: I Some organisms responsible for the malo-lactic fermentation. *Report Long Ashton Research Station for 1952*, pp. 144–150.

CARR, J. G. 1958*a* *Acetobacter estunense* nov. spec. an addition to Frateur's ten basic species. *Antonie van Leeuwenhoek* **24,** 157–160.

CARR, J. G. 1958*b* A strain of *Acetobacter aceti* giving a positive cellulose reaction. *Nature, London* **182,** 265–266.

CARR, J. G. 1968 Identification of acetic acid bacteria. In *Identification Methods for Microbiologists* Part B. eds Gibbs, B. M. & Shapton, D. A. London: Academic Press.

CARR, J. G. & SHIMWELL, J. L. 1960 Pigment producing strains of *Acetobacter aceti, Nature, London* **186,** 331–332.

CARR, J. G., COGGINS, R. A. & WHITING, G. C. 1963 Metabolism of fructose and glucose by acetic acid bacteria. *Chemistry and Industry,* p. 1279.

CHELDELIN, V. H. 1960 *Metabolic Pathways in Micro-organisms.* London and New York: John Wiley & Sons, Inc.

COCHRANE, G. C. 1975 A review of the analysis of free fatty acids [C_2–C_6] *Journal of Chromatographic Science* **13,** 440–447.

DE LEY, J. 1961 Comparative carbohydrate metabolism and a proposal for a phylogenetic relationship of the acetic acid bacteria. *Journal of General Microbiology* **24,** 31–50.

DE LEY, J. & FRATEUR, J. 1974*a* The genus *Gluconobacter.* In *Bergey's Manual of Determinative Bacteriology* 8th edn. eds Buchanan, R. E. & Gibbons, N. E. Baltimore: The Williams & Wilkins Co.

DE LEY, J. & FRATEUR, J. 1974*b* The genus *Acetobacter.* In *Bergey's Manual of Determinative Bacteriology* 8th edn. eds Buchanan, R. E. & Gibbons, N. E. Baltimore: The Williams & Wilkins Co.

FRATEUR, J. 1950 Essai sur la systématique des Acetobacters. *La Cellule* **53,** 287–392.

JAMES, A. T. & MARTIN, A. J. P. 1952 Gas liquid partition chromatography: the separation and micro-estimation of volatile fatty acids from formic acid to dodecanoic acid. *Biochemical Journal* **50,** 679–690.

JOUBERT, J. J., BAYENS, W. & DE LEY, J. 1961 The catabolism of amino-acids by acetic acid bacteria. *Antonie van Leeuwenhoek* **27,** 151–160.

LEIFSON, E. 1954 The flagellation and taxonomy of species of *Acetobacter. Antonie van Leeuwenhoek* **20,** 102–110.

MCINTOSH, A. F. 1962 A serological examination of some acetic acid bacteria. *Antonie van Leeuwenhoek* **28,** 49–62.

PASSMORE, S. M. 1973 The acetic acid bacteria: ecology, taxonomy and morphology. Ph.D. Thesis. University of Bristol.

PASSMORE, S. M. & CARR, J. G. 1975 The ecology of acetic acid bacteria with particular reference to cider manufacture. *Journal of Applied Bacteriology* **38,** 151–158.

SCOPES, A. W. 1962 The infrared spectra of some acetic acid bacteria. *Journal of General Microbiology* **28,** 69–79.

SHIMWELL, J. L. 1957 The mechanism of loss of starch production by *Acetobacter pasteurianum. Antonie van Leeuwenhoek* **23,** 235–240.

SHIMWELL, J. L. 1959 Flagella-staining acetic acid bacteria. *Journal of the Institute of Brewing* **65,** 340–341.

SHIMWELL, J. L. & CARR, J. G. 1959 The genus *Acetomonas. Antonie van Leeuwenhoek* **25,** 353–368.

SHIMWELL, J. L., CARR, J. G. & RHODES, M. E. 1960 Differentiation of *Acetomonas* and *Pseudomonas. Journal of General Microbiology* **23,** 283–286.

SNEATH, P. H. A. 1957 Some thoughts on bacterial classification. *Journal of General Microbiology* **17,** 184–200.

WIAME, J. M., HARPIGNY, R. & DOTHEY, R. G. 1959 A new type of acetobacter: *Acetobacter acidophilum* prov. sp. *Journal of General Microbiology* **20,** 165–172.

The Identification and Classification of *Rhizobium*

J. M. Vincent

Department of Microbiology, The University of Sydney,
New South Wales, Australia

P. S. Nutman and F. A. Skinner

Soil Microbiology Department, Rothamsted Experimental Station,
Harpenden, Hertfordshire, UK

The need to identify strains of rhizobia that nodulate legumes, especially those important to agriculture, has led to the grouping of strains into several species based on their ability to nodulate particular hosts. The methods available for classifying nodule bacteria in this way are outlined below. However, there are reasons for thinking that this useful grouping is unsound taxonomically and does not indicate the phylogenetic relationships of *Rhizobium*.

The Genus *Rhizobium*

Rhizobium is a genus of aerobic, heterotrophic non-sporeforming soil bacteria able to invade the roots of leguminous plants and form nodules in which atmospheric nitrogen may be fixed. This ability to form root nodules is the most important characteristic distinguishing *Rhizobium* from other bacterial genera, but the generic name can be applied to a non-invasive culture where there is other clear evidence of clonal descent or taxonomic relatedness. Exceptionally, a *Rhizobium* able to nodulate cowpea has been found nodulating the non-legume, *Trema* (*Parasponia*) *aspera* (Trinick 1973).

Characteristics in culture

Morphology

Rhizobium cells are small- to medium-sized (0·5–0·9 × 1·2–3·0 μm), Gram negative rods, occurring singly or in pairs and generally motile when young by polar, subpolar or peritrichous flagella. Young cells stain evenly with simple basic stains, but older cells contain characteristic granules of polymerized β-hydroxybutyrate (PHB) which appear as highly refractile bodies by bright field or phase contrast illumination, and remain as clear areas after the usual bacteriological staining procedures; these granules stain with Sudan Black (Burdon 1946; Schlegel 1962). PHB granules often occupy a large part of the cell and cause a banded appearance. Most strains produce abundant extra-cellular polysaccharide slime that ranges from a watery consistency to a highly tenacious gum (see section 'Fast-growing and slow-growing strains'). The composition of this exopolysaccharide varies to some extent between strains, but generally more so between species. Some strains develop a capsule-like structure (Dudman 1968) which appears to be unrelated chemically to the gum itself and could represent accumulation of cell wall antigen (Humphrey & Vincent 1969). The fine structures of wall, cytoplasm and nucleoid are typical of Gram negative bacteria (see Vincent 1974).

Types of colony

Colour of colonies ranges from almost colourless to white or cream; those of strains from *Lotononis bainseii* are pink (Norris 1958). There have been other, unsubstantiated, reports of pigmented non-invasive forms. An 'eaten', collapsed colony may be encountered; this has the appearance of phage infection but can be due to autolysis (Barnet & Vincent 1969).

Growth conditions

Rhizobia are aerobes but can grow at very low oxygen tensions (less than 0·01 atm O_2); there are no strict anaerobes. The optimal temperature range is *ca.* 25–30°C, and exposure to 40°C causes a rapid loss of viability (Bowen & Kennedy 1959); 70°C for 10 min is lethal. The rhizobia are very sensitive to drying unless protected by certain additives such as maltose or gum arabic (Fred *et al.* 1932; Vincent 1958; Vincent *et al.* 1962; Amarger *et al.* 1972).

The best growth occurs on complex media, especially those containing yeast extract, but simpler media support growth. Inorganic combined nitrogen (ammonium or nitrate) commonly suffices for growth, although care should be taken to ensure that the medium is sufficiently buffered to prevent an inhibitory drop in pH when NH_4 is utilized. Nitrates but usually not nitrites, are reduced in a dissimilatory manner (Jordan & San Clemente 1955). Some strains require growth factors, usually one or more of biotin, thiamine and pantothenic acid (Wilson 1940; Graham

1963*a*). Amino acids such as glutamic acid, may be needed for optimal growth (Jordan 1956; Bereśniewicz 1959; Bergersen 1961); urea and biuret may be used as alternative sources of organic nitrogen (Jensen & Schrøder 1965). Rhizobia are weakly proteolytic and growth on peptone agar is poor. A wide range of carbon sources is used by rhizobia (Vincent 1974), mannitol often being the preferred compound in media (see also section 'Fast- and slow-growing strains').

Major inorganic elements are usually added even to complex media; requirements for iron, magnesium, calcium, cobalt, zinc, manganese and potassium have to some extent been defined (Vincent 1974).

It is now clear that the provision of a suitable medium and an appropriate low pO_2 enables many strains in pure culture (particularly slow growers) to develop low levels of nitrogenase activity (McComb *et al.* 1975; Kurz and La Rue 1975; Pagan *et al.* 1975) and fix nitrogen (Pagan *et al.* 1975).

Inhibitory agents

Rhizobia differ widely in their resistance to antibiotics (Vintika & Vintikova 1958; Davis 1962; Graham 1963*c*; Golebiowska & Kaszubiak 1965; Škrdleta 1965; Pattison & Skinner 1974). High concentrations of amino acids can also be toxic, causing loss of viability and distortion of the cells (Skinner *et al.* 1977); selection for resistance to these, as well as to some antibiotics, may be associated with loss of symbiotic capacity (Wilson 1940; Schwinghamer 1964; Hamdi 1969*b*; Strijdom & Allen 1966, 1969; Zelazna-Kowalska 1971).

Fast-growing and slow-growing strains

The rhizobia fall into two distinctive groups which can be distinguished by relative speed of growth and change of pH on yeast extract-mannitol agar (YMA). The 'fast growers' have a mean generation time (m.g.t.) under the most favourable conditions, of *ca.* 2–4 h and produce detectable colonies on YMA (see section 'Media') in 2–3 days, and large, gummy colonies up to 5 mm diam., or more, by 5 days at 25°C. The m.g.t. of the slow growers is more likely to be 6–8 h and colonies seldom exceed 1–2 mm diam. after 10 days. Fast growers produce enough acid to be detected with bromothymol blue whereas the slow growers cause little lowering of pH or even promote a shift to the alkaline side. This distinction depends, however, on the use of YMA and the provision of adequate aeration. Some slow growers produce an acid reaction if cultivated in closed screwcapped bottles. Non-gummy mutants of fast-growing species may produce small colonies, but these are detectable after a short incubation period and in liquid culture their m.g.t. is not very different from the parent gummy form.

The fast-growing rhizobia give clear gel-diffusion reactions with antisera (containing antibodies to 'internal' antigens) to *R. trifolii* and *R. meliloti*; the slow growers with antisera to *R. japonicum* and typical cowpea rhizobia. There is little, if any, cross-reaction between the two groups. Ability to utilize sucrose as sole carbon source is general amongst the fast growers and unusual amongst the slow. They differ in their tolerance of pH on both sides of 7·0. Rhizobia show the following order of increasing tolerance of acid, and decreasing tolerance of alkaline, conditions: *R. meliloti*; *R. leguminosarum*, *R. trifolii*, *R. phaseoli*; slow growers (Graham & Parker 1964).

Cell changes within the legume host

The bacteria usually undergo morphological and biochemical changes within the cells of the nodule to form 'bacteroids'. Bacteroids are enclosed within membranes of host plant origin, singly or in groups according to the host (Dart & Mercer 1966), and during their formation lose some of the fine-structural features of the cultured cell (Mosse 1964; Dart & Mercer 1966; Goodchild & Bergersen 1966). The bacteroids of fast-growing rhizobia show considerable branching and enlargement; those of the slow growers are less altered. Fully developed bacteroids do not produce colonies on media able to support the growth of cultured rhizobia when handled in the usual way (Fred *et al.* 1932; Almon 1933). However, Gresshoff *et al.* (1977) have been able to secure growth from most isolated bacteroids when care was taken to avoid rupture of the osmotically fragile cells. In nodules that fix nitrogen the bacteroid zone is pink because of leghaemoglobin that accumulates between the bacterial cell and its enclosing membrane or in the plant cell more generally.

Identification of Rhizobia

The decision as to whether a culture is a member of the genus *Rhizobium* will depend primarily on its nodulating ability; other evidence can contribute towards a presumptive decision. Conformable characteristics and important contra-indications are given in Table 1. The history of the culture in question will affect the weight given to the various characteristics.

Abnormal morphology or staining characteristics of cells grown on the usual YMA, or abnormal cultural characteristics will indicate a non-rhizobium or a mixed culture. Cultural tests, although helpful in

Test	Conformable characteristics	Contra-indications
A. Morphology and staining	Short-medium rods (1–3 μm commonly) motile when young, often with prominent granules of polymer of β-hydroxybutyrate. Endospores absent. Gram negative	Cocci, very large rods, long chains; endospores; Gram positive
B. Cultural		
1. Growth on yeast-mannitol agar 25°–28°C	Little if any detectable growth in 24 h	Marked growth 1–2 days
	Moderate to abundant growth in 3–5 days, colourless or white, generally moderate to abundant gum ([a]*Rhizobium trifolii*, *R. leguminosarum*, *R. phaseoli*, *R. meliloti*; some rhizobia from *Lotus*, *Leucaena* and other diverse hosts) OR Little growth after 5 days; slight to moderate growth in 10 days, colourless, white or rarely pink, slight gum production (*R. lupini*, *R. japonicum*, some *Lotus* rhizobia and most cowpea rhizobia; *Rhizobium* of *Lotononis bainesii*, characteristically pink)	Colour other than white (except pink for *Lotononis* strains)
2. Growth on congo red-yeast-mannitol agar at 26°–28°C	Colonies of *Rhizobium* usually absorb very little of the dye and so remain practically colourless or only slightly pink (*R. meliloti*)	Marked absorption of the dye
3. Growth and change of pH on peptone-glucose agar, 2 days at 30°C	*Rhizobium* grows poorly on this medium and causes little change of pH	Abundant growth and marked pH change
4. Changes in litmus milk at 26°C	Slow or no change towards acid or alkaline reactions; clear 'serum' zone sometimes formed	Rapid growth and change
C. Invasiveness	Nodulation of a legume under bacteriologically controlled conditions. See Table 2 for species recognition	
D. Serology	Agglutination (flagellar and/or somatic) and development of gel-diffusion precipitation lines with *Rhizobium* antisera at 1/200 dilution is strong presumptive evidence of *Rhizobium*. Note, however, cross-reactions between *Agrobacterium* and *R. meliloti*	
E. Bacteriophage sensitivity	Similar specificities and limitations as (D)	

[a] For definition of *Rhizobium* species see Table 2.

distinguishing *Rhizobium* from most other genera, are not fully diagnostic. Useful media are given below. A culture isolated from a nodule, or with a presumed specified nodulating capacity, can be confirmed by a plant test; otherwise a more comprehensive programme may be needed to cover specific host requirements (Table 2). Good presumptive evidence can be provided by the development of internal antigen lines against antisera to representative fast and slow growing rhizobia, such as *R. trifolii* and/or *R. meliloti* and *R. japonicum* (Vincent & Humphrey 1970; Vincent *et al.* 1973).

A culture being investigated must be pure and authentic. It is not uncommon for a slow-growing rhizobium to be masked by a faster growing contaminant which, on subculture, tends to become dominant. Nodulation may be attributed to such an associated organism when in fact it is due to a slower growing, less obvious, rhizobium. Mixed cultures of rhizobia may even be recovered from the nodule itself (van Rensburg & Strijdom 1971).

Media

Yeast extract-mannitol agar (*YMA*). This is based on the medium of Fred *et al.* (1932) and is used for the routine cultivation of rhizobia. It consists of: mannitol, 10 g; K_2HPO_4, 0·5 g; $MgSO_4.7H_2O$, 0·2 g; NaCl, 0·1 g; $CaCO_3$, 4 g; aqueous extract of yeast (yeast water), 100 ml, or yeast extract (e.g. Difco, Oxoid), 0·4 g; agar, 15 g; distilled water to 1 l. The pH is adjusted to 6·8–7·0 and the medium autoclaved at 121°C for 15 min. Calcium carbonate is usually omitted from medium used for making plates or liquid medium for measuring growth turbidimetrically. The yeast water is prepared as follows: 100 g of bakers' compressed yeast is mixed with 1 l of cold water, allowed to stand at room temperature for 1–2 h and steamed for 40–60 min. The clear, supernatant fluid after settling or centrifuging is autoclaved at 121°C for 15 min and used as yeast water. Yeast extract concentrations of $> 1\%$ in media can cause cell distortion and decreased viability (Jordan & Coulter 1965; Skinner *et al.* 1977) and should not be used.

YMA with bromothymol blue. YMA may be made more useful for distinguishing fast- (acid-producing) from slow- (non-acid-producing) growing rhizobia by incorporating bromothymol blue at the rate of 5 ml of a 0·5% alcoholic solution per litre.

YMA with congo red. The detection of non-rhizobia may be facilitated by adding congo red (10 ml of sterile 1/400 aqueous solution added aseptically to each litre of melted YMA just before tubing or use to give a final concentration of 25 mg l^{-1}) (Hahn 1966). Most rhizobia absorb the dye only weakly whereas many other bacteria including agrobacteria, take it up strongly. It should be noted however that some strains of rhizobia, particularly *R. meliloti*, also become strongly coloured with the dye.

Glucose-peptone agar. A suitable formulation is: glucose, 5 g; peptone, 10 g; agar, 15 g; water, 1 l. This should be melted at 115°C for 5 min, bromocresol purple (10 ml of a 1% solution in ethanol) added, and sterilized at 115°C for 10 min. This medium is useful for detecting the presence of many non-rhizobia which grow abundantly in 24–48 h at 30°C, commonly with a marked change in pH. Some rhizobial strains show some growth but little change of pH under these conditions.

Litmus milk. Rhizobia produce slow changes in litmus milk, mainly towards slight alkalinity, though a few strains of *R. meliloti* give a slightly acid reaction. Many strains slowly digest the top layer of milk to form a 'serum' zone. Litmus milk is also useful for detecting non-rhizobia, by reason of marked or rapid change of pH or condition of the medium.

Rhizobium *and* Agrobacterium

Rhizobium is most likely to be confused with *Agrobacterium* to which it is closely related. Typical of *Agrobacterium*, and different from most rhizobia except many *R. meliloti*, are: growth on glucose-peptone agar, tolerance to 2% NaCl, production of H_2S, formation of precipitate with calcium glycerophosphate (Graham & Parker 1964), strong absorption of Congo red dye, and reduction of the dye Nile blue (Hamdi 1969*a*; Skinner 1977). Tolerance of low pH (4·5), failure to nodulate *Medicago sativa* (lucerne) and sparse sharing of internal antigens will generally distinguish *Agrobacterium* spp. from *R. meliloti*. The production of 3-ketolactose from lactose also distinguishes *A. radiobacter* from those rhizobia so far tested (Bernaerts & De Ley 1963; Clark 1969; Gaur *et al.* 1973).

Non-infective strains of *Rhizobium* that have lost the power to invade leguminous roots may be very difficult to distinguish from *Agrobacterium*. This difficulty does not arise with mutations of characterized strains if antisera are available for typing.

The Species of *Rhizobium*

The current edition of Bergey's Manual (Buchanan & Gibbons 1974) lists six species of *Rhizobium*, all defined in terms of the relatively restricted legume genera they are able to nodulate. Isolates like those from *Medicago sativa* (lucerne) nodules, which induce nodule formation on most species of *Medicago* and *Melilotus* but not, for example, on *Lupinus* spp. (lupins) or *Phaseolus* spp. (beans), are said to display cross-inoculation group specificity. Some hosts are very specific in their rhizobial requirements e.g. *Leucaena* and *Lotononis* spp. Other hosts, such as *Macroptilium atropurpureum* (siratro) nodulate with a wide range of strains.

In addition to the six species there are many other rhizobia, not yet given defined specific status, that are responsible for the nodulation of a great many legumes.

Species allocation based on nodulation specificity

Table 2 summarizes the nodulation patterns characteristic of the six named species as well as other rhizobia (fast- and slow-growing) not yet given specific status.

Reactions recorded as $\pm$ or $\underline{\underline{\pm}}$ represent, respectively, less frequent or rare cases of nodulation outside the usual range, such as *R. leguminosarum* with *Trifolium* sp. and *R. trifolii* with *Pisum* sp. (Kleczkowska *et al.* 1944). With these, however, nodulation is likely to be sparse and ineffective in nitrogen fixation. Nodulation within the specific group may also be ineffective but occurs readily and abundantly. Recognition then of a 'preferred' host will generally permit an appropriate pragmatic allocation of rhizobial strain to species or group. As a further precaution one needs to take account of possible restrictive host genotypes at the specific or varietal level. (Footnotes to Table 2 draw attention to such cases.)

Other distinctions within the groups of fast-growing and slow-growing rhizobia

The most useful cultural and serological characteristics for the four named fast-growing species and for agrobacteria are summarized in Table 4. *Rhizobium leguminosarum*, *R. trifolii* and *R. phaseoli* can be distinguished from *R. meliloti* and from the agrobacteria, but cannot be reliably separated from each other by these tests (see section 'Modern Views on Classification'). Proper definition of fast-growing rhizobia responsible for the nodulation of such hosts as *Lotus*, *Lupinus densiflorus* (Abdel-Ghaffar & Jensen 1966), *Leucaena* and *Cicer* requires more study and does not lend itself to tabulation.

There appears to be no reliable cultural or serological test to distinguish between *R. japonicum*, *R. lupini* and many slow-growing strains nodulating diverse leguminous hosts. However, rhizobia nodulating *Lotononis bainesii* have distinctive pink colonies (Norris 1958).

TABLE 2. Species and group allocation of *Rhizobium* based on nodulation specificity

Test host	Fast-growing						Slow-growing			
					Isolates from				Isolates from	
	R. leguminosarum	*R. phaseoli*	*R. trifolii*[c]	*R. meliloti*	*Lotus*	*Leucaena*	*R. japonicum*	*R. lupini*	*Lotus*	Others
Vicia sativa L.,[a] *V. hirsuta* L., S. F. Gray or *Pisum sativum* L.	+	±	±	—	—	—	—	—	—	—
Phaseolus vulgaris L.	$\underline{\pm}$	+	$\underline{\pm}$	—	—	—	±	$\underline{\pm}$	—	—
Trifolium repens L.[a]	±	$\underline{\pm}$	+	—	—	—	—	—	—	—
Medicago sativa L.[a, f]	—	—	—	+	—	$\underline{\pm}$	—	—	—	—
Lotus corniculatus L.[a]	—	—	—	—	+	—	—	—	+	—
L. uliginosus Schkuhr.[a]	—	—	—	—	+	*	—	±	+	±
Leucaena leucocephala (Lam.) De Wit	—	—	—	—	*	+	—	—	—	—
Ornithopus sativus Brot.[a]	—	—	—	—	—	—	—	+	+	±
Macroptilium atropurpureum D.C. Urb.[a, d]	—	±	—	—	*	±	±	+	+	+
Glycine max Merr.[b]	—	—	—	—	—	—	+	$\underline{\pm}$	*	±
Vinga unguiculata (L.) Walp.[e]	—	—	—	—	*	*	—	—	*	+

+: always nodulates; —: not shown to nodulate; ±: sometimes nodulates; $\underline{\pm}$: very rarely nodulates; *: information not available.

[a] Small-seeded species that will nodulate in test tube culture.

[b] *Glycine ussuriensis* can be used as a small-seeded substitute (Brockwell *et al.* 1975); non-nodulating or strain specific lines need to be avoided.

[c] Note, however, low compatibility amongst isolates from Mediterranean species of *Trifolium*, *T. ambiguum* and African clovers in respect of their non-homologous host.

[d] *Macroptilium atropurpureum* is widely susceptible to a miscellaneous collection of rhizobia, some of which are otherwise very host specific (e.g. rhizobia from *Lotononis bainesii* and from *Leucaena* spp.).

[e] *Vigna unguiculata* (cowpea) is nodulated by isolates from many genera and species of legumes and occasionally by isolates of other *Rhizobium* spp.

[f] Lucerne (*Medicago sativa*) appears to be always infectable by *R. meliloti*. Other species of *Medicago* are likely to be more strain specific.

Tests for nodule formation

The principle of testing is to grow sterile host plants in sterile nitrogen-free medium with and without the *Rhizobium* strain under test, and to observe whether nodules are formed in the presence of the bacteria. The test can also show whether or not nitrogen is fixed in the nodules. It is an advantage to use a small-seeded species, whenever possible, to enable the test to be performed under controlled bacteriological conditions. However, because of strain specificity within groups widely infectable test hosts should be used (Table 2). Better differentiation of nitrogen-fixing activity is likely to require the use of larger enclosed tubes or one of the methods of growing plants without restriction of shoot development, such as that of Gibson (1963), described by Vincent (1970).

Small-seeded plants (e.g. *Trifolium repens*) can be grown wholly enclosed within glass test tubes fitted with cotton wool plugs to protect them from contamination. Clean, uniform seed is rinsed with 95% ethanol and immersed for 3 min in 0·2% (w/v) $HgCl_2$ solution. The seed is then washed thoroughly with at least 10 changes of sterile water and planted directly or set to germinate on moist filter paper, or water agar (1·2%) in inverted petri dishes, at an appropriate temperature. Some legume seeds need temperature pretreatment before they will germinate (Crocker & Barton 1953; Mayer & Poljakoff-Mayher 1963). Hard-coated seeds can be surface-sterilized, after being dried over a desiccant, by immersion in dry concentrated sulphuric acid for 20–30 min. After draining off the excess acid the seeds are washed quickly and thoroughly with up to 10 changes of sterile water. This treatment also softens the seed coat and facilitates imbibition of water. Sterile seeds or germinated seedlings are transferred to slopes or deep layers of sterile seedling agar in suitably sized test tubes (150 × 20 mm for the smallest seeded host plants) fitted with cotton wool plugs. A suitable seedling agar (Jensen 1942) has the following composition: $CaHPO_4$, 1 g; K_2HPO_4, 0·2 g; $MgSO_4.7H_2O$, 0·2 g; NaCl, 0·2 g; $FeCl_3$, 0·1 g; distilled water, 1 l. Sufficient agar is added to give a rigid, but not too stiff, gel (8 or 15 g l^{-1} for deeps or slopes, respectively). The pH is adjusted to 6·5–7·0 and the medium sterilized at 121°C for 15 min. Fåhraeus's (1957) medium, which avoids any heavy precipitate, may be preferred: $CaCl_2.H_2O$, 0·1 g; $MgSO_4.7H_2O$, 0·12 g; KH_2PO_4, 0·1 g; $Na_2HPO_4.12H_2O$, 0·15 g; ferric citrate, 0·005 g; agar, 8–15 g; water, 1 l.

The tubes can be inoculated with bacteria either immediately or within seven days. It is convenient to arrange the tubes in racks so that the roots are shaded from direct light, and put them in a glasshouse with precautions against overheating in the sun, or in an illuminated growth cabinet. The optimal temperature for nodulation depends on the legume used, but broadly two ranges, 15–20°C and 25–30°C, are suitable for the legumes of cooler and warmer regions, respectively. Day length may need to be extended with supplementary artificial illumination according to the normal habitat and growing season of the host plant. Slopes require the periodic addition of sterile water but agar deeps can generally be left without any such addition for the period of the test.

Nodules commonly develop when the plants are 2–4 weeks old and the first true leaves are opening. Roots of some legumes, e.g. *Medicago* spp., free from rhizobia, sometimes form outgrowths that superficially resemble nodules. True root nodules can be distinguished by their structure from such pseudo-nodules (and tumours and galls produced by other bacteria and, in the field, by nematodes and insects). Nodules, except those of *Parasponia*, have a peripheral vascular system and an endodermis surrounding the infected region containing the bacteria and bacteroids; the latter stain readily with toluidine blue O (Roughley *et al.* 1976).

The seeds of larger seeded species such as *Phaseolus vulgaris*, *Lupinus luteus* and *Glycine max* can be surface-sterilized with $HgCl_2$ as described above, and sown in sterilized sand or sand and vermiculite in modified Leonard jars. The chance of contamination is reduced almost completely by using such an assembly which allows watering from below by capillary rise from a protected reservoir (Vincent 1970). Containers other than Leonard jars may also be watered through a plugged or capped tube passing through a protective layer of fine dry gravel or waxed sand overlying the sterile growth substrate; in this case, care should be exercised when watering to weight so that the water level does not rise to this layer. Plants are watered with sterile water or sterile nitrogen-free solution (e.g. the mineral salts of seedling agar at half or quarter strength), taking care, particularly with the larger legumes, that transpiration losses are adequately and regularly made good.

Strain Differences within Species

Strains of rhizobia differ in the several properties by which they are commonly characterized, notably cultural and serological properties and bacteriophage susceptibility, in their ability to form nodules on particular host species, in symbiotic nitrogen-fixing effectiveness and in other characteristics such as the number of nodules formed on the host plant. Such differences have been discussed in detail by Vincent (1974). None of these is itself an absolute basis for strain definition; groupings based solely on one property invariably break down when others are examined. Rhizobia are, to a variable degree, subject to mutational change in respect of these properties. However, when this is observed under controlled conditions, relationship of mutant substrain to the parent strain can be recognized and defined.

Serological differences

The serological reactions of *Rhizobium* strains can be studied most conveniently by using agglutination and precipitation (gel diffusion) reactions.

Agglutination may be due to heat-labile antigens located on the flagella or to heat-stable antigens on the surface of the cell itself (somatic antigens). Somatic antigens are more specific and are therefore more reliable for strain identification. Gel diffusion is a convenient alternative method for studying diffusible surface antigens (by using whole cells) and for internal antigens released from disrupted cells.

The two methods, agglutination and gel diffusion, do not give the same information; the choice of technique should depend on the kind of rhizobia under investigation and the questions to be answered. Familiarity with both methods is desirable to secure the most information and to avoid pitfalls in interpretation. Cross-agglutination between strains of *R. trifolii* is more likely than cross-precipitation with gel diffusion, possibly because the latter technique is less sensitive to lower antibody concentration and to the form of γ-globulin (IgM) likely to be associated with cross-reactive antigens (Humphrey & Vincent 1973). The gel diffusion test is therefore more convenient for strain identification within this species. Unlike *R. trifolii*, many strains of *R. meliloti* share a strong (lipopolysaccharide) antigen line, and strain recognition then depends on detecting the less conspicuous, diffuse line which coincides with the non-lipopolysaccharide surface antigen responsible for specific agglutination. Identification of *R. meliloti* strains is likely to be further complicated by loss of more specific antigens in laboratory-maintained cultures (Humphrey & Vincent 1975; Wilson *et al.* 1975).

Shared diffusion lines given by internal antigens from broken cells, though useful for revealing taxonomically significant groupings (see section 'Modern Views on Classification'), must be interpreted with caution because whole cells may leak as a result of freezing, drying or nutritional deficiency.

Bacteriophage susceptibility

Strains of *Rhizobium* differ in their susceptibility to bacteriophage

TABLE 3. Likely species allocation based on agglutination reactions[a]

Antiserum to	Fast-growing				Slow-growing			
	R. leguminosarum	*R. phaseoli*	*R. trifolii*	*R. meliloti*	*R. japonicum*	*R. lupini*	Cowpea rhizobia	Agrobacteria
R. leguminosarum	26	4	9	0	0	0	0	0
R. phaseoli	5	0	11	0	0	0	0	0
R. trifolii	8	12	31	0	0	0	0	0
R. meliloti	0	0	0	32	0	0	0	4
R. japonicum	0	0	0	0	52	8	3	0
R. lupini	0	0	0	0	12	17	5	0
Cowpea rhizobia	0	0	0	0	3	0	3	0
Agrobacteria	0	0	0	1–9	0	0	0	26–29

[a] Percentage of strains found to give agglutination (flagellar and/or somatic) with specified antisera, homologous reactions omitted (Graham 1963*b*).

(Oshima 1953; Kleczkowska 1957) and some are lysogenic (Marshall 1956; Takahashi & Quadling 1961; Schwinghamer & Reinhardt 1963; Kowalski 1966). The range of rhizobia susceptible to a particular phage may be narrowly restricted within a species, or it may cross wider taxonomic boundaries; lysis of rhizobia by bacteriophage of *Agrobacterium* has been reported (Roslycky *et al.* 1960). Differential susceptibility to particular bacteriophages may be used for strain recognition.

Modern Views on Classification

The classification of *Rhizobium* presents special problems because of its dependence on symbiotic properties. Initially only one species was recognized, *R. leguminosarum* (formerly called *Bacillus radicicola*), but as complex patterns of cross-infection were revealed, more species were proposed. The application of modern methods shows that the relationships between some of the species is closer than might be assumed from host-symbiont specificity. However, the existence of fast-growing rhizobia with peritrichous flagella and slow-growing strains with subpolar flagella indicates a fundamental lack of homogeneity within the genus.

The close relationship between *R. leguminosarum*, *R. trifolii* and *R. phaseoli*, all fast growers, is supported by cultural characteristics (see Table 4), numerical taxonomy (Graham 1964; 't Mannetje 1967), DNA hybridization studies (DeLey & Rassel 1965; Heberlein *et al.* 1967; Gibbins & Gregory 1972), analysis of internal antigens (Vincent & Humphrey 1970) and, to some extent, by esterase patterns (Murphy & Masterson 1970). It seems generally agreed that these three species have so much in common that they could be grouped to form one species, a new *R. leguminosarum*, as Graham (1964) suggested.

The same techniques show that *R. meliloti*, also a fast grower, does not have as close an affinity with the *leguminosarum* complex, but is related to *Agrobacterium*. This relationship, now well established, could lead to the merging of these groups to form a new single species of *Rhizobium*. However, there are sufficient tests (see Table 4), apart from the ability of *R. meliloti* to form root nodules, to separate this species from the agrobacteria, so it seems likely that *Agrobacterium* will remain as a separate taxon for a considerable time to come.

A third group of fast-growing rhizobia, which has not been given specific status, comprises isolates from species of *Lotus*, *Leucaena*, *Lupinus* and *Anthyllis*. Internal antigenic analysis reveals that these strains have affinities with both *R. meliloti* and the agrobacteria. As esterase analysis shows some affinity with *R. trifolii*, this group occupies

TABLE 4. Cultural and serological properties of species of fast-growing rhizobia and agrobacteria

Characteristic	*Rhizobium*				*Agrobacterium* spp.
	leguminosarum	*phaseoli*	*trifolii*	*meliloti*	
1. Tolerance of[a]					
pH 4·5	±	+	±	$\underline{\pm}$	+
pH 9·5	—	—	—	+	$\underline{\pm}$
2% NaCl	—	—	—	±	±
39°C	—	—	—	±	$\underline{\pm}$
2. Precipitate with Ca glycerophosphate[a]	—	—	—	±	+
3. Need for thiamin and/or pantothenate[a]	±	±	+	—	$\underline{\pm}$
4. Internal antigens[b] shared with *R. trifolii*	++	++	++	+	+
R. meliloti	+	+	+	+++	++
5. Reduction of Nile blue[c]	—	—	—	±	+
6. Formation of 3-Ketolactose[c]	—	—	—	±	±

[a] Adapted from Table 9.8 of Vincent (1974). +: $>$ 75%; ±: 25–75%; $\underline{\pm}$: 1–24%; —: no cases recorded.
[b] Table 9.9 (Vincent, 1974). ++: line development comparable with homologous cases; +: one or more lines, but $<$ homologous.
[c] Data from Skinner (1977).

an intermediate position. Some of these, unlike the better characterized fast-growing species, have a single subpolar flagellum.

The position of the slow growers, with subpolar flagella, is quite different. All the evidence points to their forming a separate group, antigenically distinct from all the fast growers, and with some similarities to the pseudomonads. Whether they should be regarded as forming a broad specific grouping within *Rhizobium* or a new genus, which might be outside the Eubacteriales altogether, is a matter for debate. Certainly the case for separate generic ranking for the slow growers appears at least as strong as that for *Agrobacterium*.

The pink colony isolates from *Lotononis bainesii* stand apart from all other rhizobia by their failure to react with antisera to internal antigens of fast or slow growers. Moreover, the *Lotononis* organism has a G + C content of 68–69% (Godfrey 1972), which is outside the range for all other rhizobia (59–66%).

Although the legume hosts for the fast growers and slow growers are usually distinct there are some anomalies. For example, fast-growing *R. phaseoli*, which nodulates *Phaseolus vulgaris*, occasionally invades other hosts in the 'cowpea miscellany', such as *Macroptilium lathyroides* or *M. atropurpureum*. Also, some slow-growing 'cowpea rhizobia' occasionally nodulate *P. vulgaris*.

The *Lotus* rhizobia divide into fast- and slow-growing strains each of which commonly nodulates *Lotus corniculatus*. *Anthyllis vulneraria* shows a greater compatibility with fast-growing isolates (from *L. corniculatus* and a range of other hosts); the pattern with *Lupinus densiflorus* is one that also favours fast-growing forms (Jensen 1967; Vincent 1974).

The genus *Rhizobium* as constituted at present comprises two main groups, the fast and slow growers, probably of widely-differing ancestry, but alike in one remarkable feature, the ability to form organized, nitrogen-fixing nodules on the roots of leguminous plants. The separation of the slow growers to a new genus is likely in the future; for the present the Family Rhizobiaceae comprises the two genera, *Rhizobium* and *Agrobacterium*, some members of each displaying marked similarities to each other.

Most of the work on the legume-*Rhizobium* symbiosis has been done with agricultural legumes of the temperate regions, which are not representative of the Leguminosae as a whole. This family comprises over 14 000 species, most of them tropical or sub-tropical, and many of them trees. Only about 1,200 species have been examined for the presence of nodules and not all of these nodulate (Allen & Allen 1961). An extension of this work may well reveal new relationships between the groups of rhizobia and help to classify those fast- and slow-growing rhizobia that currently lack specific status.

References

ABDEL-GHAFFAR, A. S. & JENSEN, H. L. 1966 The rhizobia of *Lupinus densiflorus* Benth., with some remarks on the classification of root nodule bacteria. *Archiv für Mikrobiologie* **54,** 393–405.

ALLEN, E. K. & ALLEN, O. N. 1961 Nitrogen fixation. The scope of nodulation in the Leguminosae. *Recent Advances in Botany* (IX *International Botanical Congress, Montreal* 1959), pp. 585–600.

ALMON, L. 1933 Concerning the reproduction of bacteroids. *Zentralblatt für Bakteriologie, Parasitenkunde, Infektionskrankheiten und Hygiene, Abt. II* **87** 289–297.

AMARGER, N., JACQUEMETTON, M. & BLOND, G. 1972 Influence de l'âge de la culture sur la survie de *Rhizobium meliloti* à la lyophilisation et à la conservation aprés lyophilisation. *Archiv für Mikrobiologie* **81,** 361–366.

BARNET, Y. M. & VINCENT, J. M. 1969 'Self-lytic' strains of *Rhizobium trifolii. Australian Journal of Science* **32,** 208.

BEREŚNIEWIČZ, K. 1959 Studies on the acceleration of the development of the slow growing strains of *Rhizobium. Acta microbiologica polonica* **8,** 333–337.

BERGERSEN, F. J. 1961 The growth of *Rhizobium* in synthetic media. *Australian Journal of Biological Sciences* **14,** 349–360.

BERNAERTS, M. J. & DE LEY, J. 1963 A biochemical test for crown gall bacteria. *Nature, London* **197,** 406–407.

BOWEN, G. D. & KENNEDY, M. M. 1959 Effect of high soil temperatures on *Rhizobium* spp. *Queensland Journal of Agricultural Science* **16,** 177–197.

BROCKWELL, J., DIATLOFF, A., GRASSIA, A. & ROBINSON, A. C. 1975 Use of wild soybean (*Glycine ussuriensis* Regel and Meack) as a test plant in dilution-nodulation frequency tests for counting *Rhizobium japonicum. Soil Biology and Biochemistry* **7,** 305.

BUCHANAN, R. E. & GIBBONS, N. E. (eds) 1974 *Bergey's Manual of Determinative Bacteriology*, 8th edn. Baltimore: Williams & Wilkins.

BURDON, K. L. 1946 Fatty material in bacteria and fungi revealed by staining dried fixed slide preparations. *Journal of Bacteriology* **52,** 665–678.

CLARK, A. G. 1969 A selective medium for the isolation of *Agrobacterium* species. *Journal of Applied Bacteriology* **32,** 348–351.

CROCKER, W. & BARTON, L. V. 1953 *Physiology of Seeds*. Waltham, Mass., U.S.A.: Chronica Botanica.

DART, P. J. & MERCER, F. V. 1966 Fine structure of bacteroids in root nodules of *Vigna sinensis, Acacia longifolia, Viminaria juncea* and *Lupinus angustifolius. Journal of Bacteriology* **91,** 1314–1319.

DAVIS, R. J. 1962 Resistance of rhizobia to antimicrobial agents. *Journal of Bacteriology* **84,** 187–188.

DE LEY, J. & RASSEL, A. 1965 DNA base composition, flagellation and taxonomy of the genus *Rhizobium. Journal of General Microbiology* **41,** 85–91.

DUDMAN, W. F. 1968 Capsulation in *Rhizobium* species. *Journal of Bacteriology* **95,** 1200–1201.

FÅHRAEUS, G. 1957 The infection of clover root hairs by nodule bacteria studied by a simple glass slide technique. *Journal of General Microbiology* **16,** 374–381.

FRED, E. B., BALDWIN, I. L. & McCOY, E. 1932 *Root Nodule Bacteria and Leguminous Plants.* Madison: University of Wisconsin.

GAUR, Y. D., SEN, A. N. & SUBBA RAO, N. S. 1973 Usefulness and limitation of ketolactose test to distinguish agrobacteria from rhizobia. *Current Science* **42,** 545–546.

GIBBINS, A. M. & GREGORY, K. F. 1972 Relatedness among *Rhizobium* and *Agrobacterium* species determined by three methods of nucleic acid hybridization. *Journal of Bacteriology* **111,** 129–141.

GIBSON, A. H. 1963 Physical environment and symbiotic fixation. I. The effect of root temperature on recently nodulated *Trifolium subterraneum* L. plants. *Australian Journal of Biological Sciences* **16,** 28–42.

GODFREY, C. A. 1972 The carotenoid pigment and deoxyribonucleic acid base ratio of a *Rhizobium* which nodulates *Lotononis bainesii* Baker. *Journal of General Microbiology* **72,** 399–402.

GOLEBIOWSKA, J. & KASZUBIAK, H. 1965 Sensitivity of *Rhizobium* to the action of thiuriam and phenylmercuric acetate. *Annales de l'Institut Pasteur, Paris, Supplement* **109,** 153–160.

GOODCHILD, D. J. & BERGERSEN, F. J. 1966 Electron microscopy of the infection and subsequent development of soybean nodule cells. *Journal of Bacteriology* **92,** 204–213.

GRAHAM, P. H. 1963*a* Vitamin requirements of root nodule bacteria. *Journal of General Microbiology* **30,** 245–248.

GRAHAM, P. H. 1963*b* Antigenic affinities of the root-nodule bacteria of legumes. *Antonie van Leeuwenhoek* **29,** 281–291.

GRAHAM, P. H. 1963*c* Antibiotic sensitivities of the root nodule bacteria. *Australian Journal of Biological Sciences* **16,** 557–559.

GRAHAM, P. H. 1964 The application of computer techniques to the taxonomy of the root nodule bacteria of legumes. *Journal of General Microbiology* **35,** 511–517.

GRAHAM, P. H. & PARKER, C. A. 1964 Diagnostic features in the characterisation of the root-nodule bacteria of legumes. *Plant and Soil* **20,** 383–396.

GRESSHOFF, P. M., SKOTNICKI, M. L., EADIE, J. F. & ROLFE, B. G. 1977 Viability of *Rhizobium trifolii* bacteroids from clover root nodules. *Plant Science Letters* **10,** 299–304.

HAHN, N. J. 1966 The congo red reaction in bacteria and its usefulness in the identification of rhizobia. *Canadian Journal of Microbiology* **12,** 725–733.

HAMDI, Y. A. 1969*a* Application of the oxidation-reduction potential indicators for characterization of *Agrobacterium* and *Rhizobium* species. *Folia Microbiologica* **14,** 92–93.

HAMDI, Y. A. 1969*b* Effect of D-, L-, and DL-methionine on growth and efficiency of *Rhizobium meliloti* strains. *Plant and Soil* **31,** 111–121.

HEBERLEIN, G. T., DE LEY, J. & TIJTGAT, R. 1967 Deoxyribonucleic acid homology and taxonomy of *Agrobacterium, Rhizobium* and *Chromobacterium. Journal of Bacteriology* **94,** 116–124.

HUMPHREY, B. & VINCENT, J. M. 1969 The somatic antigens of two strains of *Rhizobium trifolii. Journal of General Microbiology* **59,** 411–425.

HUMPHREY, B. & VINCENT, J. M. 1973 Differential antibody response to the somatic antigenic determinants of *Rhizobium trifolii. Microbios* **7,** 87–93.

HUMPHREY, B. A. & VINCENT, J. M. 1975 Specific and shared antigens in *Rhizobium meliloti. Microbios* **13,** 71–76.

JENSEN, H. L. 1942 Nitrogen fixation in leguminous plants. I. General characters of root-nodule bacteria isolated from species of *Medicago* and *Trifolium* in Australia. *Proceedings of the Linnean Society of New South Wales* **67,** 98–108.

JENSEN, H. L. 1967 Mutual host plant relationships in two groups of legume root nodule bacteria (*Rhizobium* spp.). *Archiv für Mikrobiologie* **59,** 174–179.

JENSEN, H. L. & SCHRØDER, M. 1965 Urea and biuret as nitrogen sources for *Rhizobium* spp. *Journal of Applied Bacteriology* **28,** 473–478.

JORDAN, D. C. 1956 Observations on the enzymatic degradation and conversion of certain L- and D-amino acids by an effective strain of *Rhizobium meliloti. Canadian Journal of Microbiology* **1,** 743–748.

JORDAN, D. C. & COULTER, W. H. 1965 On the cytology and synthetic capacities of natural and artificially produced bacteroids of *Rhizobium leguminosarum. Canadian Journal of Microbiology* **11,** 709–720.

JORDAN, D. C. & SAN CLEMENTE, C. L. 1955 The utilization of purines, pyrimidines and inorganic nitrogenous compounds by effective and ineffective strains of *Rhizobium meliloti. Canadian Journal of Microbiology* **1,** 668–674.

KLECZKOWSKA, J. 1957 A study of the distribution and the effects of bacteriophage of root nodule bacteria in the soil. *Canadian Journal of Microbiology* **3,** 171–180.

KLECZKOWSKA, J., NUTMAN, P. S. & BOND, G. 1944 Note on the ability of certain strains of rhizobia from peas and clover to infect each other's host plants. *Journal of Bacteriology* **48,** 673–675.

KOWALSKI, M. 1966 Lysogeny in *Rhizobium meliloti. Acta microbiologica polonica* **25,** 119.

KURZ, W. G. W. & LA RUE, T. A. 1975 Nitrogenase activity in rhizobia in absence of plant host. *Nature, London* **256,** 407–409.

MARSHALL, K. C. 1956 A lysogenic strain of *Rhizobium trifolii. Nature, London* **177,** 92.

MAYER, A. M. & POLJAKOFF-MAYHER, A. 1963 *The Germination of Seeds.* Oxford: Pergamon Press.

MCCOMB, J. A., ELLIOTT, J. & DILWORTH, M. J. 1975 Acetylene reduction by *Rhizobium* in pure culture. *Nature, London* **256,** 409–410.

MOSSE, B. 1964 Electron-microscope studies of nodule development in some clover species. *Journal of General Microbiology* **36,** 49–66.

MURPHY, P. M. & MASTERSON, C. L. 1970 Determination of multiple forms of esterases in *Rhizobium* by paper electrophoresis. *Journal of General Microbiology* **61,** 121–129.

NORRIS, D. O. 1958 A red strain of *Rhizobium* from *Lotononis bainesii* Baker. *Australian Journal of Agricultural Research* **9,** 629–632.

OSHIMA, H. 1953 Studies on the application of bacteriophage to the identification of the root nodule bacteria (Japanese). *Bulletin of the National Institute of Agricultural Sciences, Tokyo Series B.* **2,** 15–48.

PAGAN, J. D., CHILD, J. J., SCOWCROFT, W. R. & GIBSON, A. H. 1975 Nitrogen fixation by *Rhizobium* cultured on a defined medium. *Nature, London* **256,** 406–407.

PATTISON, A. C. & SKINNER, F. A. 1974 The effects of antimicrobial substances on *Rhizobium* spp. and their use in selective media. *Journal of Applied Bacteriology* **37,** 239–250.

ROSLYCKY, E. B., ALLEN, O. N. & MCCOY, E. 1960 Certain properties of bacteriophages of *Agrobacterium radiobacter*. *Bacteriological Proceedings* 28–29.

ROUGHLEY, R. J., DART, P. J. & DAY, J. M. 1976 The structure and development of *Trifolium subterraneum* L. root nodules. I. In plants grown at optimal root temperatures. *Journal of Experimental Botany* **27,** 431–440.

SCHLEGEL, H. G. 1962 Die Isolierung von Poly-8-hydroxybuttersäure aus Wurzelknöllchen von Leguminosen. *Flora, Jena* **152,** 236.

SCHWINGHAMER, E. A. 1964 Association between antibiotic resistance and ineffectiveness in mutant strains of *Rhizobium* spp. *Canadian Journal of Microbiology* **10,** 221–233.

SCHWINGHAMER, E. A. & REINHARDT, D. T. 1963 Lysogeny in *Rhizobium leguminosarum* and *R. trifolii*. *Australian Journal of Biological Sciences* **16,** 597–605.

SKINNER, F. A. 1977 An evaluation of the nile blue test for differentiating rhizobia from agrobacteria. *Journal of Applied Bacteriology* **43,** 91–98.

SKINNER, F. A., ROUGHLEY, R. J. & CHANDLER, M. R. 1977 Effect of yeast extract concentration on viability and cell distortion in *Rhizobium* spp. *Journal of Applied Bacteriology* **43,** 287–297.

ŠKRDLETA, V. 1965 The sensitiveness of *Rhizobium japonicum* to some antibiotics and sulphadimidine. *Vědecké Práce Űstředníko Výzkumného Űstavu Rostlinné Výroby v Praze-Ruzyni* **9,** 177–183.

STRIJDOM, B. W. & ALLEN, O. N. 1966 Medium supplementation with L- and D-amino acids relative to growth and efficiency of *Rhizobium meliloti*. *Canadian Journal of Microbiology* **12,** 275–283.

STRIJDOM, B. W. & ALLEN, O. N. 1969 Properties of strains of *Rhizobium trifolii* after cultivation on media supplemented with amino-acids. *Phytophylactica* **1,** 147–151.

TAKAHASHI, I. & QUADLING, C. 1961 Lysogeny in *Rhizobium trifolii*. *Canadian Journal of Microbiology* **7,** 455–465.

't MANNETJE, L. 1967 A re-examination of the taxonomy of the genus *Rhizobium* and related genera using numerical analysis. *Antonie van Leeuwenhoek* **33,** 477–491.

TRINICK, M. J. 1973 Symbiosis between *Rhizobium* and the non-legume, *Trema aspera*. *Nature, London* **244,** 459–460.

VAN RENSBURG, H. J. & STRIJDOM, B. W. 1971 Stability of infectivity in strains of *Rhizobium japonicum*. *Phytophylactica* **3,** 125–130.

VINCENT, J. M. 1958 Survival of the root-nodule bacteria. In *Nutrition of the Legumes*, ed. Hallsworth, E. G. London: Butterworths.

VINCENT, J. M. 1970 *A Manual for the Practical Study of the Root-nodule Bacteria*. IBP Handbook No. 15. Oxford: Blackwell Scientific Publications.

VINCENT, J. M. 1974 Root-nodule symbioses with *Rhizobium*. In *The Biology of Nitrogen Fixation*, ed. Quispel, A. North-Holland Research Monographs, Frontiers of Biology, Vol. 33. Amsterdam: North-Holland.

VINCENT, J. M. & HUMPHREY, B. 1970 Taxonomically significant group antigens in *Rhizobium*. *Journal of General Microbiology* **63,** 379–382.

VINCENT, J. M., HUMPHREY, B. & ŠKRDLETA, V. 1973 Group antigens in slow-growing rhizobia. *Archiv für Mikrobiologie* **89,** 79–82.

VINCENT, J. M., THOMPSON, J. A. & DONOVAN, K. O. 1962 Death of root-nodule bacteria on drying. *Australian Journal of Agricultural Research* **13,** 258–270.

VINTIKA, J. & VINTIKOVA, H. 1958 Effect of antibiotics on rhizobia. Za sotsialisticheskuyu sel'skokhozyaistvennuyu nauku, Prague **7,** 349–360.

WILSON, M. H. M., HUMPHREY, B. A. & VINCENT, J. M. 1975 Loss of agglutinating specificity in stock cultures of *Rhizobium meliloti*. *Archives of Microbiology* **103,** 151–154.

WILSON, P. W. 1940 *Biochemistry of Symbiotic Nitrogen Fixation*. Madison, U.S.A.: University of Wisconsin Press.

ZELAZNA-KOWALSKA, I. 1971 Correlation between streptomycin resistance and infectiveness in *Rhizobium trifolii*. *Plant and Soil*. (Special Volume) 67–71.

Techniques in the Identification and Classification of *Brucella* species

M. J. CORBEL, C. D. BRACEWELL, E. L. THOMAS AND K. P. W. GILL

Ministry of Agriculture, Fisheries and Food, Central Veterinary Laboratory, Weybridge, Surrey, UK

The genus *Brucella*, as originally defined by Meyer & Shaw (1920), comprised the two species *Brucella abortus* (Schmidt & Weis 1901) Meyer & Shaw 1920 and *B. melitensis* (Hughes 1893) Meyer & Shaw 1920. Subsequently the brucella organisms isolated from porcine infections were included as a third species, *B. suis* Huddleson 1929.

This list has been extended in recent years to include *B. ovis* Buddle 1956, *B. neotomae* Stoenner & Lackman 1957 and *B. canis* Carmichael & Bruner 1968.

Procedures for the identification and differentiation of the three 'classical' species, *B. abortus*, *B. melitensis* and *B. suis*, were developed several decades ago, largely as the result of work by Huddleson (1929) and Wilson & Miles (1932). This allowed the species to be distinguished on the basis of cultural characteristics, including CO_2 requirement, production of H_2S from sulphur-containing amino acids, relative sensitivity to azo dyes and antigenic structure revealed by agglutination reactions with monospecific antisera.

This scheme provided an effective means of differentiating isolates on a practical basis which correlated closely with the principal natural hosts of the three main species. Subsequently, however, modifications have had to be introduced to deal with isolates which differed in some respects from the specifications laid down for the original species. The introduction of new taxonomic methods, particularly phage typing and the measurement of oxidative metabolism of specific substrates, also led to a reconsideration of the criteria used for speciation within the genus. Details of these changes have been discussed in a number of reviews, including those by Huddleson (1943; 1961); Renoux (1958); Stableforth (1959); Biberstein & Cameron (1961); Morgan (1964; 1970); Morgan &

Gower (1966) and Meyer (1974). Consequent revisions of the recommendations for classification of the genus *Brucella* have been made by the Subcommittee on Taxonomy of the genus *Brucella* of the International Committee for Systematic Bacteriology (ICSB) (Stableforth & Jones 1963; Jones 1967; Jones & Wundt 1971; Wundt & Morgan 1975).

Recently, concern has been expressed about the difficulties involved in the identification of atypical non-smooth isolates, particularly at the genus level. This has led to delays in deciding the generic status of *Brucella* isolates, such as *B. canis* and *B. ovis* for which no corresponding smooth phase has been observed and has also occasionally resulted in the erroneous inclusion of superficially similar organisms in the genus (Renoux & Philippon 1969; Corbel 1973*a*). To obviate such difficulties attention has been directed towards the adoption of standard methods for identification at the genus level (Corbel & Morgan 1975; Morgan & Corbel 1976).

Other advances since the first edition of this chapter by Morgan & Gower (1966) have included the extension of the phage typing scheme by the introduction of phages lytic for all smooth *Brucella* spp. and some rough strains (Moreira-Jacob 1968; Morris & Corbel 1973; Morris *et al.* 1973; Corbel & Thomas 1976*a*, *b*; Douglas & Elberg 1976; Corbel 1977; Thomas & Corbel 1977). Methods for differentiating the classical species, *B. abortus*, *B. melitensis* and *B. suis*, into biotypes have also been established, largely through the efforts of the ICSB Subcommittee on Taxonomy of *Brucella*.

The procedures recommended for the performance of these techniques have been added to those described by Morgan & Gower (1966) in revising this chapter to accord with the current status of knowledge of the genus *Brucella*.

Definitions of Genus and Species

Specifications for the identification of cultures at genus, species and biotype level are incorporated in the proposal prepared on behalf of the ICSB Subcommittee on Taxonomy of *Brucella* for minimal standards for descriptions of new species and biotypes of the genus *Brucella* (Corbel & Morgan 1975). These are based on recommendations made by the Subcommittee over the past 15 years. The genus and the six species of *Brucella* currently recognized are defined as follows:

Brucella

Small, non-motile, non-sporing, Gram negative cocci, coccobacilli or short rods, 0·5–0·7 μm × 0·5–1·5 μm occurring singly, in pairs or short chains. Aerobic, no growth under strictly anaerobic conditions. Carboxyphilic. Metabolism is mainly oxidative. Usually show little fermentative action on carbohydrates in conventional media. Multiple amino acids, thiamin, biotin and nicotinamide are required. Growth of many strains is improved by calcium pantothenate and *meso*-erythritol. Haemin (X factor) and nicotinamide adenine dinucleotide (V factor) are not required. Catalase positive. Usually oxidase positive. Reduce nitrates. Produce H_2S and hydrolyse urea to a variable extent. Do not produce indole. Do not liquefy gelatin or lyse blood. Do not produce acetylmethylcarbinol (Voges-Proskauer test). Do not utilize citrate. Give a negative methyl red reaction. Do not release *o*-nitrophenol from *o*-nitrophenol-β-D-galactoside. Do not change litmus milk or may render it alkaline. Optimum growth temperature 37°C, range 20–40°C. Optimum pH for growth between pH 6·6 and 7·4. Contain cytochrome c. The G + C content of DNA is 55–58 mol %. Show > 90% homology in DNA hybridization tests. Possess characteristic internal antigens. Intracellular parasites producing characteristic infections in animals, transmissible to man.

Brucella abortus

Catalase and oxidase positive. Usually require supplementary CO_2 for growth, especially on primary isolation. Usually produce H_2S from sulphur-containing amino acids or proteins. Usually hydrolyse urea but some strains may not. Usually grow in the presence of basic fuchsin, methyl violet, pyronin and safranin O but not thionin, at standard concentrations. Reduce nitrates to nitrites and may also reduce nitrites. Smooth strains may have A, M or A and M surface antigens reactive in tests with monospecific antisera, depending upon biotype. Oxidize L-alanine, D-alanine, L-asparagine, L-glutamic acid, D-galactose, D-glucose, D-ribose and *meso*-erythritol. Do not oxidize D-xylose, L-arginine, DL-citrulline, DL-ornithine or L-lysine. Cultures in the smooth or smooth-intermediate phase are lysed by brucella-phages of the Tbilisi (Tb), Weybridge (Wb), Firenze (Fz), M51-S708, Berkeley (Bk), MC/75 and D groups at routine test dilution (RTD) and 10^4 RTD. Non-smooth cultures are lysed by brucella-phage R at RTD. Usually pathogenic for cattle, causing abortion, but may also infect other species including horses and man.

Nine biotypes are recognized. The reference/neotype strains are: *B. abortus* 544 (biotype 1), 86/8/59 (biotype 2), Tulya (biotype 3), 292 (biotype 4), B3196 (biotype 5), 870 (biotype 6), 63/75 (biotype 7),

C68 (biotype 9). No reference strain exists for *B. abortus* biotype 8.

Brucella melitensis

Catalase and oxidase positive. Do not require supplementary CO_2 for growth. Do not produce H_2S, or no more than a trace, when grown on recommended media. Usually hydrolyse urea but some strains may not. Usually grow in the presence of basic fuchsin, thionin, methyl violet, pyronin and thionin blue at the standard concentrations. Reduce nitrates to nitrites and may also reduce nitrites. Smooth cultures may have the A, M or both A and M surface antigens reactive in tests with monospecific sera. Oxidize D-glucose, *meso*-erythritol, L-alanine, D-alanine, L-asparagine and L-glutamic acid. Do not oxidize L-arabinose, D-galactose, D-ribose, D-xylose, L-arginine, DL-citrulline, DL-ornithine or L-lysine. Cultures in the smooth phase are usually susceptible to lysis by Bk phage at RTD and 10^4 RTD, but are not lysed by brucella-phages of the Tb, Wb, Fz, M51-S708, MC/75, D or R groups at RTD or 10^4 RTD. Non-smooth cultures are resistant to lysis by all of these phages. Usually pathogenic for sheep and goats but may also infect other species, including man.

Reference strains are: *B. melitensis* 16M (biotype 1), 63/9 (biotype 2), Ether (biotype 3).

Brucella suis

Catalase and oxidase positive. Supplementary CO_2 is not required for growth. Produce large amounts of H_2S (biotype 1) or none at all (other biotypes). Usually hydrolyse urea rapidly. Usually grow in the presence of thionin but most strains are inhibited by basic fuchsin, methyl violet and pyronin, and safranin O at standard concentrations. Smooth cultures have the A surface antigen predominant except for those of biotype 4 which react equally with antisera monospecific for the A and M surface antigens. Usually oxidize D-ribose, D-glucose, *meso*-erythritol, D-xylose, L-arginine, DL-citrulline, DL-ornithine and L-lysine. Oxidation of L-asparagine, L-glutamic acid, L-arabinose and D-galactose varies with the biotype. Do not oxidize L-alanine or D-alanine. Smooth or smooth-intermediate cultures are lysed by brucella-phages of the Wb, M51-S708, Bk, MC/75 and D groups at RTD and 10^4 RTD. Phages of the Fz group produce partial lysis at RTD, those of the Tb group do not produce lysis at RTD but are lytic at 10^4 RTD. Non-smooth cultures are not lysed by any of these phages or by phage R at RTD or 10^4 RTD.

Usually pathogenic for pigs, except biotype 4 which is usually pathogenic for reindeer. Other species may be infected however, including hares and man.

Four biotypes are recognized. The reference strains are: *B. suis* 1330 (biotype 1), Thomsen (biotype 2), 686 (biotype 3), 40 (biotype 4).

Brucella neotomae

Catalase positive, oxidase negative. Do not require supplementary CO_2 for growth. Usually produce H_2S in large amounts and hydrolyse urea rapidly. Do not grow in the presence of basic fuchsin even at 1/150 000, nor in the presence of safranin O at 1/10 000 or thionin blue at 1/500 000, but will grow in the presence of thionin at 1/150 000. Reduce nitrates to nitrites. In smooth cultures the A surface antigen is predominant. Oxidize L-asparagine, L-glutamic acid, L-arabinose, D-galactose, D-glucose, *meso*-erythritol and D-xylose. Do not oxidize L-alanine, D-alanine, L-arginine, DL-citrulline, DL-ornithine or L-lysine. Oxidation of D-ribose may be variable. In peptone media, ferment D-glucose, D-galactose, L-arabinose and D-xylose. Smooth or smooth-intermediate cultures are lysed by brucella-phages of the Wb, M51-S708, Fz, Bk, MC/75 and D groups at RTD. Phages of the Tb group produce partial lysis with few very small plaques at RTD, complete lysis at 10^4 RTD. Rough or mucoid cultures are not lysed by these phages or brucella-phage R, at RTD or 10^4 RTD. Pathogenic for the desert wood rat (*Neotoma lepida* Thomas). Authenticated natural infections of other species have not been reported. No biotypes are recognized.

Reference/neotype strain: *B. neotomae* 5K33.

Brucella ovis

Catalase positive, oxidase negative. Require supplementary CO_2 for growth. Does not produce H_2S or hydrolyse urea. Inhibited by methyl violet but grows in the presence of basic fuchsin and thionin at standard concentrations. Does not produce nitrite from nitrate. A true smooth phase does not occur, cultures are always in the rough phase on primary isolation. Rough (R)—specific surface antigens cross-reacting with other non-smooth brucellae are predominant. Oxidize L-alanine, D-alanine, L-asparagine, D-asparagine, L-glutamic acid, DL-serine and adonitol. Do not oxidize L-arabinose, D-galactose, D-glucose, D-ribose, *meso*-erythritol, D-xylose, L-arginine, DL-citrulline, DL-ornithine or L-lysine. The cultures are not lysed by brucella-phages of the Tb, Wb, M51-S708, Fz, Bk, MC/75, D or R groups at RTD or 10^4 RTD. Pathogenic for sheep, causing epididymitis in rams and sometimes abortion in ewes. Natural infections in other species have not been reported. No biotypes are recognized.

Reference/neotype strain: *B. ovis* 63/290.

Brucella canis

Catalase positive, oxidase positive. Do not require supplementary CO_2 for growth. Do not produce H_2S. Hydrolyse urea very rapidly. Reduce nitrates to nitrites. Growth is inhibited by basic fuchsin but not by thionin at the standard concentrations. A smooth phase is not known, cultures are always in the rough or mucoid phase on primary isolation. Form a mucoid sediment in liquid media. Rough-(R)-specific surface antigens cross-reacting with other species of *Brucella* are predominant. Oxidize D-ribose, D-glucose, L-arginine, DL-citrulline, DL-ornithine, L-lysine. Oxidation of *meso*-erythritol is variable. Do not oxidize L-alanine, D-alanine, L-asparagine, L-glutamic acid, L-arabinose, D-galactose. They are not lysed by brucella-phages of the Tb, M51-S708, Wb, Fz, Bk, Mc/75, D or R groups at RTD or 10^4 RTD. Pathogenic for dogs, causing abortion in pregnant females and epididymo-orchitis in males. Occasionally transmitted to man. No biotypes are recognized.

Reference/neotype strain: *B. canis* RM6/66.

Other species

A number of *Brucella* isolates have been described whose properties do not closely agree with the descriptions for recognized species. The status of most of these strains has yet to be finally determined and it is possible that some or all of them will be found eventually to correspond to atypical cultures of existing species or biotypes (Meyer 1976). Included among these are the proposed species '*Brucella murium*' (Korol & Parnas 1967) and the rodent isolates of Dushina *et al.* (1964), Cook *et al.* (1966) and Taran *et al.* (1966).

In addition to these, strains of *B. abortus*, *B. suis* and other species have been isolated recently which differ significantly from the standards currently recognized and it is likely that the present scheme may have to be extended to include new biotypes.

Basic Materials and Methods

Culture media

Serum dextrose agar (SDA)

This medium is recommended for making subcultures and for primary isolation of *Brucella* from uncontaminated sources such as guinea pig spleen. Its composition is: Oxoid Blood Agar Base, 40 g; distilled water, 1 l; equine serum, inactivated at 56°C for 30 min, 50 ml; 25% (w/v) glucose autoclaved at 121°C for 15 min, 40 ml.

The Oxoid base is added to the water and the mixture allowed to stand for 10 min to prevent the powder from caking. Gentle heat is applied to aid solution before autoclaving at 121°C for 15 min. The base is cooled to 56°C before adding the serum and dextrose solution. Plates or slopes are then poured immediately.

The SDA medium can be made selective for *Brucella* by the addition of antibiotics and antimicrobial agents. The formulation developed by Farrell (1974) is recommended for the isolation of *Brucella* from contaminated material.

Farrell's modified SDA

This is prepared from SDA basal medium by the addition of bacitracin, polymyxin B, actidione, vancomycin, nalidixic acid and nystatin. Stock solutions of these are prepared as follows:

Bacitracin (Burroughs Wellcome and Co., London). The sterile powder is dissolved in sterile distilled water to a concentration of 2000 u ml^{-1} and stored at 4°C.

Polymyxin B (Burroughs Wellcome and Co., London). The sterile powder is dissolved in sterile distilled water to give a concentration of 5000 u ml^{-1} and stored frozen at –20°C. Unused solution should be discarded and not re-frozen.

Actidione (Upjohn and Co., Michigan, USA). A stock solution containing 10 000 μg ml^{-1} is prepared by dissolving 1 g actidione in 5 ml acetone and diluting 1:20 in distilled water. This may be stored at 4°C.

Vancomycin (Lilly and Co., Indianapolis, USA). A stock solution containing 50 mg ml^{-1} is made by dissolving the sterile powder in sterile distilled water immediately before use.

Nalidixic acid (Winthrop Laboratories, Edgefield Avenue, Newcastle). A 5% (w/v) stock solution is prepared in 0·5 mol l^{-1} NaOH and stored at 4°C. Immediately before use this is diluted 1:10 in distilled water.

Nystatin (E. R. Squibb, New York, USA). Immediately before use the sterile powder is suspended in sufficient sterile distilled water to give a concentration of 50 000 u ml^{-1}.

The volumes of each of these stock solutions to be added to a 1 litre volume of melted SDA base are given in Table 1.

TABLE 1. Antibiotic composition of Farrell's modified SDA medium

Antibiotic	Volume of stock solution l^{-1} basal medium (ml)	Final concentration (ml^{-1})
Bacitracin	12·5 ml	25 u ml^{-1}
Polymyxin B	1·0 ml	5 u ml^{-1}
Actidione	10·0 ml	100 μg ml^{-1}
Vancomycin	0·4 ml	20 μg ml^{-1}
Nalidixic acid	1·0 ml	5 μg ml^{-1}
Nystatin	2·0 ml	100 u ml^{-1}

Glycerol dextrose agar (GDA)

This medium is used for examining cultures for dissociation. It is prepared by dissolving 40 g of Oxoid Blood Agar Base in 1 l of distilled water and autoclaving at 126°C for 20 min. Batches of 250 ml of this melted basal medium are added to a solution of 25 g glucose and 50 ml glycerol in 50 ml of distilled water, previously sterilized by autoclaving at 108°C for 15 min.

Albimi Brucella broth

This medium is used, *inter alia*, in the propagation of brucella-phages. Its composition (gl^{-1}) is: peptone M, 20; glucose, 1; sodium bisulphite, 0·1; yeast autolysate, 2; sodium chloride, 5.

This medium is available as a dehydrated powder from Albimi Laboratories Inc., Brooklyn 2, New York or from V. A. Howe and Co. Ltd, Peterborough Road, London SW6. For use, 28 g is added to 1 l of distilled water and sterilized by autoclaving at 126°C for 20 min.

Modified Thayer Martin medium

This medium is a modification of that developed by Brown *et al.* (1971) and has been used successfully for the isolation of *B. ovis* from ram semen. It is less inhibitory than other selective media and is particularly recommended for the isolation of fastidious or antibiotic-sensitive brucella strains from contaminated material. It is prepared from SDA base by addition of VCN inhibitor and furadantin. The VCN inhibitor is obtainable from Baltimore Biological Laboratories, Division of Becton-

Dickinson, Cockeysville, Md., USA in vials each containing: vancomycin, 3000 μg; colistimethate, 7500 μg; nystatin, 12 500 u.

To prepare the medium, the contents of one vial of VCN inhibitor are dissolved in 10 ml of sterile distilled water and added to 1 l of melted SDA base to give final concentrations of vancomycin 3 $\mu g\ ml^{-1}$, colistimethate 7·5 $\mu g\ ml^{-1}$ and nystatin 12·5 $u\ ml^{-1}$. To this mixture is added 1 ml of a 1% (w/v) solution of furadantin in 0·1 mol l^{-1} NaOH. This medium will usually suppress the growth of most contaminating organisms but where swarming bacteria pose a particular problem, 2% (w/v) Ionagar No. 2 or Oxoid agar No. 1 may be added to the medium.

Dye sensitivity test media

These are prepared by the addition of the relevant dye to a basal medium. The inhibitory activity of each dye to be used varies with its source and batch and also with the basal medium employed. Dyes obtained from the National Aniline Division, Allied Chemical and Dye Company, New York, N.Y., USA are recommended for this purpose. Each new batch must be tested to determine the optimum dye concentration required to inhibit *Brucella* reference strains growing on the basal medium used. As it will support the growth of all *Brucella* spp. and biotypes, SDA is recommended for this purpose. To prepare the media, the dyes are dissolved in distilled water to produce stock solutions of the concentrations shown in Table 2. These are sterilized by steaming for one hour. The volume of dye solution to produce the required final concentration is then added to the SDA basal medium. Plates or slopes are poured from this immediately. A similar procedure is used to prepare plates containing *meso*-erythritol, which is previously sterilized by filtration.

The concentrations of these reagents which will usually give satisfactory results in inhibition tests are shown in Table 2.

Bacto-glutamate medium

This is used for suspending *Brucella* strains for preservation at liquid nitrogen temperature. It is prepared by dissolving 25 g of Bacto-casitone in 1 l of sterile distilled water and autoclaving at 115°C for 20 min. To the solution, 50 g of sucrose and 10 g of monosodium glutamate are added and dissolved by steaming for 10 min. The medium is pumped through a Seitz EKS filter into sterile flasks and finally autoclaved at 106°C for 15 min.

TABLE 2. Concentrations used in dye/erythritol sensitivity test media

Inhibitor	Stock solution (% w/v)	Volume added to 1 l (ml)	Final concentration
Basic fuchsin	0·2% (w/v)	10 ml	1/50 000
Thionin	0·2% (w/v)	10 ml	1/50 000
Safranin O	0·5% (w/v)	20 ml	1/10 000
Thionin blue	0·01% (w/v)	20 ml	1/500 000
Methyl violet	0·0125% (w/v)	10 ml	1/800 000
Pyronin	0·01% (w/v)	20 ml	1/500 000
meso-Erythritol	10% (w/v)	10 ml	1 mg ml^{-1}
meso-Erythritol	10% (w/v)	20 ml	2 mg ml^{-1}

H_2S test strips

Sheets of filter paper are soaked in a 10% (w/v) solution of lead acetate and hung up to dry at 37°C in air free of pollution with H_2S. When dry, the sheets are cut into strips of 8 mm × 100 mm, wrapped in grease-proof paper in bundles of 50, and vacuum-autoclaved at 126°C for 20 min.

Preservation of strains

Brucella strains may be stored for short periods by streaking on SDA slopes, incubating for 72 h in air + 10% (v/v) CO_2 (if required), then sealing the slope and storing at 4°C. The procedure should be repeated every 6 to 8 weeks. This technique is not very satisfactory, however, since each subculture increases the chance of changes in the characteristics of the strain, and of contamination of the slope.

Vacuum drying

This is the method most commonly used at this laboratory. It requires relatively little apparatus other than a vacuum pump, compressor and a manifold to which tubes may be attached for sealing.

Brucella strains are grown on SDA slopes at 37°C in air + 10% (v/v) CO_2 (if required) for 72 h. The growth is then washed off the slopes and suspended in sterile rabbit serum to give a concentration of *ca.* 10^{10} organisms ml^{-1}. (For preserving non-smooth strains horse serum may be substituted for rabbit serum.) The suspension is distributed in 0·1 ml volumes into sterile 63 mm × 8 mm tubes which are sealed with cotton wool plugs.

The tubes are placed in a desiccator containing calcium chloride or phosphorus pentoxide, and air is evacuated until small bubbles appear in the liquid. The desiccator is then sealed, and stored at 4°C overnight.

The following day, the air in the desiccator is evacuated for 10 min and the material replaced at 4°C. This procedure is repeated for at least one week, or until a pressure of 0·05 mm Hg can be obtained on evacuation.

When drying is complete, the tubes are placed inside 150 mm × 12 mm test tubes, which have a cotton wool plug in the bottom covering a layer of silica gel + $CoCl_2$ dryness indicator. A second plug is placed over the inner tube, and a narrow neck is drawn above this by means of a gas/air blowtorch.

When cool, the tubes are placed on the manifold, evacuated to a pressure of 0·05 mm Hg and the neck is sealed. The ampoules should then be tested for retention of vacuum, and stored at 4°C. Provided the vacuum is maintained, cultures should retain their viability for many years.

Storage in liquid nitrogen

This technique has been shown to cause a smaller decrease in viability than either freeze drying or vacuum drying (see Davies *et al.* 1973; Lapage *et al.* 1970). The method used will vary according to the apparatus available, hence only general indications can be given. At this laboratory, cultures are grown for 72 h on SDA slopes at 37°C in air + 10% (v/v) CO_2 (if required), and suspended in single strength Bacto-glutamate to the desired concentration. The suspension is left undisturbed at 4°C for one week, after which volumes of up to 1 ml are placed in sterile half-dram glass screw-capped vials. The vials are allowed to equilibrate in the vapour phase of liquid nitrogen for 10–15 min, and are then immersed in the liquid for storage.

Freeze drying

This technique is also dependent on the apparatus available. General indications are given by Lapage *et al.* (1970), and a description of the technique as applied to *Brucella* has been given by Boyce & Edgar (1966).

Antisera

Antisera to the antigens specific for smooth and rough *Brucella* strains are prepared in healthy 6 month old male or female rabbits of a large breed. Before use the animals must be shown by serum agglutination tests to be free of non-specific agglutinins and anti-brucella antibodies.

A and M monospecific antisera

(a) Prepare suspensions of smooth *B. abortus* biotype 1 strain 544 and

B. melitensis biotype 1 strain 16M containing *ca.* 10^9 viable organisms ml^{-1} PBS.

(b) Inject single doses of 0·5 ml of these intravenously into rabbits.

(c) Bleed the animals daily from the fifth day onwards and titrate the sera for brucella agglutinins. By the ninth or tenth day most animals will have serum agglutination titres of at least 1:1280.

(d) Exsanguinate those animals with high agglutinin titres and prepare separate pools of antisera against *B. abortus* and *B. melitensis*.

(e) Titrate each serum pool in the serum agglutination test against standard *B. abortus* and *B. melitensis* suspensions. A titre of at least 1:1280 against the homologous antigen is desirable.

(f) Determine the quantity of packed brucella cells of the heterologous species required to remove cross-reacting agglutinins from each batch of serum. This should be done by trial runs with 5 ml volumes. Usually about 1 ml of packed cells will be required for each 10 ml of antiserum.

(g) Absorb each batch of serum by mixing with the appropriate volume of packed cells and incubating at 37°C for 2 h. Recover the antiserum by centrifugation and again titrate against the homologous and heterologous brucella antigens. After satisfactory absorption it will agglutinate the homologous antigen at a dilution of at least 1:160 but will not agglutinate the heterologous antigen even at a 1:10 dilution. If necessary the absorption process must be repeated until these conditions are met.

(h) Dilute the absorbed serum 1:5 in phenol-saline (0·5% (w/v) phenol in 0·15 mol l^{-1} NaCl) and check its reaction to *B. abortus* and *B. melitensis* in slide agglutination tests. Satisfactory monospecific serum will rapidly agglutinate the homologous organism but have no visible effect on the heterologous species.

(i) The diluted serum should be stored at 4°C until required but allowed to warm to room temperature before use.

Anti-R specific serum

Antiserum reacting with rough strains of *Brucella* is prepared using the R antigen of *B. ovis* as antigen.

(a) Grow *B. ovis* strain 63/290 on SDA slopes incubated at 37°C in air + 10% (v/v) CO_2 for 4 days.

(b) Check the slopes for purity and resuspend the growth in 0·15 mol l^{-1} NaCl to give a concentration of *ca.* 10^{10} organisms ml^{-1}. Use 3 ml volumes of this to inoculate Roux flasks.

(c) Incubate the Roux flasks at 37°C in air + 10% (v/v) CO_2 for 4 days.

(d) Check the cultures for purity and harvest the growth from each in 10–15 ml of 0·15 mol l^{-1} NaCl by shaking with a few glass beads.

(e) Strain the suspension through 2 layers of muslin and centrifuge at 3000 **g** for 30 min. Discard the supernatant fluid and wash the deposited cells twice by centrifugation in 0·15 mol l^{-1} NaCl.

(f) Heat the organisms resuspended in 0·15 mol l^{-1} NaCl, at 80°C for 2 h.

(g) Centrifuge out the cells at 3000 **g** for 30 min.

(h) Collect the supernatant liquid from (g) and centrifuge out the antigen at 60 000 **g** for 6 h. Discard the supernatant liquid.

(i) Resuspend the deposit from (h) in 20 volumes of 0·15 mol l^{-1} NaCl. This can be aided by repeated freezing and thawing. Store the R antigen solution at 4°C until required.

(j) Homogenize the R antigen solution with an equal volume of Freund's incomplete adjuvant.

(k) Inject 6 rabbits (more if a larger volume of serum is required) each with 2 ml of antigen/adjuvant homogenate spread over 4 subcutaneous sites on the left flank. Repeat the injections 7 days later on the right flank.

(l) Four weeks later exsanguinate the rabbits and collect the sera individually.

(m) Titrate each serum for agglutinins to rough brucella cells using a modification of the standard serum agglutination test. For this, serum dilutions are made in Menzel's buffer (0·5% (w/w) NaCl + 0·5% (w/w) phenol in a mixture of 1·5 volumes of 0·05 mol l^{-1} Na_2CO_3 and 8·5 volumes of 0·1 mol l^{-1} $NaHCO_3$, final pH 8·9). Equal volumes of an antigen consisting of a suspension of *B. ovis* 63/290 suspended in the same buffer to an O.D. of 3·30 at 450 nm are added to each tube. Incubation and reading of the test is as for the standard procedure using smooth antigen. The sera are also titrated against smooth *B. abortus* antigen in the standard test. Satisfactory sera should have a titre of at least 1:640 against the rough antigen and 1:20 or less against smooth antigen.

(n) Remove smooth agglutinins by absorbing the sera with smooth *B. abortus* cells. This is done by adding 1 volume of packed *B. abortus* 544 or S99 cells to 20 volumes of serum and incubating at room temperature overnight. The serum is recovered by centrifugation.

(o) Test each batch of serum again for agglutinins for rough and smooth brucella strains in the tube and slide agglutination tests. Satisfactory samples will agglutinate rough organisms to high titre but produce no reaction with smooth brucella cells even at the lowest dilution. Batches of similar titre are pooled and freeze-dried in 0·5 ml volumes then sealed under nitrogen in ampoules and stored at 4°C. For use the serum is reconstituted in 0·5 ml volumes of distilled water and diluted 1:5 in brucella-negative rabbit serum. The diluent is used in parallel as a negative control.

Tests for dissociation

Direct observation using obliquely transmitted light (*Henry* 1933)

This requires a microscope lamp with a blue filter, a stereoscopic microscope with a clear glass stage, capable of giving magnification of × 15 to × 25, and a concave mirror placed between the lamp and microscope (Fig. 1). Ideally this should be screened to reduce interference from incidental light.

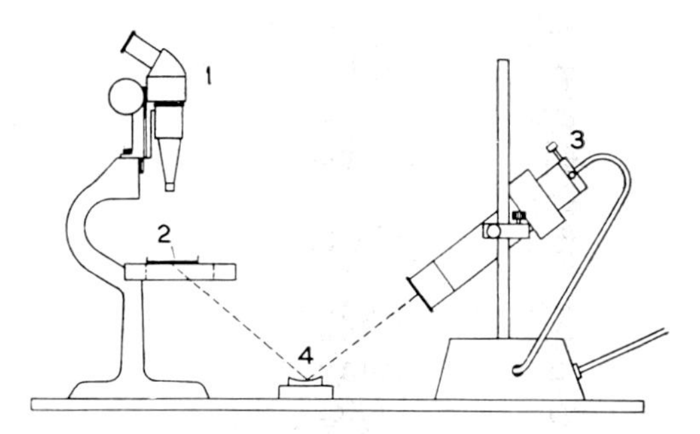

FIG. 1. Position of microscope, mirror and lamp for oblique-light observation of dissociation in *Brucella* cultures. 1: microscope (× 21); 2: glass stage for culture plate; 3: lamp giving a focused, concentrated light source; 4: mirror.

Plates containing GDA are streaked with the culture so as to produce areas of dense growth as well as isolated colonies. It is important that the medium should be clear and of even depth. After 4 days of incubation, the growth is examined under the microscope. Smooth colonies of *Brucella* spp. are convex, 2 mm diam., with an entire edge, and a blue, glistening and homogeneous appearance. Rough colonies are yellow, opaque, granular and break up when touched with a needle. Intermediate colonies are the most difficult to classify and have an appearance intermediate between smooth and rough, i.e. are slightly more opaque and granular than the smooth colonies. Mucoid colonies are glistening, greyish in colour and have a slimy consistency when touched with a needle.

Agglutinability in acriflavine

A 0·1% (w/v) solution of neutral acriflavine is used. The test is easily done by placing a drop of the acriflavine solution on a clean slide. A colony is then touched with a straight wire and emulsified in the acriflavine. Smooth colonies emulsify easily to produce a homogeneous suspension of cells undergoing Brownian movement. Rough colonies are difficult to emulsify and/or are agglutinated immediately; mucoid colonies form threads. Intermediate colonies behave like smooth colonies or may produce a fine, granular agglutination.

Staining of colonies with crystal volet (White & Wilson 1951)

The stain is prepared as follows: Solution A, 2 g crystal violet in 20 ml of 95% alcohol; Solution B, 0·8 g ammonium oxalate in 80 ml of distilled water. The two solutions are mixed to form a stock solution which is then diluted 1/40 in distilled water for use.

The plate is flooded with the stain and allowed to act for 15 to 20 s, after which it is drained into disinfectant. Under the stereoscopic microscope, smooth colonies are pale yellow; rough colonies are red with a coarse, granular appearance; other dissociated colonies are stained various shades of purple and blue (Fig. 2).

Agglutination by anti-Rough serum

Antiserum prepared in rabbits against the R antigen of *B. ovis*, as described above, is used; normal rabbit serum known to be free of antibodies to *Brucella* being used as a control. The reconstituted antiserum is diluted for use 1:5 in normal rabbit serum. A drop of the serum is placed on a clean microscope slide, and a colony is touched with a straight wire and emulsified in the serum. Agglutination with the antiserum but not with the control serum indicates that the organism is probably a rough *Brucella*. Agglutination with acriflavine but not with anti-Rough serum indicates that it probably is not *Brucella*.

R phage

Rough strains of *B. abortus* (but not other species) are lysed by phage R, as described in the section below on phage typing.

Fluorescent antibody test

This uses monospecific antiserum prepared against the R antigen of *B. ovis*, or *B. abortus* strain 45/20, as described above, conjugated so as to give a specific, direct fluorescence with rough strains of *Brucella* (Corbel 1973*b*). It is regarded as a supplementary test which could be valuable for isolates that give inconclusive results on the other tests. A smear of the colony on a microscope slide is dried, fixed by 95% (v/v) ethanol for 15 min, and stained and examined by u.v. light (Corbel 1973*b*).

Thermoagglutination test

This test consists of heating a saline suspension of the organism at 80°C or 100°C for 2 h (Burnet 1928; Wilson & Miles 1975). *Brucella* strains which have undergone dissociation auto-agglutinate under these conditions. The test may be useful as a supplement to the other procedures described.

Precautions to be taken when working with Brucella *cultures*

Pathogenicity for man

Infection with *Brucella* is very readily acquired by laboratory workers. A survey in the USA by Sulkin & Pike (1951) showed brucellosis to be the most common of all laboratory-acquired infections. The routes of infection include skin contact (probably mainly through cuts and abrasions), inhalation of aerosols, contact with the eyes and mouth, and accidental inoculation by needle and syringe. Most strains are capable of causing disease in man, and no safe and effective method of immunization is available.

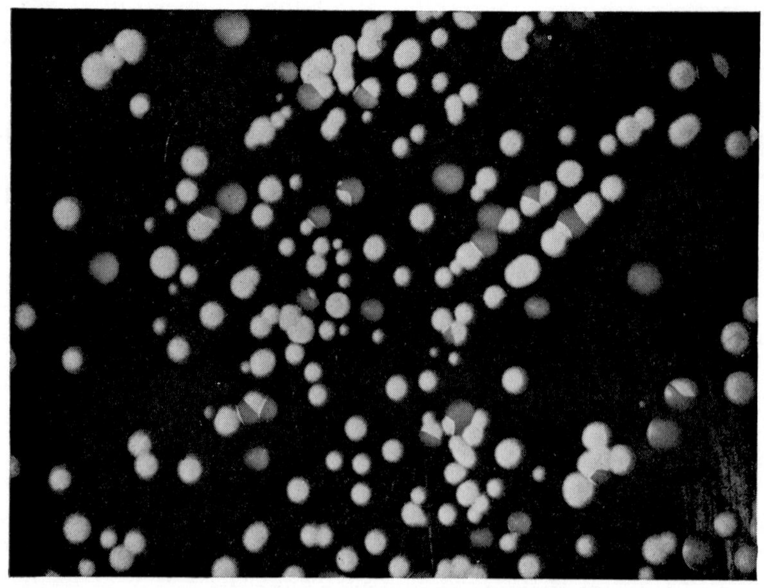

FIG. 2. Appearance of a *B. abortus* culture undergoing dissociation after 3 days' growth on glycerol-dextrose agar at 37°C. The plate was examined under oblique illumination after flooding with ammonium oxalate-crystal violet for 15–20 s. The smooth colonies appear pale and glistening whereas those undergoing dissociation are dark and granular.

Many strains of *B. melitensis* and *B. suis* are liable to cause especially severe disease in man. The handling of such strains, or those of unknown virulence, should be restricted to experienced staff working in an area set aside for the purpose. Whenever possible, less virulent strains, such as *B. abortus* strain 19 or 99 or *B. melitensis* Rev 1, should be substituted.

Safety equipment and clothing

Exposure of live cultures on the open bench should not be permitted. Safety cabinets are required, of the type now known as 'exhaust protective' in order to distinguish them from 'sterile' cabinets that are designed to protect the operation rather than the operator. Such cabinets are now recommended for handling all human pathogens (Darlow 1969). There is a trend in design towards total enclosure of all stages of the work, by placing all apparatus inside one large cabinet, or a series of connected cabinets (Rutter 1977). Because of the large volume of work that is usual in *Brucella* laboratories, it is particularly desirable that their cabinets should have convenient transfer devices, such as ports, air locks or liquid traps (dunk tanks) for getting materials in and out, and also facilities for *in situ* fumigation such as a paraformaldehyde vaporizer.

Centrifuges must either be totally enclosed in exhaust protective cabinets, or must use aerosol-proof, sealed rotors and buckets, which remain sealed under the stresses of centrifugation. Even with such precautions, breaking of containers is a serious hazard. Glass containers should not be used for centrifuging infective materials.

The choice of clothing will be influenced by local conditions, but as a guide it is suggested that it could be composed of two layers. Firstly, a comfortable, non-waterproof, but closely-woven suit and cap, which can either cover or replace the normal clothing; and secondly, a waterproof gown or apron, boots and gloves. The first layer and the boots would be worn when entering the biohazard area for any purpose. The gown or apron and the gloves would be put on when about to handle infected materials outside cabinets, as a protection against accidental exposure.

Planning procedures to reduce hazards

Many of the procedures described in this chapter need to be well planned from the safety angle. Even when an operation is carried out entirely within a cabinet, it is better to do it in a way that avoids contamination of the outsides of containers, the gloves, and the cabinet itself. Some examples are given below.

The extraction of *Brucella* from tissues by pestle and mortar, Griffiths (ground glass) tube or homogenizer with rotating blades, is not recommended, because of the aerosols these methods produce. The Colworth stomacher is considered less dangerous, but must only be used in a completely closed cabinet.

Ultrasonication or freeze-pressing of living *Brucella* suspensions may also produce infective aerosols, and must be done in a cabinet.

Catalase tests should not be done by dropping the hydrogen peroxide on to exposed surface colonies, because the effervescence produces minute splashes and aerosols. Instead these tests should be done in small test tubes, plugged with cotton wool. The hydrogen peroxide is added, by inserting the Pasteur pipette through the cotton wool, to a small volume of organisms suspended in saline.

Slide agglutination tests with live *Brucella* also produce minute splashes and aerosols of infective material, and must not be performed or examined outside a cabinet.

Pipetting by mouth must be totally prohibited in a *Brucella* laboratory, even when the material is thought to be non-infective. Instead, rubber teats, syringes or other mechanical pipetting aids can be used.

The use of disposable plastic petri dishes, pipettes and loops assists the ease of working in a cabinet, and reduces the risks from fire and explosion and from transporting infected glassware to the autoclave for sterilization. Plastic articles are easily and safely disposed of by incineration while sealed in polythene bags.

The Warburg manometric apparatus and the Gilson respirometer have to be operated from above as well as from the front, making their use within a cabinet particularly difficult. The fitting and removal of the glass flasks containing *Brucella* suspensions involve serious hazards. If no suitable way of enclosing them can be found and the work must be done, then staff must be protected by every other means possible, including waterproof clothing covering the whole body, enclosure of the whole head and neck in a hood supplied with sterile air, discarding of all clothing (for sterilization) on leaving the room, taking of a shower bath, and isolation of the room by an air lock. All the preliminary stages, including the filling of the flasks with the *Brucella* suspensions, must take place in a cabinet.

Disinfection

Personal hygiene is important. Washing facilities should include sinks with elbow taps and paper towel dispensers, and a shower. Hands and forearms should be washed on leaving the biohazard room, before touching the normal clothing. The shower is needed in case of an accident involving contamination of clothing. No eating, drinking or smoking should of course be allowed in any laboratory.

The main methods of disinfection of discarded cultures and infected materials and instruments are incineration and autoclaving. Boiling is not recommended, because the bubbling may force infective material out of the container. Formaldehyde fumigation can be used for clothing and books and papers, and ethylene oxide for electrical apparatus, respirators and hoods.

Procedures for dealing with accidental exposure of cultures outside cabinets should be worked out and memorized, and the equipment required should always be ready at hand. Workers should either not work alone, or be able to summon help quickly by telephone. Most accidents give rise to an aerosol hazard, so a hood, covering the whole head and neck, and supplied with sterile air, should be kept ready. A disinfectant spray can be used for floors, walls, furniture and clothing. A solution of 2·5% (v/v) Lysol B.P. is effective for this purpose. However, spillages should be covered with cotton wool or lint before being sprayed, to avoid the risk of dispersing the infected material in all directions.

Fumigation of the whole room should be carried out after any spillage of cultures outside the cabinets. This may be done as follows. All ventilation is switched off and all openings except the exit from the room are sealed up. By boiling water in open containers or by other means, the air temperature is raised to 30°C and the relative humidity to 75% r.h. After vacating the room, formalin is sprayed through an aerosol generator at the rate of 20 ml m^{-3} of air. The object is to maintain a concentration of 1 g of formaldehyde gas m^{-3} over a period of 3 h. The generator is set to produce the finest possible spray but as only a small proportion of the formaldehyde will become truly gaseous, an apparently large excess of formalin must be used.

The effectiveness of fumigation can be tested by exposing *Brucella* test strips. These consist of strips of filter paper that have been soaked in a suspension of *B. abortus* strain 19, then dried at 37°C and stored at 4°C. After exposure, they are laid on an SDA plate, left at room temperature for 1 h, and then removed. The plate is incubated at 37°C, and examined after 7 days. Unexposed strips are tested for their viability as controls at the same time.

Official obligations

Cultures sent by post must conform with the regulations of all the post offices that will handle the package. Use should be made of reliable, waterproof containers, and wrappings such as sealed polythene bags. These should be surrounded by enough absorbent packing material to prevent movement and absorb any leaking fluids, and enclosed in a strong, rigid, outer container of metal or wood. Unpacking of cultures should be done in a cabinet by trained staff.

Officially approved procedures for handling pathogens such as *Brucella* are becoming more clearly defined. In the United Kingdom, many detailed recommendations are contained in the Report of the Working Party on the Laboratory Use of Dangerous Pathogens (Chairman: Sir George Godber) (Anon 1975). Also relevant are the legal obligations of employers and employees as stated in the Health and Safety at Work, Etc. Act 1974.

Methods for Identification at the Genus Level

General methods

In most instances the application of conventional staining and cultural procedures, in conjunction with slide agglutination tests with specific antisera, will suffice to establish the identity of an organism as a member of the *Brucella* genus. For atypical or doubtful cultures, supplementary procedures are available which give more specific information on generic status. Unless stated to the contrary, it is recommended that all tests should be conducted on cultures which have been grown for not more than 5 days on a recommended solid medium incubated at 37°C in air or air + supplementary CO_2.

In the first instance, it is recommended that a heat-fixed smear should be examined after staining by Hucker's modification of Gram's method as described by Cowan (1974). The organisms should be Gram negative with a morphology consistent with that of brucellae. Examination for the presence of capsules by the wet Indian ink method (Duguid 1951) and the method of Howie & Kirkpatrick (1934); for flagella by the silver-staining method of Rhodes (1958); for spores according to Ashby (1938); for bipolar staining by Wayson's method and for acid fastness by the Ziehl-Neelsen method (Cowan 1974) should then give negative results. It should be noted, however, that although some brucella cultures may appear to show small capsules with special staining procedures, these are probably a continuation of the lipopolysaccharide-protein layer of the cell envelope.

At this stage, the culture should be examined for agglutination with unabsorbed *Brucella* antisera. One drop of a suspension of the culture is first mixed with an equal volume of 0·1% (w/v) acriflavine and examined for auto-agglutination. If no auto-agglutination occurs the culture is tested against unabsorbed antiserum to a smooth *Brucella* strain. If the culture auto-agglutinates with acriflavine, it is tested against antiserum to a rough *Brucella* strain. The absence of agglutination with the appropriate *Brucella* antiserum will enable the isolate to be dismissed as a member of the genus. A positive agglutination reaction on the other hand, although suggestive of *Brucella*, may also be given by other organisms and further proof of identity is required.

Broth cultures incubated at 22°C and 37°C should be examined for motility by the hanging drop method (Cowan 1974). A positive motility test excludes *Brucella*.

The culture should be streaked out on nutrient agar, sheep blood agar, MacConkey or deoxycholate-citrate agar and SDA plates and examined for separate colonies after incubation at 37°C. *Brucella* cultures will grow the most rapidly on the SDA plates and usually produce little or no growth on bile salt media, even after prolonged incubation. Even on suitable media growth is slow and well-developed colonies are unlikely to be seen in under 48 h. Growth tends to be most vigorous with *B. suis* strains and slowest with *B. melitensis* and *B. ovis*; *B. abortus* strains usually show an intermediate growth rate. On SDA, round, convex colonies, 2–3 mm diam., bluish-white in reflected light, with a smooth glistening surface, are produced after 2–3 days at 37°C. In transmitted light, the colonies are transparent with a pale honey colour. Rough strains produce opaque off-white colonies of a similar size and shape but with a more granular surface (see section 'Dissociation'). It should be noted that other organisms may produce similar colonies on SDA but their growth is usually much more rapid than that of brucellae.

On blood agar, growth is slower with the production of non-haemolytic greyish-white, glistening colonies *ca.* 0·5 mm diam. after 2–3 days at 37°C. Growth on nutrient agar is similar. The development of rapidly growing, haemolytic or lactose-fermenting colonies on the appropriate media is inconsistent with *Brucella*.

Subsequently the cultures are tested for the production of catalase (Cowan 1974). A negative result excludes *Brucella*. Catalase positive cultures are also tested for cytochrome c oxidase using the method of Ewing & Johnson (1960). All *Brucella* cultures, except those of *B. ovis* and *B. neotomae* are oxidase positive, although some may give weak reactions.

All *Brucella* cultures, with the exception of *B. neotomae* strains, do not produce acid from glucose in peptone water when tested according to Cowan (1974). In general, a positive glucose fermentation reaction will exclude *Brucella*.

Additional tests which should be done include those for *o*-nitrophenol-β-D-galactosidase activity (Lowe 1962), acetylmethylcarbinol production (Clarke & Cowan 1952), the methyl red reaction (Cowan 1974), indole production (Arnold & Weaver 1948), Koser citrate utilization (Cowan 1974), nitrate reduction (Method 1; Cowan 1974) and gelatin liquefaction (Method 2; Cowan 1974).

Positive reactions to any of these tests, with the exception of nitrate reduction, will exclude *Brucella*. All *Brucella* cultures, with the exception of *B. ovis*, will reduce nitrates to nitrite. Most *Brucella* cultures will also split urea in Christensen's medium (Cowan 1974), a negative result does not exclude *Brucella* however.

Specific methods

Collectively, the application of the above general tests will enable the generic status of *Brucella* cultures to be identified in the vast majority of cases. Difficulty may be experienced none the less, particularly with isolates that give atypical reactions to one or more of the conventional tests; those that are highly dissociated; those that have been grown on unfavourable media, for example bile salt agar; those that have been deliberately or inadvertently exposed to brucella-phages. Supplementary procedures suitable for determining the generic status of such cultures include:

(1) Measurement of the guanine + cytosine content of their DNA.

(2) Measurement of the hybridization of their DNA with the DNA of *Brucella* reference strains.

(3) Comparison of the electrophoretic properties of their acid-phenol soluble proteins with those of *Brucella* reference strains.

(4) Examination of the antigenic relationship between their intracellular antigens and those of *Brucella* reference strains.

(5) Examination of their cytochrome a and cytochrome c absorption spectra.

(6) Examination of the elution profiles produced by gas-liquid chromatography of their fatty acid methyl esters.

Procedures (1), (5) and (6) do not yield information which is totally specific for *Brucella* but give results which are unlikely to be duplicated by those organisms most liable to be confused with brucellae.

Determination of DNA base ratios (modified from Hoyer & McCullough 1968*a*)

The test organism and at least one *Brucella* reference strain are grown on Trypticase Soy Agar (Baltimore Biological Laboratories, Division of Becton-Dickinson, Cockeysville Md., USA) layers in Roux flasks incubated in air at 37°C for 48 h, with supplementary CO_2 if required. After checking each culture for purity, the organisms are harvested by washing the agar layers with phosphate buffered saline (0·15 mol l^{-1} NaCl in 0·01 mol l^{-1} phosphate buffer, pH 7·2) and sedimented by centrifugation. The DNA is extracted by the following procedure:

(a) Resuspend the packed cells in 50 volumes of tris-EDTA buffered saline (0·03 mol l^{-1} tris (hydroxymethyl) amino methane-HCl, pH 8·2; 0·05 ml l^{-1} ethylene diamine tetra-acetic acid, disodium salt; 0·1 mol l^{-1} NaCl).

(b) Add crystalline lysozyme to a concentration of 50 μg ml^{-1} and incubate at 37°C for 1 h.

(c) Add pronase (Calbiochem) to a concentration of 50 μg ml^{-1} followed by sodium lauryl sulphate to a final concentration of 5 mg ml^{-1} and incubate the mixture overnight at 37°C.

(d) Increase the concentration of sodium lauryl sulphate to 10 mg ml^{-1} and add an equal volume of pure water-saturated phenol. Shake the mixture steadily for 15 min and then collect the aqueous phase after centrifugation at 17 500 g for 30 min to break the emulsion.

(e) Extract the aqueous phase from (d) with an equal volume of water-saturated phenol under the same conditions. After centrifugation, collect the aqueous phase and chill to between 0 and 4°C.

(f) Slowly, with gentle mixing add 2 volumes of ethanol at 0°C to the aqueous phase. As the DNA precipitates, collect it by spooling it on to a glass rod. Rinse the spooled DNA repeatedly with cold ethanol and resuspend with gentle agitation in 0·15 mol l^{-1} NaCl containing 0·015 mol l^{-1} sodium citrate.

(g) Dilute the DNA suspension in 10 volumes of distilled water, add pancreatic ribonuclease to a final concentration of 50 μg ml^{-1} and incubate at 60°C for 1 h.

(h) Re-precipitate the DNA by slow addition of 2 volumes of cold ethanol and spooling on a glass rod as in (f).

(i) Resuspend the spooled DNA in 0·02 ml l^{-1} tris (hydroxymethyl) amino-methane-HCL buffer, pH 8·2 and repeat the digestion with pronase and sodium lauryl sulphate, phenol extraction and ethanol precipitation described in (c), (d), (e) and (f).

(j) Resuspend the spooled DNA in 0·015 mol l^{-1} NaCl containing 0·0015 mol l^{-1} sodium citrate, saturate with chloroform and store at 4°C.

DNA prepared in this way is suitable both for determination of G + C ratios and for use in hybridization experiments.

The G + C ratios may be determined by buoyant density measurements in the analytical ultracentrifuge according to Schildkraut *et al.* (1962), by the melting point method of Marmur & Doty (1962) or by the spectrophotometric procedure of Skidmore & Duggan (1971). Modifications of these procedures may also be used but whichever method is chosen, DNA from a *Brucella* reference strain must be examined in parallel with the unidentified culture.

The method of Bendich (1957) is described here as it requires only readily available apparatus.

(k) Centrifuge the DNA preparation from (j) at 10 000 **g** for 20 min and discard the supernatant liquid. Dry the residue over P_2O_5 to constant weight.

(l) Weigh out 5 mg (or proportionately more) of the dried DNA and suspend in 0·1 ml of 70% perchloric acid. Heat in a glass stoppered tube or sealed hard glass ampoule at 100°C for 1 h.

(m) On cooling, dilute to 0·5 ml with distilled water and grind up residue with a glass rod. Centrifuge out insoluble material and collect the supernatant hydrolyzate.

(n) Apply volumes of 10 to 20 μl of hydrolysate to an origin line marked across a 150 mm $\times$ 500 mm strip of Whatman No. 1 paper. At the same time apply similar volumes of perchloric acid as a control and perchloric acid solutions of known concentrations of adenine, thymine, cytosine and guanine at other points on the line. Allow the samples to dry and chromatograph by downward development in a solvent mixture of freshly mixed 12 mol l^{-1} HCl 16·7 ml, *iso*propanol A.R., 65 ml and water to 100 ml.

(o) After development for about 15 h at room temperature, dry the paper strip by hanging it upside down in a current of warm air. Locate the bases by scanning with a u.v. lamp fitted with a filter transmitting only radiation of 254 nm wavelength. Under these conditions the bases form dark patches on a weakly fluorescent background. Mark the position of each patch with a pencil.

(p) Cut out each marked area together with a non-absorbing patch of similar area from the negative control lane. Elute each piece of paper with 5 ml of 0·1 ml l^{-1} HCl for 2 h at room temperature. Then centrifuge the eluates to remove paper fibres.

(q) Scan the eluate from each absorbing area, including those corresponding to reference standards, against the control eluate in a spectrophotometer reading continuously from 230 nm to 310 nm. Determine the molar extinction coefficient (ϵ) for each reference base and for its corresponding component in the hydrolysate at its absorption maximum.

(r) Calculate the molar concentration of each base from the formula:

$$\text{Molarity} = \frac{\text{OD}}{\epsilon}$$

where OD is defined as the optical density at the absorption maximum of the eluate (or reference standard) multiplied by the total volume in millilitres. The values for ϵ at their respective absorption maxima are adenine, 12·6 $\times$ 10^3 (262·5 nm); thymine, 7·95 $\times$ 10^3 (265 nm); cytosine, 10 $\times$ 10^3 (276 nm) and guanine, 11·1 $\times$ 10^3 (249 nm). In practice only the molar concentrations of guanine + cytosine need be determined. These are likely to be more accurate than estimations of adenine + thymine because of partial destruction of thymine by the hydrolysis procedure.

DNA hybridization studies (Hoyer & McCullough 1968*a*, *b*)

Measurement of the extent of hybridization of the DNA of the unidentified isolate with that of a *Brucella* reference strain would be expected to give the most definitive information on their genetic relationship. However, the method is technically difficult and only suited to specialist laboratories. The authors have no personal experience of the procedure and those seeking further details are referred to the papers of Hoyer & McCullough (1968*a*, *b*).

Disc electrophoresis of acid-phenol soluble proteins

This procedure gives indirect information on the genetic constitution of organisms and is easily applied to the comparison of strains. The technique described here is essentially that developed by Morris (1973) for the characterization of *Brucella* isolates. All members of the genus produce very similar electrophoretograms under the conditions specified and the method is very useful for determining the generic status of atypical strains. It should be stressed, however, that standard conditions should be rigidly adhered to in applying the technique and comparison should only be made between electrophoretograms of extracts of identical protein concentration run simultaneously.

The organisms for examination, including at least one *Brucella* reference strain, are grown on slopes of trypticase soy agar or SDA incubated aerobically at 37°C for 48–72 h, with supplementary CO_2 if required. The cells are harvested in phosphate buffered saline (0·15 mol l^{-1} NaCl, 0·01 mol l^{-1} phosphate buffer, pH 7·0) at 4°C and washed once by centrifugation in this medium. The harvest from at least two slopes will be required for each batch of extract which is prepared as follows:

(a) Suspend about 100 mg wet weight of washed packed cells in 100 μl of phenol + acetic acid + water (4:2:1 (w/w)) and incubate in the dark at 18°–20°C overnight.

(b) Deposit residual cellular material by centrifugation at 40 000 **g** for 30 min and collect the clear supernatant liquid. From *Brucella* strains this is usually a dark straw colour. If not required for immediate use, the extracts may be stored at –20°C in sealed containers.

(c) For the protein separation prepare cylinders of polyacrylamide gel containing 7·5% (w/v) acrylamide, 5 mol l^{-1} urea and 35% (v/v) aqueous acetic acid. To do this, prepare a stock solution of acrylamide (Eastman Kodak Ltd, London) 7·5 g, N,N′-methylene-bis-acrylamide (Eastman

Kodak Ltd, London) 0·2 g and urea A.R. 30 g, in aqueous acetic acid A.R. 35% (v/v) to 100 ml. This may be stored in the dark at 4°C and kept for up to 4 weeks.

(d) For use, add 10 mg of fresh ammonium persulphate A.R. to each 12 ml of stock solution and degas the mixture by exposure to a vacuum until no more air is evolved. Then add immediately 60 μl of N,N,N′,N′-tetramethylethylenediamine and dispense the mixture into clean, grease-free glass running tubes (7 mm internal diam. × 85 mm length) to a height of about 72 mm. Overlay the mixture with distilled water to a depth of 5 mm, using a 1 ml tuberculin syringe fitted with a 25G needle.

(e) Allow polymerization to proceed at 37°C for 90 min. Tip the tubes after this period to decant the water overlay and to check that a rigid transparent gel has formed. If the gel has not set, a fresh batch of reagents must be prepared and particular attention paid to the de-gassing step.

(f) Position the gels vertically between the upper and lower reservoirs of an analytical polyacrylamide electrophoresis apparatus. Fill both reservoirs with 10% (v/v) aqueous acetic acid.

(g) Using a syringe and needle or micropipette, layer between 50 μl and 75 μl of acid-phenol extract directly on to the top of each gel cylinder, through the acetic acid in the top reservoir. The quantity of protein applied to each gel should be between 200 μg and 300 μg. This should be checked by subjecting samples of the extracts to nitrogen estimation by the micro-Kjeldahl method (Meynell & Meynell 1970).

(h) Connect the upper reservoir electrode as anode and the lower reservoir as cathode and apply a constant current equivalent to 3 mA per gel cylinder from a stabilized direct current power supply unit for 15 min. Then increase the current to 5 mA per gel for a further 60 min. A trace of methyl green in 10% (v/v) acetic acid added to the top reservoir electrolyte will enable the progress of electrophoresis to be monitored visually.

(i) Stop the electrophoresis when the green marker dye disc has travelled to within about 10 mm of the bottom of each tube. Remove each gel cylinder from its tube and fix and stain the protein zones by immersion in 1·0% (w/v) Amido Black 10B in 7% (v/v) aqueous acetic acid for 30 min. Remove unbound stain by repeated washing in 2 l volumes of 30% (v/v) aqueous acetic acid.

(j) Evaluate patterns visually by comparison with those produced by reference strains run simultaneously (Fig. 3). There should be coincidence of nearly all of the protein bands for cultures under comparison to be considered to be of the same genus. It should be noted that some of the reagents used, particularly acrylamide, methylene-bis-acrylamide and

N,N,N′,N′-tetramethylethylenediamine, are toxic by a variety of routes and should be handled with suitable precautions.

Examination of intracellular antigens by immunodiffusion and immuno-electrophoresis

The members of the genus *Brucella* form an antigenically coherent and close-knit group. Nevertheless, serological cross-reactions with organisms of other genera do occur (Francis & Evans 1926; Cioglia 1950; Wundt 1959; Ahvonen *et al.* 1969; Feeley 1969; Corbel 1975). These are attributable to antigenic similarities between the surface antigens of the organisms involved, in particular the lipopolysaccharide components of the O agglutinogens. On the other hand, cross-reactions between the intracellular antigens of brucellae and those of unrelated organisms occur rarely. Furthermore, many of the intracellular antigens are common to all *Brucella* strains, irrespective of species, biotype or stage of dissociation (Diaz *et al.* 1967; 1968; Bhonghbhibhat *et al.* 1970; Freeman *et al.* 1970;

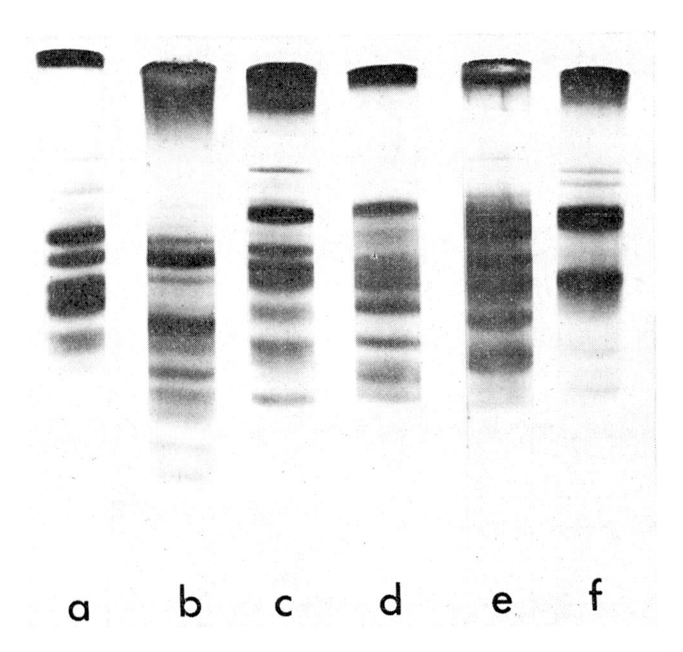

FIG. 3. Comparison of the phenol-acetic acid-water soluble proteins of *Brucella* and other Gram negative bacterial strains after polyacrylamide disc electrophoresis. The gels were stained with 1·0% (w/v) Amido Black 10B in 7% (v/v) aqueous acetic acid. (a) *B. abortus* 544, (b) *Yersinia enterocolitica* 0:9 296/68, (c) *Acinetobacter anitratus*, (d) *A. lwoffii*, (e) *Alkaligenes bronchiseptica*, (f) *Moraxella non-liquefaciens*.

Corbel 1975). Thus, direct comparison of the intracellular antigens of an unidentified culture with those of authentic *Brucella* strains provides a valid and effective means of determining its relationship to the genus. Antigenic comparison is most easily made using the agar immunodiffusion technique of Ouchterlony (1953). The procedure recommended is as follows.

Preparation of bacterial antigens for immunodiffusion

Procedures suitable for the release of soluble, diffusible antigens from *Brucella* cells include cold acetone treatment, followed by slow hypertonic saline extraction (Bhonghbhibhat *et al.* 1970); ultrasonic disintegration (Hinsdill & Berman 1967; Corbel & Cullen 1970; McGhee & Freeman 1970; Freeman *et al.* 1970); cell wall rupture by means of a Hughes or Edebo X-press or French pressure cell (Hughes *et al.* 1971); extraction with sodium dodecyl sulphate (Ellwood *et al.* 1967); extraction with diethyl ether—water (Ribi *et al.* 1959). Largely on the grounds of simplicity the first two procedures are recommended.

For both methods the organisms for examination, including a *Brucella* reference strain, are grown and harvested as described for DNA preparation.

Cold acetone/hypertonic saline extraction

(a) Wash packed cells 3 times by centrifugation in 20 volumes of phosphate buffered saline (0·15 mol l^{-1} NaCl, 0·01 mol l^{-1} phosphate, pH 7·0) at 4°C and resuspend in this medium to a concentration of 10^{10}–10^{11} organisms ml^{-1} as determined turbidimetrically.

(b) Pour this mixture into 2 volumes of acetone at –20°C and allow to stand at 4°C for 16 h.

(c) Decant the bulk of the acetone and recover the organisms by centrifugation at 4°C.

(d) Wash the deposited cells 3 times with acetone pre-cooled to –20°C and dry over calcium chloride in a desiccator under vacuum.

(e) Resuspend the dried organisms to 5% (w/v) in 2·5% (w/w) aqueous sodium chloride at 4°C. Agitate the mixture on a magnetic stirrer at 4°C for 3 days.

(f) Centrifuge out the cells at 20 000 **g** for 30 min at 4°C. Collect the supernatant fluid and discard the deposit.

(g) Add 3 volumes of ethanol at 0°C to the supernatant liquid, with constant stirring. Allow the mixture to stand at 4°C overnight.

(h) Discard the ethanolic supernatant liquid and collect the deposit after centrifugation at 20 000 **g** for 15 min.

(i) Redissolve the deposit in the minimum volume of distilled water and dialyse against distilled water at 4°C to remove traces of ethanol.

(j) Clarify the dialysed solution by centrifugation at 20 000 **g** for 20 min at 4°C and freeze-dry.

(k) For use, reconstitute the dried extract in 0·15 mol l^{-1} NaCl to a concentration of 50 mg ml^{-1}. Alternatively, omit the freeze-drying step and adjust the protein concentration to 25 mg ml^{-1} as determined by the Folin biuret method (Lowry *et al.* 1951). These preparations may be used in immunodiffusion and immunoelectrophoresis experiments.

Ultrasonic disruption

(a) Prepare dried acetone-extracted organisms as described in the previous section (a)–(d). Live organisms may be used but this makes the procedure unnecessarily hazardous for routine purposes.

(b) Suspend the bacteria to a concentration of 20% (w/w) in sterile distilled water containing 0·001 mol l^{-1} EDTA and 0·01 mol l^{-1} cysteine hydrochloride.

(c) Place the suspension in a sonication vessel, preferably with continuously circulating coolant, and subject to ultrasonic vibration with the apparatus tuned to maximum energy output until satisfactory cellular disintegration has been produced. This period has to be determined for each type of apparatus but for the Branson Soniprobe type 1130A (Dawe Instruments Ltd, London) two 20-min periods of sonication, separated by an interval for cooling, will produce effective breakage of *Brucella* cells. Close attention must be paid to the temperature of the suspension and this should not be allowed to rise above 30°C or loss of protein antigen through thermal denaturation will occur. Aerosols are produced by this procedure and even with non-viable preparations may provoke allergic reactions in sensitized individuals. It is recommended that this process should be conducted in an exhaust protective cabinet.

(d) Centrifuge the disrupted cell suspension at 20 000 **g** for 20 min and discard the deposit.

(e) Determine the protein concentration of the supernatant fluid by the method of Lowry *et al.* (1951) and adjust to 25 mg ml^{-1} either by dilution or by concentration by freeze-drying or ultra-filtration.

(f) Add sodium azide to 0·1% (w/v) and store frozen at –20°C or, for short periods, at 4°C.

Comparison of antigenic composition by immunodiffusion

Immunodiffusion tests are performed in a medium composed of 1% (w/v) Oxoid No. 1 agar or 0·9% (w/v) agarose in 0·15 mol l^{-1} NaCl and 0·1% (w/v) sodium azide. Uniform layers of gel 3 mm deep are formed in scratch-free polystyrene petri dishes. Reagent wells 5 mm diam. are cut in isometric patterns with centres 7 mm apart. The central well is filled with antiserum prepared by repeated subcutaneous injection of rabbits with acetone-killed *Brucella* organisms emulsified in Freund incomplete adjuvant. Soluble antigen preparations of standard concentration, made from the test organisms and *Brucella* reference strains, are placed in adjacent peripheral wells (Fig. 4). The plates are incubated in a water-saturated atmosphere for 1–3 days and inspected daily. Incubation need not be prolonged after the precipitation patterns produced by the reference antigens have clearly developed.

The precipitation line formed by the lipopolysaccharide-protein agglutinogen is easily recognized as a broad crescent formed close to the antigen well in extracts prepared from smooth strains of *B. abortus*, *B. suis* or *B. neotomae*. That produced by *B. melitensis* strains of biotype 1 tends to migrate further from the antigen well. Organisms of other genera, particularly *Yersinia enterocolitica* 0:9, may contain antigens giving reactions of partial identity with this component in immunodiffusion tests. The other antigens of *Brucella* extracts, corresponding

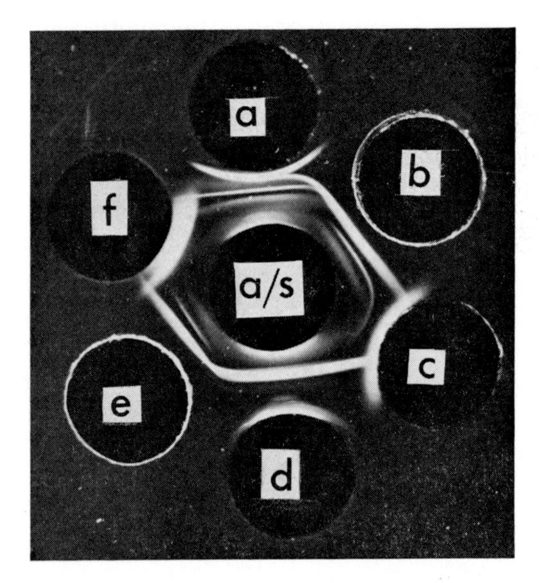

FIG. 4. Immunodiffusion of ultrasonically disrupted cells of (a) *B. abortus* 544, (b) *B. abortus* 45/20, (c) *B. melitensis* 16M, (d) *B. canis* RM6/66, (e) *B. ovis* 63/290, (f) *Yersinia enterocolitica* 0:9 296/68. Precipitin lines corresponding to the smooth lipopolysaccharide-protein agglutinogen have formed adjacent to the antigen well in (a), (c) and (f). Precipitin lines corresponding to genus-specific antigens have formed against the extracts in antigen wells (a), (b), (c), (d) and (e) but not (f). The antiserum used in the central well (a/s) was prepared against *B. abortus* 544.

mainly to the intracellular antigens, do not cross-react with those of organisms of other genera but give reactions of identity with similar components of other *Brucella* strains (Fig. 4).

The failure of an extract of an unidentified organism to cross-react in the immunodiffusion test with *Brucella* intracellular antigens would be sufficient for the organism to be discounted as a *Brucella* strain.

If considered necessary, the immunodiffusion reactions may be confirmed by immunoelectrophoresis under the conditions described by Hinsdill & Berman (1967).

Examination of cytochrome absorption spectra

The procedure to be used is that described by Dranovskaya & Kushnarev (1968). Briefly, killed suspensions of cells in phosphate buffer are examined spectrophotometrically for the absorption characteristics of cytochrome a and cytochrome c. All *Brucella* strains show strong cytochrome c absorption maxima at 522–524 nm (β) and 552–554 nm (α). Most strains of the classical species, *B. abortus*, *B. melitensis* and *B. suis*, show a weak cytochrome a absorption maximum at 620–630 nm. Strains of the other *Brucella* spp., *B. canis*, *B. neotomae*, *B. ovis* and those proposed as '*B. murium*' show a stronger cytochrome a absorption maximum at 620–630 nm.

The absence of a typical cytochrome absorption spectrum is of value in excluding an isolate from the genus *Brucella* but a positive result, in isolation, should be regarded as no more than supporting evidence.

Gas-liquid chromatography of fatty acids and their methyl esters

A method for examining bacterial metabolites by gas-liquid chromatography (g.l.c.) has been described by Mitruka & Alexander (1970). This permitted the differentiation of *Brucella* strains from other bacteria and incidentally differentiated *B. canis* from the other species. A method dependent upon the elution profile produced by the methyl esters of the whole cell fatty acids has been described by Tanaka *et al.* (1977). This has the advantage of using small quantities of formalin-killed cells. Although the short chain fatty acids of *Brucella* strains do not differ markedly from those of other bacteria, the longer chain fatty acids are distinct and the elution profile produced by each species is characteristic. This method promises to be of considerable value in the identification of organisms as members of the genus but requires further evaluation before it can be recommended as a definitive method without the support of ancillary procedures. For further details of the techniques of g.l.c. applicable to bacterial taxonomy the reader is referred to Drucker (1976).

Methods for Identification at the Species and Biotype Levels

Phage typing

Description of phages

The phages in use at this laboratory may be divided into five groups according to their host range. Group 1 includes phages Tb, A422 and 22/XIV. Group 2 contains phages Wb, M51, S708, MC/75 and D. Group 3 contains the Fz phages, Group 4 the Bk phage and Group 5, phage R. Although the phages within each group have the same host range, other properties such as efficiency of plating and plaque morphology may vary (cf. Morris *et al.* 1973; Morris & Corbel 1973; Corbel & Thomas 1976*a*, *b*; Thomas & Corbel 1977; Corbel 1977). The host range of the phage groups is shown in Table 3, and examples are given in Fig. 5.

For routine typing of *Brucella* spp., the phages Tb, Wb, Bk and R are used. The methods of propagation of these phages are modifications of those described by Adams (1959).

Propagation

The host strain used for propagation of phages Tb and Wb is smooth *B. abortus* strain 19. For phage Bk smooth *B. melitensis* strain 16M has been used, and for phage R, rough *B. abortus* strain 45/20. The propagating strain should be grown on SDA (see section 'Media') for 24–36 h at 37°C in air. It should then be suspended in ABB (see section 'Media') to give a concentration of *ca.* 10^{10} organisms ml^{-1}. The phage suspension should be diluted to a little less than RTD (see titration of phage stocks for full definition) i.e. between 10^4 and 10^5 plaque forming units (p.f.u.) ml^{-1}. If necessary a preliminary titration should be carried out to determine RTD of the phage stock.

Method 1. This procedure is that most commonly used at this laboratory, and may be employed for all brucella-phages. The yield is about 1·5 ml of high-titre phage stock per 9 cm petri-dish culture used.

(a) Place 0·1 ml of the suspension of propagating strain, and 0·1 ml of the phage stock described above, in the centre of an SDA plate. Spread the liquid uniformly over the surface of the plate, using a sterile glass or stainless steel spreader.

(b) Invert the plates, and incubate in air at 37°C for 24–36 h.

(c) Inspect the plates and select those showing confluent or near-confluent lysis with no bacterial contamination.

(d) Flood the surface of each plate with about 3 ml of sterile ABB. Leave the plates undisturbed on a level surface at room temperature for 20 min.

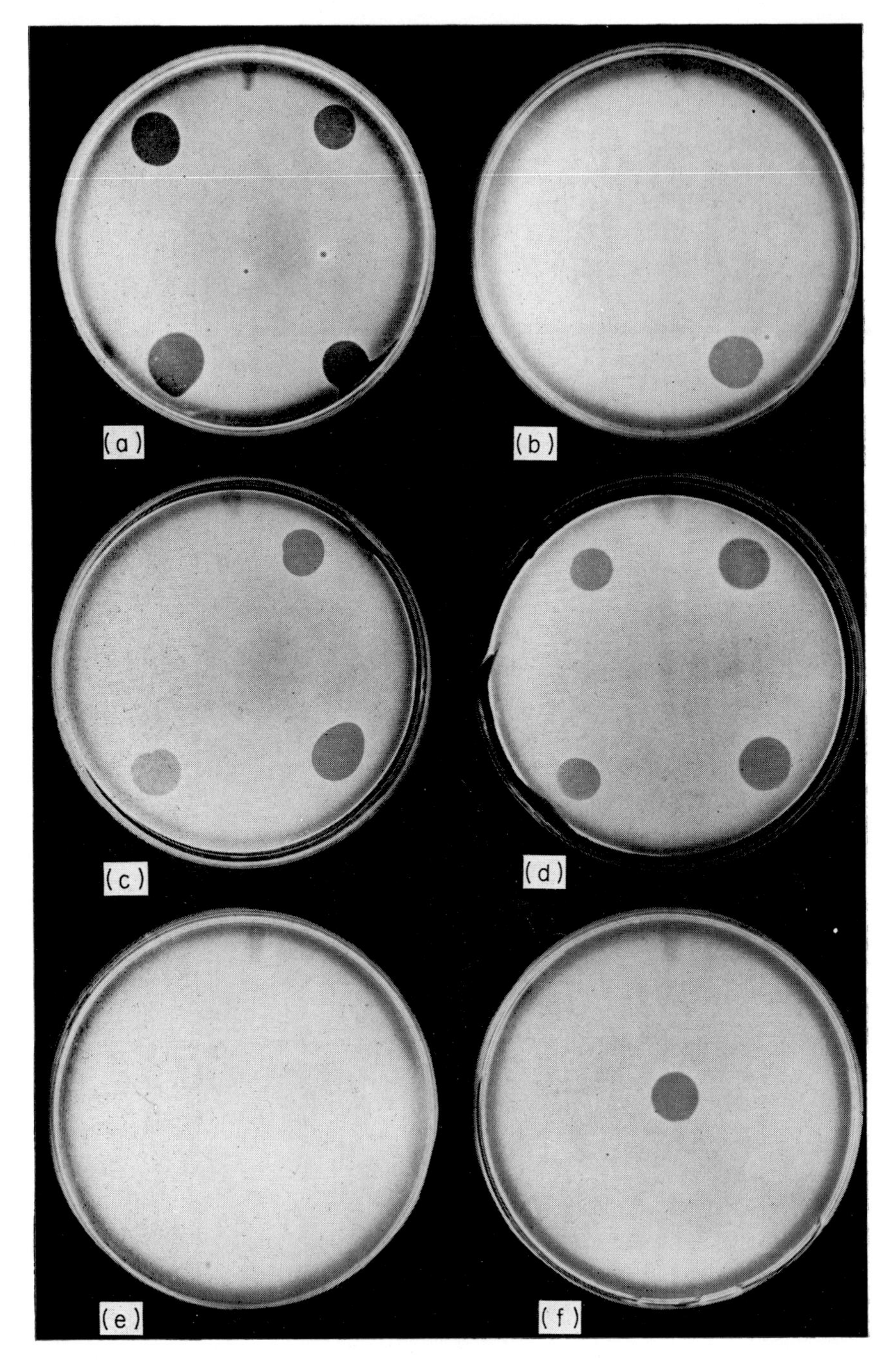

FIG. 5. Patterns of lysis produced by brucella-phages. (On each plate, reading anti-clockwise (1 to 4) from top L.H. corner, with no. 5 centred.) (1) Tb, (2) Wb, (3) Bk, (4) Fz and (5) R on cultures of (a) smooth *B. abortus* 544, (b) *B. melitensis* 16M, (c) *B. suis* 1330, (d) *B. neotomae* 5K33, (e) *B. ovis* 63/290 and (f) rough *B. abortus* 45/20. *B. canis* produces a result identical with (e).

TABLE 3. Host ranges of brucella-phages (tested on cultures grown under standard conditions on SDA medium)

			B. abortus		*B. suis*		*B. melitensis*		*B. neotomae*		*B. canis*	*B. ovis*
Group[a]	Phage	Titre	Smooth	Non-smooth	Smooth	Non-smooth	Smooth	Non-smooth	Smooth	Non-smooth	Non-smooth	Non-smooth
1	Tbilisi A422	RTD[a]	L	NL	NL	NL	NL	NL	PL	NL	NL	NL
	22/XIV	RTD $\times 10^4$	L	NL	L	NL	NL	NL	L	NL	NL	NL
2	Weybridge M51 S708	RTD	L	NL	L	NL	NL	NL	L	NL	NL	NL
	MC/75 D	RTD $\times 10^4$	L	NL	L	NL	NL	NL	L	NL	NL	NL
3	Firenze	RTD	L	NL	PL	NL	NL	NL	L	NL	NL	NL
		RTD $\times 10^4$	L	NL	L	NL	NL	NL	L	NL	NL	NL
4	Berkeley[b]	RTD	L	NL	L	NL	L	NL	L	NL	NL	NL
		RTD $\times 10^4$	L	NL	L	NL	L	NL	L	NL	NL	NL
5	R	RTD	NL	L	NL	NL	NL	NL	NL	NL	NL	NL
		RTD $\times 10^4$	L	L	L	NL	NL	NL	L	NL	NL	NL

L: confluent lysis; PL: partial lysis; NL: no lysis.
[a] See text for definitions.
[b] Results as described by Douglas and Elberg (1976).

(e) Decant the liquid into a sterile container. This constitutes the crude phage stock.

(f) Clarify the stock by centrifugation at 500 g for 10–15 min. Discard any deposit. If possible, the supernatant liquid should be further purified by passage through a membrane filter of 0·45 μm average pore diameter. Finally, the stock may be saturated with toluene to maintain bacterial sterility.

(g) The purified stock should be titrated, and stored at 4°C, at which temperature it will maintain its titre for up to 2 years. It should not be frozen, since this often greatly reduces the titre.

Method 2. For some phages, this method has been reported to produce more consistent results than Method 1 (Adams 1959). The technique was used for propagation of phage Bk by Douglas & Elberg (1976). The yield is similar to that of Method 1.

(a) To about 3 ml of ABB + 0·4% (w/v) Oxoid agar No. 1 (or its equivalent), which has been melted, and is maintained at a temperature of 47°C, is added 0·1 ml of propagating strain and 0·1 ml of phage suspension, with mixing.

(b) Pour the mixture on to the surface of an SDA plate and keep undisturbed on a level surface until the agar has set.

(c) Invert the plate and incubate at 37°C in air for 24–36 h.

(d) Scrape off the soft agar layer into a sterile container, and add to this 3 ml sterile ABB per plate used. Leave overnight at 4°C. This constitutes the crude phage stock.

(e) Clarify and filter the crude stock as described in Method 1, (f) and (g).

Method 3. This method is most suitable for the phages of Group 2. It does not consistently produce high titre preparations, although large volumes of phage stock are easily obtained.

(a) To 250 ml sterile ABB in a 1 l Ehrlenmeyer flask, add 1 ml of the propagating strain and 1 ml of the phage suspension described above.

(b) Incubate for 48–72 h at 37°C in air. The culture should be shaken or stirred vigorously, to ensure efficient aeration. After incubation, the suspension constitutes the crude phage stock.

(c) Clarify and filter the crude stock as described in Method 1, (f) and (g).

Titration of phage stocks

Method 1. For most culture typing purposes, brucella-phages are used at the Routine Test Dilution (RTD). The RTD is the highest dilution of a phage suspension which will produce confluent lysis on a lawn of the propagating bacterial strain.

(a) Prepare serial tenfold dilutions between 10^{-1} and 10^{-9} of the phage stock, in sterile ABB. A separate pipette should be used in preparing each dilution, to ensure that no carry-over occurs.

(b) Place 0·1 ml of the suspension of propagating strain on an SDA plate, and spread it evenly over the surface using a sterile glass or stainless steel spreader. Allow the plate to dry at room temperature for about 1 h.

(c) Pipette standard drops of the phage dilutions on to the agar surface. The drops should be discrete, and hence the number that can be contained on one plate will vary with the size of drop. The position and degree of dilution of each drop should be noted.

(d) Allow the plates to remain undisturbed at room temperature for 1 h, then invert and incubate at 37°C in air for 48 h.

(e) Examine the plates, and determine the highest dilution at which confluent lysis occurs. This is the RTD.

Method 2. In cases where a more accurate knowledge of the number of active phage particles in a suspension is required, a plaque assay should be performed. The count is expressed in p.f.u.

(a) Prepare dilutions of the phage stock as in Method 1 (a).

(b) Add 0·1 ml of each phage dilution and 0·1 ml of propagating strain to 3 ml volumes of ABB + 0·4% (w/v) Oxoid No. 1 agar which have been melted and then maintained at 47°C. Mix, and spread the suspensions on the surface of SDA plates. Leave undisturbed on a level surface until the agar has set. Repeat this procedure with the other dilutions. The test should preferably be performed in duplicate.

(c) Invert the plates and incubate in air at 37°C until individual plaques can easily be seen. Count the number of plaques present on those plates showing discrete lysis, and from these counts determine the number of p.f.u. per unit volume of the original stock.

Host range tests

Before use, the phage should be tested at RTD, RTD $\times 10^2$, and RTD $\times 10^4$ against authentic rough and smooth strains of *B. abortus*, *B. suis* and *B. melitensis* and preferably also, strains of *B. neotomae*, *B. canis* and *B. ovis*. Such a test will determine whether any host-range mutants are present in the phage stock. This is particularly important in the case of phage R, which has a high rate of mutation to a variant having a host range similar to those of the phages of Group 2 (see Table 3). In order to minimize any effect caused by the presence of this smooth-specific variant, the RTD of the rough-specific phage should not contain more than 10 p.f.u. ml^{-1} of this variant.

Testing the host range of a phage is most simply done by repeating the procedure for determination of RTD using the three dilutions described above and authentic *Brucella* reference strains. The pattern of lysis should then be compared with those of Table 3.

If the relative titres of any host-range mutants are required, a series of plaque assays should be performed in parallel, using the relevant *Brucella* strains.

Any phage stock having an anomalous host range should be discarded.

Phage typing of an organism known to be *Brucella* is conveniently performed in conjunction with biotyping procedures. The practical details will be found under 'Routine Identification Procedures'.

Oxidative metabolism tests

Patterns of utilization of amino acid and carbohydrate substrates are determined for standardized resting cell suspensions by a Warburg constant volume respirometer (Umbreit *et al.* 1959; Morgan & Gower 1966) or a Gilson differential respirometer and are expressed as quotients representing the rate of oxygen consumption in μl^{-1} mg^{-1} of cell nitrogen h^{-1} (Qo_2N).

The quotients obtained by either method are compared with the expected values for each species shown in Table 4 (from Meyer 1969*a*, *b*). In practice, the figures obtained by the Gilson procedure are somewhat lower than those given by Warburg manometry.

The following substrates are normally used:

Group 1: L-alanine; L-asparagine; L-glutamic acid.

Group 2: L-arginine; DL-ornithine: DL-citrulline and L-lysine.

Group 3: L-arabinose; D-galactose; D-glucose; D-ribose; D-xylose; *meso*-erythritol.

Other substrates, including D-amino acids, may be used in addition to these. They are prepared as 1% (w/v) solutions in Sorensen phosphate buffer, sterilized by filtration and stored at –20°C. It may be necessary to adjust the pH of some of these solutions by adding dilute KOH.

Routine identification procedures for Brucella *cultures*

When typing organisms suspected to be *Brucella*, it is essential to include in all tests at least the *Brucella* reference strains *B. abortus* 544, *B. melitensis* 16M and *B. suis* 1330, as a check on media and methods. If the strains undergoing typing are non-smooth and/or are thought to belong to one of the other *Brucella* spp., then at least one strain typical of these should be included in the tests.

Most cultures received for typing will have been subjected to a pre-

TABLE 4. Range of QO_2N values for *Brucella* strains on various substrates

	L-alanine	L-asparagine	L-glutamic acid	L-arabinose	D-galactose	D-ribose	D-xylose	L-arginine	DL-citrulline	DL-ornithine	L-lysine	*meso*-erythritol	Dextrose
B. abortus 1–9	65–172	80–198	140–420	45–187	50–230	103–349	10–93	15–48	0–50	0–50	0–25	210–462	108–430
B. melitensis 1–3	61–216	60–222	101–450	0–41	0–42	0–52	0–53	0–40	0–40	0–46	0–50	125–400	100–570
B. suis 1	21–101	0–30	0–50	167–524	101–526	200–509	80–500	55–150	60–219	75–264	56–150	185–472	144–466
2	0–38	0–119	50–177	107–412	37–172	200–433	63–180	51–111	86–187	93–278	0–34	154–280	194–351
3	10–106	0–22	40–379	20–54	0–60	204–401	28–83	45–108	65–123	104–216	46–145	156–241	136–413
4	0–30	0–26	25–260	0–29	0–37	130–385	0–41	93–241	68–184	129–330	54–144	142–424	110–400
B. neotomae	0–107	70–116	139–469	194–363	274–403	0–431	0–102	0–10	0–15	0–26	38–90	126–387	201–474
B. ovis	20–140	85–272	110–328	0–51	0–13	0–11	0–10	0–16	0–13	0–21	0–14	0–16	0–10
B. canis	14–115	0–56	63–280	0–254	9–191	331–515	0–56	82–284	106–228	138–361	72–134	0–395	216–366

liminary examination to exclude other organisms. In these cases it is usually only necessary to check that the Gram reaction, morphology, colonial appearance and reaction with *Brucella* antisera are consistent with *Brucella* before proceeding with typing. If the genus identification procedures have not been carried out or if atypical results are encountered in these screening tests, then the full range of procedures described in the first part of the section 'Methods for identification at the genus level' should be performed.

(a) As a first step in typing, the culture should be streaked on GDA to give isolated colonies and confluent growth on the same plate, and incubated for 48–72 h at 37°C in air + 10% (v/v) CO_2. The plate is then examined in obliquely transmitted light (cf. section 'Tests for dissociation').

(b) If all colonies are smooth or a mixture of smooth and non-smooth colonial forms is present a smooth colony is picked off and subcultured on three SDA slopes. If only non-smooth colonies are present then a typical colony is picked off and subcultured on three SDA slopes. In either case, one of the slopes is incubated at 37°C in air and the other two in air + 10% (v/v) CO_2. A strip of lead acetate paper is placed in the mouth of one of the latter two tubes but without allowing it to come into contact with the agar.

(c) Examine the lead acetate papers daily for the next 4 days. Record the presence or absence of any blackening and change the strips each day. See 'Notes' and Fig. 6 for interpretation of the results.

(d) After 72 h incubation examine the slope incubated in air and record any growth as a negative CO_2 requirement. Absence of growth on this slope but growth on the slopes incubated in the CO_2 supplemented atmosphere indicates a positive CO_2 requirement.

(e) After 48 h incubation, suspend the growth from one of the slopes incubated in air + 10% (v/v) CO_2 in *ca.* 0·5 ml sterile saline to give an opacity equivalent to approximately 10^{10} organisms ml^{-1}. This suspension is used for the procedures described below.

(f) Dye plates which have been incubated overnight as a sterility check are quartered as shown in Fig. 7(a) permitting four samples to be tested on one plate. One loopful of suspension is used to inoculate each quarter with 5 consecutive streaks without recharging the loop. The loop is sterilized between each sample. After drying, the plates are incubated for 48 h in air + 10% (v/v) CO_2 (see 'Notes'). Normally incubation for 48 h will be sufficient but for slow growing strains it may be necessary to incubate for up to 4 days.

(g) Place drops of A, M and R antisera and 0·1% (w/v) acriflavine on a clean microscope slide. To each drop add a similar drop of the bacterial

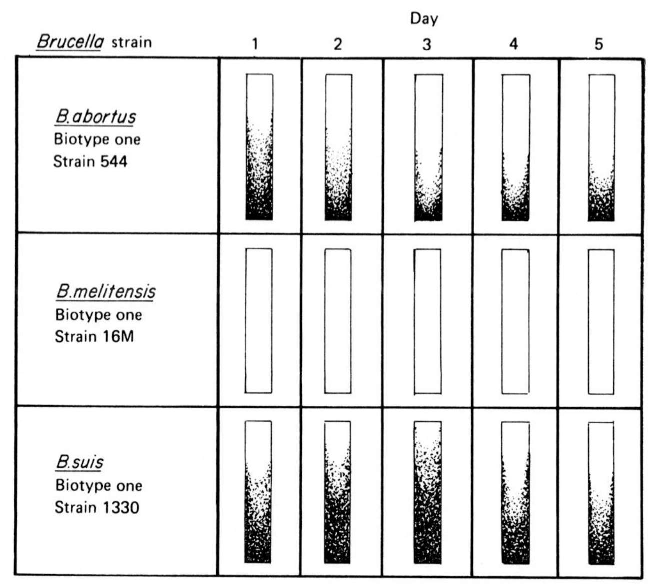

FIG. 6. Examples of hydrogen sulphide production by the various *Brucella* spp. over a period of 5 days. Strains of *B. ovis* and *B. canis* do not produce hydrogen sulphide. Production in *B. neotomae* strains is similar to that of *B. abortus*.

suspension prepared for (e). Gently rock the slide for about 0·5–1 min and examine each drop for agglutination. Record the presence or absence of agglutination with each reagent.

(h) Inoculate a sector of an SDA plate with a loopful of suspension from (e) as shown in Fig. 7(b). Carefully place discrete drops of phage suspension on each inoculated area as shown. Using this arrangement 4 *Brucella* strains can be examined for sensitivity to the 3 phages, Tb, Wb and Bk, on a single plate. The plate should be kept undisturbed on a flat surface for 1 h to allow the drops of phage suspension to absorb into the agar. It is then incubated at 37°C in air + 10% (v/v) CO_2 for 48 h before being examined for lysis.

(i) The suspension from (e) may be used at this stage to inoculate SDA Roux flasks to prepare suspensions for oxidative metabolism tests if required.

The pattern of results obtained in these tests may be compared with those shown in Table 5. If the pattern does not conform with any of those in the table, the procedures described in section 'Methods for identification at the genus level' should be followed through carefully to check that the organism is a member of the genus *Brucella*.

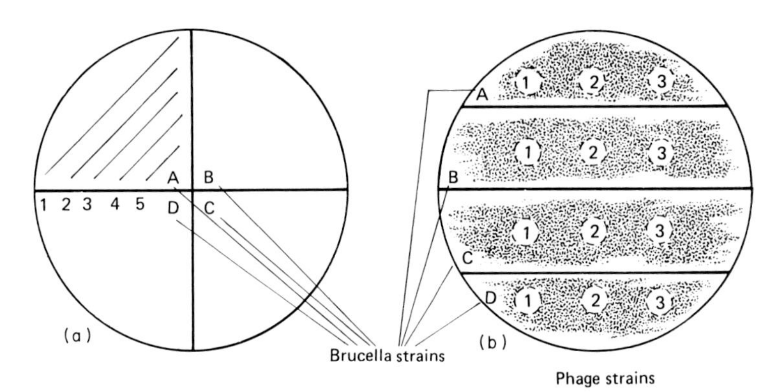

FIG. 7(a). Method for inoculating dye plates with *Brucella* cultures. The plate is divided into 4 marked quadrants A, B, C and D. Each *Brucella* strain is used to inoculate one quadrant by streaking a loopful of culture suspension five times in sequence as shown. (b) Method for inoculating *Brucella* cultures with phage. Four cultures A, B, C and D, are streaked laterally across the plate to produce zones of confluent growth. The phage strains 1, 2 and 3, corresponding to Tb, Wb and Bk for routine use, are then deposited as rows of discrete drops across each inoculated zone.

Notes on the interpretation of results

Production of hydrogen sulphide. The papers should be changed each day whether blackening has occurred or not, to exclude the possibility of false positive or negative results.

In general, blackening of the paper on one day only is not regarded as significant at this laboratory. It should be noted that strains may lose the ability to produce hydrogen sulphide on prolonged storage without subculture.

Agglutination reactions. If the strain under test is found to be non-smooth (i.e. agglutinates in acriflavine and/or in anti-R serum), then any positive results obtained with A and M antisera should be ignored, since they may be due to auto-agglutination. The identity of such strains should be checked by other means, since agglutination in acriflavine is not specific to *Brucella*, and, on rare occasions agglutination of organisms other than *Brucella* has been observed in anti-R serum.

Dye plates. Care should be taken that the suspension used for the inoculation of the dye plates (e) contains not more than about 10^{10} organisms ml^{-1}, since false positive results may otherwise be obtained. In general, the appearance of growth on the first two streaks only is not regarded as significant at this laboratory. If growth appears on three or more streaks, the organism is regarded as resistant to inhibition by the dye.

TABLE 5. Characters used in the differentiation of *Brucella* spp. and biotypes

Species	Biotype	CO_2 req't	H_2S prod'n	Growth on media containing thionin[a]	basic fuchsin[a]	Agglutination with monospecific antisera A	M	R	Lysis by phage[b] at RTD Tb	Wb	Bk
B. abortus	1	(+)	+	—	+	+	—	—	L	L	L
	2	(+)	+	—	—	+	—	—	L	L	L
	3[c]	(+)	+	+	+	+	—	—	L	L	L
	4	(+)	+	—	+[d]	—	+	—	L	L	L
	5	—	—	+	+	—	+	—	L	L	L
	6[c]	—	(—)	+	+	+	—	—	L	L	L
	7	—	(+)	+	+	+	+	—	L	L	L
	8	+	—	+	+	—	+	—	L	L	L
	9	—	+	+	+	—	+	—	L	L	L
B. suis	1	—	+	+	—	+	—	—	NL	L	L
	2	—	—	+	—	+	—	—	NL	L	L
	3	—	—	+	+	+	—	—	NL	L	L
	4	—	—	+	(—)	+	+	—	NL	L	L
B. melitensis	1	—	—	+	+	—	+	—	NL	NL	L
	2	—	—	+	+	+	—	—	NL	NL	L
	3	—	—	+	+	+	+	—	NL	NL	L
B. ovis	/	+	—	+	(+)	—	—	+	NL	NL	NL
B. canis	/	—	—	+	—	—	—	+	NL	NL	NL
B. neotomae	/	—	+	—	—	+	—	—	NL or PL	L	L

L: confluent lysis; PL: partial lysis; NL: no lysis; (+): most strains positive; (—): most strains negative.
[a] Concentration is 1/50 000 (w/v).
[b] Phage R will lyse non-smooth *B. abortus* at RTD.
[c] For more certain differentiation of *B. abortus* Type 3 and Type 6 thionin at 1/25 000 (w/v) is used in addition. Type 3: +; Type 6: —.
[d] Some strains of this biotype are inhibited by basic fuchsin.

Phage typing patterns. The pattern of lysis produced on a *Brucella* strain by brucella-phages is very dependent upon the dissociation status of the strain. If anomalous results are obtained with a phage, the dissociation status of the culture on the phage plate should be compared with that of the GDA plate, to ensure that no change has taken place during subculture. If a strain is found to be non-smooth, and does not undergo lysis with phage R, its identity should be confirmed by the techniques described previously.

Cultures in the intermediate phase may show resistance to some phages at RTD but still be lysed by phage at 10^4 RTD. In such cases tests with the higher phage concentration may be of value. Rarely, smooth cultures of *B. abortus* and other species may be encountered which are resistant to their respective phage or to all phages, even at 10^4 RTD. The identity of such isolates can usually be established from the results of the serological and cultural tests, but where possible, the oxidative metabolism patterns should be determined as confirmatory evidence.

Urease test. In cases where there is doubt about the species of *Brucella* under test, the semi-quantitative NCTC micro-method (Cowan 1974) may be useful or Bauer's method as described by Alton *et al.* (1975) may be employed. Suspensions prepared as described under (e) are used for inoculating the medium which is incubated at 37°C and inspected at intervals of 15 min for up to 4 h. In general, *B. abortus* strains will give a positive reaction, registered as a deep red colour by the phenol red indicator, after 1–2 h. Strains of *B. canis*, *B. suis* and *B. neotomae* usually produce a positive reaction almost immediately or at least within 30 min. The behaviour of *B. melitensis* strains is variable, with many strains giving negative reactions at 4 h. *Brucella ovis* always gives a negative reaction. In assessing the results of this test it should be remembered that atypical strains are quite frequently encountered.

Identification of vaccine strains

Brucella abortus *strain 19*

Using the routine identification procedures described previously, *B. abortus* strain 19 cannot be distinguished from other CO_2-independent strains of *B. abortus* biotype 1. As a first step in differentiation of this strain isolates should be examined for the characteristics described in Table 6. The techniques used for determining inhibition of growth by thionin blue or *meso*-erythritol are identical with those used in the dye sensitivity tests of the routine identification procedures (see (e)). Penicillin sensitivity is assayed by a modification of the method described for determining phage typing (see section 'Routine identification procedures' (h)) but substituting sensitivity test discs containing 5 u and 10 u of benzylpenicillin for the drops of phage suspension.

It has been shown by Morgan (1961) that some field isolates of *B. abortus* biotype 1 are sensitive to thionin blue (see Table 6). Although no fully virulent field isolate has to date shown all the cultural properties of *B. abortus* strain 19, it appears that some isolates can mimic these properties very closely. Therefore, before a suspect organism is designated as *B. abortus* strain 19, tests should be performed to ensure that its virulence is compatible with that strain. Virulence may be assessed by the following methods.

Intramuscular inoculation of guinea-pigs with a dose of 5×10^9 *viable organisms.* After 11 days the animals are killed. The spleen of each animal should contain not more than 5×10^5 viable strain 19 organisms g^{-1} on cultural examination and the serum not more than 1000 i.u. when tested by the serum agglutination test (Hulse & Carnaghan 1968).

Determination of the duration of bacteraemia following inoculation with a large number of viable organisms (Cruickshank 1957). Guinea-pigs are inoculated with a dose of $1 \cdot 8 \times 10^{10}$ viable organisms by the subcutaneous route, and blood samples taken by cardiac puncture at weekly intervals. These samples should be cultured on solid and liquid media. *Brucella abortus* strain 19 should be cleared from the blood within 2 weeks, whereas positive cultures may be obtained for 4 weeks or more from animals inoculated with fully virulent field strains.

When performing either of the above tests, a strain of known virulence, e.g. *B. abortus* strain 544, and a known sample of *B. abortus* strain 19 should be included as controls.

Brucella abortus *strain 45/20*

This is a non-smooth strain, originally derived from a smooth, CO_2-independent strain of *B. abortus* biotype 1 (McEwen & Roberts 1936). It may be distinguished from smooth strains of *B. abortus* biotype 1 by the characters shown in Table 7. There is at present no certain means of distinguishing the strain from other rough variants of biotype 1 and therefore great care should be taken in the storage and subculture of strain 45/20 for vaccine production.

The strain has been reported to revert to the smooth state *in vivo* (McEwen 1940), and hence is used as a heat-killed vaccine in adjuvant. The vaccine should contain no smooth cells. Methods for detecting the presence of small numbers of smooth cells in a suspension are given by Roerink (1966). The first method involves dilution of the suspension, followed by plating out, incubation and direct observation of individual colonies. The second involves storage of the suspension at 4°C for 24 h,

TABLE 6. Differential characters of *B. abortus* strain 19

B. abortus strain	CO_2 req't	Growth on media containing			Growth in the presence of benzylpenicillin sensitivity discs	
		thionin blue	*meso*-erythritol			
		1/500 000 (w/v)	1 mg ml^{-1}	2 mg ml^{-1}	5 iu	10 iu
Biotype 1	(+)	(+)	+	+	(+)	(+)
Variant field isolates	(+)	—	+	+	+	+
Strain 19	—	—	(—)	(—)	—	—

(+): most strains positive; (—): most strains negative.

TABLE 7. Properties of *B. abortus* strain 45/20

	CO_2	Agglutination with monospecific antisera			Lysis by phage at RTD				Minimum inhibitory concentration (MIC) of antibiotics[a] in $\mu g\ ml^{-1}$			
B. abortus strain	req't	A	M	R	Tb	Wb	Bk	R	Erythromycin	Tetracycline	Streptomycin	Ampicillin
Biotype 1 strain 544	+	+	—	—	L	L	L	NL	100	1·6	12·5	12·5
Biotype 1 field isolates	(+)	+	—	—	L	L	L	NL	100	1·6	6·2	6·2
Biotype 1 strain 19	—	+	—	—	L	L	L	NL	50	0·2	6·2	1·6
Strain 45/20	—	—	—	+	NL	NL	NL	L	⩾ 100	0·8	12·5	12·5
Rough field isolates	(+)	—	—	+	NL	NL	NL	L	⩾ 50	1·6	6·2	6·2

L: confluent lysis; NL: no lysis.
[a] MIC determined by serial dilution method in Albimi brucella broth using a standard inoculum of 10^6 viable organisms ml^{-1}.

followed by dilution, plating out and incubation of the supernatant. The dissociation status of individual colonies may then be observed. This method is said to detect smooth colonies up to a ratio (smooth/rough) of $1:10^5$.

Brucella melitensis *strain Rev 1*

This strain cannot be differentiated from field isolates of *B. melitensis* biotype 1 using the techniques described previously for routine identification. The additional characteristics given in Table 8 are useful for this purpose. The strain is of reduced virulence for guinea-pigs, doses of 10^3 viable organisms being cleared from the tissues in up to 5 months. Further details of the characteristics of strain Rev 1 are given by Alton & Elberg (1967).

TABLE 8. Differential characters of *B. melitensis* strain Rev 1

B. melitensis strain	Growth[a] on media containing basic fuchsin[b]	thionin[b]	Colony size (mm)	Growth in the presence of sensitivity discs: Penicillin 5 iu	Streptomycin 5 μg ml^{-1}
Biotype 1/16M	+	+	3–4	+	—
Rev 1	—	—	1–2	—	+

[a] Incubation in air without supplementary CO_2 for 48 h.
[b] 1/50 000 (w/v).

The authors wish to thank Dr W. J. Brinley Morgan for his encouragement in producing this chapter and for permitting the use of some of the material incorporated in the original edition, Miss E. Bain for typing the manuscript and Messrs R. Sayer and R. Cooper for the photographs.

References

ADAMS, M. H. 1959 In *Bacteriophages*. New York: Interscience.

AHVONEN, P., JANSSON, E. & AHO, K. 1969 Marked cross-agglutination between brucellae and a sub-type of *Yersinia enterocolitica*. *Acta pathologica et microbiologica scandinavica* **75**, 291–295.

ALTON, G. G. & ELBERG, S. S. 1967 Rev 1 *Brucella melitensis* vaccine. A review of ten years' study. *Veterinary Bulletin* **37**, 793–800.

ALTON, G. G., JONES, L. M. & PIETZ, D. E. 1975 *Laboratory Techniques in Brucellosis*, 2nd edn. Geneva: World Health Organization.

ANON. 1928 *International Critical Tables*. National Research Council Vol. III. New York: McGraw-Hill.

ANON. 1975 *Report of the Working Party on the Laboratory Use of Dangerous Pathogens*. (Cmnd 6054) London: Her Majesty's Stationery Office.

ARNOLD, W. M. JR. & WEAVER, R. H. 1948 Quick microtechniques for the identification of cultures. I. Indole production. *Journal of Laboratory and Clinical Medicine* **33,** 1334–1337.

ASHBY, G. K. 1938 Simplified Schaeffer spore stain. *Science* **87,** 443.

BENDICH, A. 1957 Methods for characterization of nucleic acids by base composition. In *Methods in Enzymology Vol. 3*, pp. 715–723, eds Colowick, S. P. & Kaplan, N. O. New York: Academic Press.

BHONGHBHIBHAT, N., ELBERG, S. S. & CHEN, T. H. 1970 Characterization of *Brucella* skin test antigens. *Journal of Infectious Diseases* **122,** 70–82.

BIBERSTEIN, E. L. & CAMERON, H. S. 1961 The family Brucellaceae in veterinary research. *Annual Review of Microbiology* **15,** 93–118.

BOYCE, K. J. & EDGAR, A. W. 1966 Production of freeze-dried *Brucella abortus* Strain 19 vaccine using cells produced by continuous culture. *Journal of Applied Bacteriology* **29,** 401–408.

BROWN, G. M., RANGER, C. R. & KELLEY, D. J. 1971 Selective media for the isolation of *Brucella ovis. Cornell Veterinarian* **61,** 265–280.

BUDDLE, M. B. 1956 Studies on *Brucella ovis* (N.sp.), a cause of genital disease of sheep in New Zealand and Australia. *Journal of Hygiene, Cambridge* **54,** 351–364.

BURNET, E. 1928 La thermo-agglutination et l'evaluation des espèces de *Brucella. Archives de l'Institut Pasteur de Tunis* **17,** 128–146.

CIOGLIA, L. 1950 Antigeni communi a brucelle e salmonelle. *Giornale di Batteriologia i Immunologia* **42,** 81–90.

CLARKE, P. H. & COWAN, S. T. 1952 Biochemical methods for bacteriology. *Journal of General Microbiology* **6,** 187–197.

COOK, I., CAMPBELL, R. W. & BARROW, G. 1966 Brucellosis in North Queensland rodents. *Australian Veterinary Journal* **42,** 5–8.

CORBEL, M. J. 1973*a* Examination of two bacterial strains designated '*Brucella suis* biotype 5'. *Journal of Hygiene, Cambridge* **71,** 271–282.

CORBEL, M. J. 1973*b* The direct fluorescent antibody test for detection of *Brucella abortus* in bovine abortion material. *Journal of Hygiene, Cambridge* **71,** 123–129.

CORBEL, M. J. 1975 The serological relationship between *Brucella spp., Yersinia enterocolitica* serotype IX and *Salmonella* serotypes of Kauffmann-White group N. *Journal of Hygiene, Cambridge* **75,** 151–171.

CORBEL, M. J. 1977 Production of a phage variant lytic for non-smooth *Brucella* strains. *Annali Sclavo* **19,** 99–108.

CORBEL, M. J. & CULLEN, C. A. 1970 Differentiation of the serological response to *Yersinia enterocolitica* and *Brucella abortus* in cattle. *Journal of Hygiene, Cambridge* **68,** 519–530.

CORBEL, M. J. & MORGAN, W. J. B. 1975 Proposal for minimal standards for descriptions of new species and biotypes of the genus *Brucella. International Journal of Systematic Bacteriology* **25,** 83–89.

CORBEL, M. J. & THOMAS, E. L. 1976*a* Description of a new phage lytic for several *Brucella* species. *Journal of Biological Standardisation* **4,** 195–201.

CORBEL, M. J. & THOMAS, E. L. 1976*b* Properties of some new *Brucella* phage isolates; evidence for lysogeny within the genus. International Symposium on Brucellosis (II) Rabat 1975. *Developments in Biological Standardisation* **31,** 38–45. Basel: S. Karger.

COWAN, S. T. 1974 In *Cowan and Steel's Manual for the Identification of Medical Bacteria*, 2nd edn. London: Cambridge University Press.

CRUICKSHANK, J. C. 1957 The duration of bacteraemia in relation to the virulence of *Brucella* strains. *Journal of Hygiene, Cambridge* **55,** 140–147.

DARLOW, H. M. 1969 Safety in the microbiological laboratory. In *Methods in Microbiology* 1, eds Norris, J. R. & Ribbons, D. W. London: Academic Press.

DAVIES, G., HERBERT, N. C. & CASEY, A. 1973 Preservation of *Brucella abortus* (strain 544) in liquid nitrogen and its virulence when subsequently used as a challenge. *Journal of Biological Standardisation* **1,** 165–170.

DIAZ, R., JONES, L. M. & WILSON, J. B. 1967 Antigenic relationships of *Brucella ovis* and *Brucella melitensis*. *Journal of Bacteriology* **93,** 1262–1268.

DIAZ, R., JONES, L. M. & WILSON, J. B. 1968 Antigenic relationship of the gram-negative organism causing canine abortion to smooth and rough brucellae. *Journal of Bacteriology* **95,** 618–624.

DOUGLAS, J. T. & ELBERG, S. S. 1976 Isolation of *Brucella melitensis* phage of broad biotype and species specificity. *Infection and Immunity* **14,** 306–308.

DRANOVSKAYA, E. A. & KUSHNAREV, V. M. 1968 Tsitokhromy brutsell. *Zhurnal Mikrobiologii, Epidemiologii i Immunobiologii* **45,** (12) 3–5.

DRUCKER, D. B. 1976 Gas-liquid chromatographic chemotaxonomy. In *Methods in Microbiology* **9,** ed. Norris, J. R. London: Academic Press.

DUGUID, J. P. 1951 The demonstration of bacterial capsules and slime. *Journal of Pathology and Bacteriology* **63,** 673–685.

DUSHINA, O. P., MITROFANOVA, L. I., CHUDENTSOVA, E. N. & SAUCHENKO, N. T. 1964 Isolation of atypical *Brucella* from murine rodents in the Checheno-Ingush Autonomic Republic. (In Russian.) *Zhurnal Mikrobiologii, Epidemiologii i Immunobiologii* **41,** (3) 143–144.

ELLWOOD, D. C., KEPPIE, J. & SMITH, H. 1967 The chemical basis of the virulence of *Brucella abortus*. VIII. The identity of purified immunogenic material from culture filtrate and from the cell wall of *Brucella abortus* grown *in vitro*. *British Journal of Experimental Pathology* **48,** 28–39.

EWING, W. H. & JOHNSON, J. G. 1960 The differentiation of *Aeromonas* and C27 cultures from Enterobacteriaceae. *International Bulletin of Bacteriological Nomenclature and Taxonomy* **10,** 223–230.

FARRELL, I. D. 1974 The development of a new selective medium for the isolation of *Brucella abortus* from contaminated sources. *Research in Veterinary Science* **16,** 280–286.

FEELEY, J. C. 1969 Somatic O antigen relationship of *Brucella* and *Vibrio cholerae*. *Journal of Bacteriology* **99,** 645–649.

FRANCIS, E. & EVANS, A. C. 1926 Agglutination, cross-agglutination and agglutinin absorption in tularaemia. *Public Health Reports, Washington D.C.* **41,** 1273–1295.

FREEMAN, B. A., MCGHEE, J. R. & BAUGHN, R. E. 1970 Some physical, chemical and taxonomic features of the soluble antigens of the *Brucellae*. *Journal of Infectious Diseases* **121,** 522–527.

HENRY, B. S. 1933 Dissociation in the genus *Brucella*. *Journal of Infectious Diseases* **52,** 374–402.

HINSDILL, R. D. & BERMAN, D. T. 1967 Antigens of *Brucella abortus*. 1. Chemical and immunoelectrophoretic characterization. *Journal of Bacteriology* **93,** 544–549.

HOWIE, J. W. & KIRKPATRICK, J. 1934 Observations on bacterial capsules as demonstrated by a simple method. *Journal of Pathology and Bacteriology* **39,** 165–169.

HOYER, B. H. & MCCULLOUGH, N. B. 1968*a* Polynucleotide homologies of *Brucella* deoxyribonucleic acids. *Journal of Bacteriology* **95,** 444–448.

HOYER, B. H. & MCCULLOUGH, N. B. 1968*b* Homologies of deoxyribonucleic acids from *Brucella ovis*, canine abortion organisms, and other *Brucella* species. *Journal of Bacteriology* **96,** 1783–1790.

HUDDLESON, I. F. 1929 The differentiation of the species of the genus *Brucella. Michigan State College Agricultural Experimental Station, Technical Bulletin* No. 100.

HUDDLESON, I. F. 1943 *Brucellosis in Man and Animals.* Revised edn. New York: Commonwealth Fund.

HUDDLESON, I. F. 1961 Emergence during growth of *Brucella* strains on dye-agar media of cells that show changes in sulfur metabolism. *Bulletin of the World Health Organization* **24,** 91–102.

HUGHES, D. E., WIMPENNY, J. W. T. & LLOYD, D. 1971 The disintegration of micro-organisms. In *Methods in Microbiology* **5B,** eds Norris, J. R. & Ribbons, D. W. London: Academic Press.

HULSE, E. C. & CARNAGHAN, R. B. A. 1968 Requirements for the production control of '*Brucella abortus*' (Strain 19) vaccine. International Symposium on Brucellosis, Tunis 1968; *Symposium Series, Immunobiological Standardisation* **12,** 1–10. Basel: S. Karger.

JONES, L. M. 1967 Report to the International Committee on Nomenclature of Bacteria by the Subcommittee on Taxonomy of *Brucellae*. Minutes of meeting, July 1966. *International Journal of Systematic Bacteriology* **17,** 371–375.

JONES, L. M. & WUNDT, W. 1971 International Committee on Nomenclature of Bacteria. Subcommittee on the Taxonomy of *Brucella*. Minutes of meeting, 7 August 1970. *International Journal of Systematic Bacteriology* **21,** 126–128.

KOROL, A. G. & PARNAS, J. 1967 Ein neue Serobiotyp von *Brucella—Brucella murium* (Korol). *Zeitschrift für die gesamte Hygiene und ihre Grenzgebiete* **13,** 799–800.

LAPAGE, S. P., SHELTON, J. E., MITCHELL, T. G. & MACKENZIE, A. R. 1970 Culture collections and the preservation of bacteria. In *Methods in Microbiology* **3A,** eds Norris, J. R. & Ribbons, D. W. London: Academic Press.

LOWE, G. H. 1962 The rapid detection of lactose fermentation in paracolon organisms by the demonstration of β-D-galactosidase. *Journal of Medical Laboratory Technology* **19,** 21–25.

LOWRY, O. H., ROSEBROUGH, N. J., FARR, A. L. & RANDALL, R. J. 1951 Protein measurements with the Folin phenol reagent. *Journal of Biological Chemistry* **193,** 265–275.

MARMUR, J. & DOTY, P. 1962 Determination of the base composition of deoxyribonucleic acid from its thermal denaturation temperature. *Journal of Molecular Biology* **5,** 109–118.

MCEWEN, A. D. 1940 The virulence of *Br. abortus* for laboratory animals and pregnant cattle. *Veterinary Record* **52,** 97–106.

MCEWEN, A. D. & ROBERTS, R. S. 1936 Bovine contagious abortion. The use of guinea-pigs in immunisation studies. *Journal of Comparative Pathology and Therapeutics* **49,** 97–117.

MCGHEE, J. R. & FREEMAN, B. A. 1970 Separation of soluble *Brucella* antigens by gel-filtration chromatography. *Infection and Immunity* **2,** 48–53.

MEYER, K. F. & SHAW, E. B. 1920 A comparison of the morphologic, cultural and biochemical characteristics of *Br. abortus* and *Br. melitensis*. *Journal of Infectious Diseases* **27,** 173–184.

Meyer, M. E. 1969*a* *Brucella* organisms isolated from dogs: comparison of characteristics of members of the genus *Brucella*. *American Journal of Veterinary Research* **30,** 1751–1756.

Meyer, M. E. 1969*b* Phenotypic comparison of *Brucella ovis* to the DNA-homologous *Brucella* species. *American Journal of Veterinary Research* **30,** 1757–1764.

Meyer, M. E. 1974 Advances in research on brucellosis, 1957–1972. *Advances in Veterinary Science and Comparative Medicine* **18,** 231–250.

Meyer, M. E. 1976 Evolution and taxonomy in the genus *Brucella*: brucellosis of rodents. *Theriogenology* **6,** 263–269.

Meynell, G. G. & Meynell, E. 1970 In *Theory and Practice in Experimental Bacteriology* Ch. 1 pp. 10–11. London: Cambridge University Press.

Mitruka, B. M. & Alexander, M. 1970 Differentiation of *Brucella canis* from other brucellae by gas chromatography. *Applied Microbiology* **20,** 649–650.

Moreira-Jacob, M. 1968 New group of virulent bacteriophages showing differential affinity for brucella species. *Nature, London* **219,** 752–753.

Morgan, W. J. B. 1961 The use of the thionin blue sensitivity test in the examination of brucella. *Journal of General Microbiology* **25,** 135–139.

Morgan, W. J. B. 1964 Reviews of the progress of dairy science. Section E. Diseases of dairy cattle. Brucellosis. *Journal of Dairy Research* **31,** 315–359.

Morgan, W. J. B. 1970 Reviews of the progress of dairy science. Section E. Diseases of dairy cattle. Brucellosis. *Journal of Dairy Research* **37,** 303–360.

Morgan, W. J. B. & Gower, S. G. M. 1966 Techniques in the Identification and Classification of *Brucella*. In *Identification Methods for Microbiologists* Part A, eds Gibbs, B. M. & Skinner, F. A. Society for Applied Bacteriology Technical Series No. 1. London & New York: Academic Press.

Morgan, W. J. B. & Corbel, M. J. 1976 Recommendations for the description of species and biotypes of the genus *Brucella*. *Developments in Biological Standardization* **31,** 27–37.

Morris J. A. 1973 The use of polyacrylamide gel electrophoresis in taxonomy of *Brucella*. *Journal of General Microbiology* **76,** 231–237.

Morris, J. A. & Corbel, M. J. 1973 Properties of a new phage lytic for *Brucella suis*. *Journal of General Virology* **21,** 539–544.

Morris, J. A., Corbel, M. J. & Phillip, J. I. H. 1973 Characterisation of three phages lytic for *Brucella* species. *Journal of General Virology* **20,** 63–73.

Ouchterlony, O. 1953 Antigen-antibody reactions in gels. *Acta pathologica et microbiologica scandinavica* **32,** 231–240.

Renoux, G. 1958 La notion d'espèce dans le genre *Brucella*. *Annales de l'Institut Pasteur, Paris* **94,** 179–206.

Renoux, G. & Philippon, A. 1969 Position taxonomique dans le genre *Brucella* de bactéries isoleés de brebis et de vaches. *Annales de l'Institut Pasteur, Paris* **117,** 524–528.

Rhodes, M. E. 1958 The cytology of *Pseudomonas* spp. as revealed by a silver-plating staining method. *Journal of General Microbiology* **18,** 639–648.

Ribi, E., Milner, K. C. & Perrine, T. D. 1959 Endotoxic and antigenic fractions from the cell wall of *Salmonella enteritidis*. Methods for separation and some biologic activities. *Journal of Immunology* **82,** 75–84.

ROERINK, J. H. G. 1966 Development of a non-agglutinogenic killed *Brucella abortus* adjuvant vaccine and its applicability in the control of bovine brucellosis. Thesis, Ryksuniversiteit te Utrecht.

RUTTER, D. A. 1977. Safety cabinet for use in laboratory studies on hazardous infectious diseases. *British Medical Journal* **2,** 24.

SCHILDKRAUT, C. L., MARMUR, J. & DOTY, P. 1962 Determination of the base composition of deoxyribonucleic acid from its buoyant density in CsCl. *Journal of Molecular Biology* **4,** 430–443.

SKIDMORE, W. D. & DUGGAN, E. L. 1971 Base composition of nucleic acids. In *Methods in Microbiology* **5B,** eds Norris, J. R. & Ribbons, D.W. London: Academic Press.

STABLEFORTH, A. W. 1959 Species of *Brucella* and their general distribution. In *Infectious Diseases of Domestic Animals, Diseases due to Bacteria*, eds Stableforth, A. W. & Galloway, I. A. London: Butterworths.

STABLEFORTH, A. W. & JONES, L. M. 1963 Report of the Subcommittee on Taxonomy of the Genus *Brucella. International Bulletin of Bacteriological Nomenclature and Taxonomy* **13,** 145–158.

SULKIN, S. E. & PIKE, R. M. 1951 Survey of laboratory-acquired infections. *American Journal of Public Health* **41,** 769–781.

TANAKA, S., SUTO, T., ISAYAMA, Y., AZUMA, R. & HATAKEYAMA, H. 1977 Chemotaxonomical studies on fatty acids of *Brucella* spp. *Annali Sclavo* **19,** 67–82.

TARAN, I. F., POGORELOV, N. A., KULIKOVA, G. G., KUTSEMAKINA, A. Z., RUDNEV, M. M., NELYAPIN, N. M., RUDNEVA, V. A. & SUVAROVA, A. E. 1966 Izucheniye brutsellyeznykh kultur, vȳdelennȳkh ot mȳshevidnȳkh grȳzunov i ikh ektoparazitov. *Zhurnal Mikrobiologii, Epidemiologii i Immunobiologii* **43,** (9) 70–74.

THOMAS, E. L. & CORBEL, M. J. 1977 Isolation of a phage lytic for several *Brucella* species following propagation of Tbilisi phage in the presence of Mitomycin C. *Archives of Virology* **54,** 259–261.

UMBREIT, W. W., BURRIS, R. H. & STAUFFER, J. F. 1959 In *Manometric Techniques* Ch. 4, pp. 46–63. Minneapolis: Burgess Publishing Company.

WHITE, P. G. & WILSON, J. B. 1951 Differentiation of smooth and non-smooth colonies of *Brucellae. Journal of Bacteriology* **61,** 239–240.

WILSON, G. S. & MILES, A. A. 1932 The serological differentiation of smooth strains of the *Brucella* group. *British Journal of Experimental Pathology* **13,** 1–13.

WILSON, G. S. & MILES, A. A. 1975 In *Topley and Wilson's Principles of Bacteriology, Virology and Immunity*. 6th edn., Vol. 1, Ch. 33, pp. 1054–1078. London: Edward Arnold.

WUNDT, W. 1959 Zur Frage der Antigengemeinschaften zwischen Brucellen und Bakterien anderer Gattungen. *Zeitschrift für Hygiene und Infektionskrankheiten* **145,** 556–563.

WUNDT, W. & MORGAN, W. J. B. 1975 International Committee on Systematic Bacteriology. Subcommittee on Taxonomy of *Brucella*. Minutes of the meeting, 3 September 1974. *International Journal of Systematic Bacteriology* **25,** 235–236.

Biochemical Identification of Enterobacteriaceae

S. P. Lapage, B. Rowe, B. Holmes and R. J. Gross

Central Public Health Laboratory, Colindale, London NW9 5HT, UK

Though serological methods are of great importance in the routine identification of the major pathogenic groups of the Enterobacteriaceae, only biochemical methods are given here. Initial, accurate, biochemical identification may be supplemented by further serological, bacteriophage and similar studies which may lead to finer divisions within the species.

There are many purposes for which a set of tests may be needed for screening isolates, e.g. the isolation of *Salmonella* and *Shigella* from patients, the screening of urinary isolates or of food and water bacteria. The reader is therefore left to select suitable screening sets from the tests used for generic differentiation in Tables 2 and 3. Sets of tests for screening purposes can be derived from a series of single tests, from composite media of which many commercial multiple media are now available, and from diagnostic kits of various kinds.

Methods

Biochemical tests and media

The temperature of incubation is 37°C unless otherwise stated. In all the media, Oxoid peptone L37 has been substituted for other peptones described in the original formulations unless otherwise stated.

Acetate test medium. This is described in Edwards & Ewing (1972) who state on page 342: "This medium is the same as Simmons' citrate agar except that 0·25% sodium acetate is used instead of citrate. It is inoculated, incubated, and interpreted in the same manner as Simmons' agar."

β-galactosidase (ONPG) test. The test of Lowe (1962) is used (Lab M No. 1 peptone).

Catalase production. Hydrogen peroxide, 1 ml of 6% (v/v), in water is run down an 18–24 h agar slope culture and the presence of bubbles is taken as a positive result.

Citrate utilization, Christensen's. The medium of Christensen (1949) is used. See also Anon. (1958*a*).

Citrate utilization, Simmons'. The medium of Simmons (1926) is used.

Decarboxylase production. The method of Møller (1955) is used.

Fermentation tests. Carbohydrates 0·5% (w/v) in 1% (w/v) peptone water with Andrade's indicator are used, and the reactions observed daily.

Gelatin liquefaction, stab culture. Nutrient gelatin stab cultures (Difco Bacto-Peptone) are grown at room temperature and observed daily; alternatively, they are incubated at 37°C and observed daily after cooling to about 18°C.

Gelatinase production, plate method. The method of Frazier (1926) is used.

Gluconate test. The method of Shaw & Clarke (1955) is used but modified (Carpenter 1961) by substituting 'Clinitest' Reagent Tablets (Ames Company, Stoke Poges, Slough, Bucks.) for Benedict qualitative solution.

H_2S production, triple sugar iron agar (TSI). The method of Sulkin & Willett (1940) is used. This is described by Edwards & Ewing (1972) on page 346. The use of lead acetate papers to detect H_2S production is not described since different results are obtained with different media; and different interpretations of weak or doubtful positive reactions are possible.

Indole production. 2% (w/v) peptone water cultures (Lab M No. 1 peptone), after incubation for 24 h, are tested with Kovács' (1928) reagent (an occasional weakly positive strain may require 48 h incubation).

KCN tolerance. The method of Møller (1954) is used, modified by using bijou bottles with caps very tightly screwed. Positive and negative control strains are always used and all strains must be tested for their ability to grow in the basal medium.

Malonate test. The medium and method is that of Shaw & Clarke (1955). This medium is also used for the PPA test, q.v.

Methyl red test (MR). Cultures grown in buffered glucose-phosphate broth are tested after incubation for 5 days at 30°C or 2 days at 37°C.

Nitrate reduction. Cultures in 0·1% (w/v) nitrate broth are tested, after incubation for 5 days, by the Griess-Ilosvay method (Wilson & Miles 1964) with replacement of α-naphthylamine by 1-naphthylamine-7-sulphonic acid, i.e. Cleve's acid (Crosby 1967).

Organic acids. Tartrates, mucate and sodium citrate tests (Difco Bacto-Peptone). These tests are described in Kauffman (1966).

Oxidase production. The method of Kovács' (1956) is used. The reagent is either 1% (w/v) *p*-aminodimethylaniline oxalate (Difco) or tetramethyl-*p*-phenylenediamine hydrochloride in distilled water. A platinum (not nichrome) wire or glass rod must be used for this test.

Oxidation/fermentation test. The medium is that of Hugh & Leifson (1953).
Phenylpyruvic acid production (PPA). The production of PPA from phenylalanine is detected either in the medium of Shaw & Clarke (1955), which is also used for the malonate utilization test; or on phenylalanine agar (Test 18, Anon. 1958*a*).
Urease production. Christensen (1946) medium is used.
Voges-Proskauer (VP). The medium described under Methyl Red is employed, and the VP test of O'Meara (1931) is used.

TABLE 1. Definition of the family, Enterobacteriaceae

Gram negative rods; motile by peritrichous flagella or non-motile
Capsulated or non-capsulated
Not sporeforming; not acid fast
Aerobic and facultatively anaerobic
Grow readily on simple nutrient and bile salt media
Chemo-organotrophic; metabolism respiratory and fermentative
Acid is produced from the fermentation of glucose, other carbohydrates and alcohols; usually aerogenic but anaerogenic groups and mutants occur
Oxidase negative
Catalase positive, with the exception of *Shigella dysenteriae* type 1
Nitrates are reduced to nitrites, with the exception of one biotype of *Yersinia*
G + C content of the DNA is 39–59 mol. %. Values for individual genera are listed below:[a]

Arizona	50–53	*Klebsiella*	52–56
Citrobacter	52–52·6[b, c]	*Proteus*	38–50
Edwardsiella	Unpublished[c]	*Providencia*	41·5[3]
Enterobacter	52–59	*Salmonella*	50–53
Erwinia[d]	52·6–57·7	*Serratia*	53–59
Escherichia	50–51	*Shigella*	49–53[b]
Hafnia	52–57	*Yersinia*	45·8–46·8

[a] From Buchanan & Gibbons (1974) unless otherwise indicated.
[b] From Normore (1973), for *C. freundii* only.
[c] As the DNA base composition (mol. % G + C) of certain genera and of some species within genera was not known, the mol. % G + C was estimated in the NCTC by R. J. Owen from the thermal denaturation temperature (T*m*) determined by the method of Marmur & Doty (1962). The mol. % G + C was calculated from the T*m* determined in 0·015 mol. l^{-1} NaCl buffered with 0·0015 mol. l^{-1} trisodium citrate at pH 7·0 by the equation: % G + C = 2·08 T*m* − 106·4 (Owen & Lapage 1976). The % G + C values below were expressed relative to a value of 50·9 mol. % G + C for *Escherichia coli* strain B (NCTC 10537) which was used as a DNA reference strain.
The results for the following strains were:

Citrobacter (Levinea) amalonaticus	NCTC 10805 = 54·0
Citrobacter koseri	NCTC 10786 = 54·2
Edwardsiella tarda	NCTC 10396 = 56·7
Escherichia adecarboxylata	NCTC 10599 = 55·1
Escherichia blattae	NCTC 10965 = 55·9
Providencia alcalifaciens	NCTC 10286 = 42·0
Providencia stuartii	NCTC 1038 = 42·0

[d] *E. herbicola* group only.

Biochemical Reactions of the Genera

Key to tables

A	Acid production from carbohydrate fermentation after overnight incubation.
G	Gas, $\frac{1}{4}$ or more volume of Durham tube; recorded for glucose only, except in Table 6.
g	Gas less than $\frac{1}{4}$ volume of Durham tube; recorded for glucose only.
+	Acid production from carbohydrate fermentation or a positive result in other tests after overnight incubation except where otherwise specified under 'Methods' or in the tables.
[+]	Acid production from carbohydrate fermentation or a positive reaction in other tests after incubation for 48 h or longer.
D	Different species, biotypes or serotypes give different reactions.
d	Different strains give different reactions.
D/d	Different species, biotypes or serotypes usually give different reactions but occasional variants occur.
/	The symbol before the '/' is the commoner result and that after is a rare finding e.g. +/—; —/+; +/[+].
—	Negative reaction in the test specified.

The reactions shown in the tables are those given by a majority of strains. Inevitably there are exceptions. Allowance has been made for the isolation of rare species from clinical material, and for strains giving an unusual result for a species. Tables 2 and 3 therefore represent a compromise between typical findings and allowance for rarer results.

Some Characters of the Genus *Salmonella*

Motile with peritrichous flagella, non-motile flagellate or non-flagellate strains may occur; two species, *Salmonella gallinarum* and *S. pullorum* are non-flagellate. Some salmonellae possess fimbriae.

Metabolism. Salmonellae show a mixed acid fermentation, the methyl red test is positive and acetylmethylcarbinol is not produced. Indole is not formed, nor is urea hydrolysed. Glucose is fermented with the production of acid, and gas is produced by most strains, except members of the species *S. gallinarum* and *S. typhi*. Mannitol is fermented. Maltose is fermented by the majority of species, but for a few, e.g. *S. pullorum*, the absence of maltose fermentation is characteristic. Dulcitol is fermented by most species, but a few medically important species attack it late or do not ferment it, e.g. *S. choleraesuis*, *S. paratyphi* A, *S. typhi*. Lactose, sucrose and adonitol are not fermented. Occasional biochemically aberrant strains occur, but the vast majority yield the typical reactions given above. See Tables 8 and 9.

TABLE 2. Fermentation reactions of genera[a]

Genus	Glucose	Adonitol	Dulcitol	Inositol	Lactose	Mannitol	Salicin	Sucrose	Xylose
Arizona	AG	—	—	—	+/—	+	—	—	+
Citrobacter	AG	D	d	—	+/—	+	d	d	+
Edwardsiella	AG	—	—	—	—	—	—	—	—
Enterobacter	AG	d	d	D	+	+	+	+	+
Erwinia	A	—	—	—	d	+	+	+/—	+/—
Escherichia	AG/A	—	d	—	+/—	+	d	d	+
Hafnia	AG	—	—	—	—/+	+	—/+	—/+	+
Klebsiella[b]	AG/A	+/—	d	+/—	+/[+]	+	+	+	+
Proteus	AG/A	D	—	D	—	D	d	d	D
Providencia	A/Ag	D	—	D	—	—/+	—	[+]/—	—
Salmonella	AG/A	—	D	—/+	—	+	—/+	—	+/—
Serratia	Ag/A	D	—	+/—	d	+	+	+	d
Shigella	A	—	D	—	—/[+]	D	—	—/[+]	D
Yersinia	A	—	—	—/[+]	—	+	d	D	D

[a] See detailed tables listed in the footnote to Table 3 for species differentiation.
[b] Excluding *K. rhinoscleromatis* (see Table 12 for the biochemical reactions of this species).

TABLE 3. Other biochemical reactions of genera

Genera	β-galactosidase	Simmons' citrate	Decarboxylases: Arginine	Decarboxylases: Lysine	Decarboxylases: Ornithine	Gelatin[a]	Gluconate
Arizona[c]	+/—	+	[+][b]	+	+	[+]	—
Citrobacter[d]	+/—	+/—	+/[+]	—	d	—	—
Edwardsiella	—	—	—	+	+	—	—
Enterobacter[e]	+	+	D	D	+	[+]/—	+
Erwinia[f]	+	d	—	—	—	[+]/+	d
Escherichia[g, h]	+/—	—	d	+/—	d	—	—
Hafnia[e, i]	+	+	—	+	+	—	+
Klebsiella[e, j]	+	+/—	—	+	—	—/[+]	D
Proteus[k]	—	D	—	—	D	D	—/+
Providencia[k]	—	+/—	—	—	—	—	—
Salmonella[c, l]	—/+	D	[+][b]	+	+	—/[+]	—
Serratia[e, m]	+	+	—	+/—	D	+/[+]	+
Shigella[g, n]	D	—	—/[+]	—	D	—	—
Yersinia[o]	+	—	—	—	D	—	—

[a] Rapid or slow gelatin liquefiers (tube test) should give a positive plate result within 5 days. The symbol [+] for this test thus implies a positive plate test.
[b] Characteristically positive after incubation for 48 h.
[c] See Table 8 for the diagnostic biochemical reactions of *Salmonella* subgenera.
[d] See Table 10 for the diagnostic biochemical reactions of the genus *Citrobacter*.
[e] See Table 11 for the diagnostic biochemical reactions of the genera *Enterobacter*, *Hafnia*, *Klebsiella* and *Serratia*. Note that *Enterobacter liquefaciens* has been transferred to the genus *Serratia* (see Table 13).
[f] *Erwinia: E. herbicola* group only; this organism is referred to by Ewing & Fife (1972) as *Enterobacter agglomerans*.
[g] See Table 7 for the differentiation of *Escherichia coli* and *Shigella*.
[h] *Escherichia*: *E. adecarboxylata* (Leclerc 1962) gives the opposite results to typical *E. coli* in the following tests: Christensen's citrate, lysine, adonitol, KCN and malonate. Ewing & Fife (1972) include *E. adecarboxylata* in *Enterobacter agglomerans* but these two species give opposite results to each other in the following tests: indole, gas from glucose and fermentation of adonitol and dulcitol. *E. blattae* (Burgess *et al.* 1973) gives the opposite results to typical *E. coli* in the following tests: β-galactosidase, mannitol, gluconate and indole. Both these species have been placed in the genus *Escherichia* for convenience rather than on their resemblance to *E. coli*.

H_2S (TSI)	Indole	KCN	Malonate	MR	PPA	Urease	VP	Motility
+	—	—	+	+	—	—	—	+
D	D	D	D/d	+	—	d	—	+
+	+	—	—	+	—	—	—	+
—	—	+	+/—	—/+	—	d	+/—	+
—	—	—/+	+/—	+/—	—	—	d	+
—	+/—	—	—	+	—	—	—	d
—	—	+/—	d	—	—	—	+	+
—	—/+	D	D	D	—	+/—	D	—
D	D	+	—	+/—	+	+	—/+	+
—	+	+	—	+	+	—	—	+
+/—	—	—/+	—/+	+	—	—	—	+/—
—	—	+	—/+	d	—	—/+	+/—	+
—	d	—	—	+	—	—	—	—
—	D	—/+	—	+/—	—	D	D/d	D

[i] *Hafnia*: MR, VP and citrate reactions are those given after incubation at 30°C. Some strains may give positive MR test and negative VP test results and fail to grow on Simmons' citrate if incubated at 37°C.
[j] See Table 12 for the diagnostic biochemical reactions of the genus *Klebsiella*; *K. rhinoscleromatis* is not included in the *Klebsiella* row in this Table; some strains of *K. ozaenae* give a delayed or negative reaction in the lysine test.
[k] See Table 14 for the diagnostic biochemical reactions of the genera *Proteus* and *Providencia*.
[l] See Table 9 for the differentiation of some aberrant biotypes of *Salmonella*.
[m] See Table 13 for the diagnostic biochemical reactions of the genus *Serratia*.
[n] See Table 4 for the classification and nomenclature of *Shigella*, Table 5 for the diagnostic biochemical reactions of *Shigella* subgroups, and Table 6 for biotypes of *S. flexneri* 6.
[o] See Table 15 for the diagnostic biochemical reactions of the genus *Yersinia*.

TABLE 4. Classification and nomenclature of *Shigella*

Characters	Species and serotypes	Main synonyms
Subgroup A	*S. dysenteriae* 1	*S. shigae*
	S. dysenteriae 2	*S. schmitzii*, *S. ambigua*
Non-mannitol fermenters; each serologically distinct	*S. dysenteriae* 3	*S. largei* Q771; *S. arabinotarda* A
	S. dysenteriae 4	*S. largei* Q1167; *S. arabinotarda* B
	S. dysenteriae 5	*S. largei* Q1030
	S. dysenteriae 6	*S. largei* Q454
	S. dysenteriae 7	*S. largei* Q902
	S. dysenteriae 8	Serotype 599-52
	S. dysenteriae 9	Serotype 58
	S. dysenteriae 10	Serotype 2050-50
Subgroup B	*S. flexneri* 1a	V
	S. flexneri 1b	VZ
Usually mannitol fermenters; members serologically related to each other	*S. flexneri* 2a	W
	S. flexneri 2b	WX
	S. flexneri 3a	Z
	S. flexneri 3b	
	S. flexneri 3c	
	S. flexneri 4a	103[a]
	S. flexneri 4b	103Z
	S. flexneri 5	P119
	S. flexneri 6	Newcastle, Manchester or Boyd 88 bacillus[b]
	S. flexneri X variant	X
	S. flexneri Y variant	Y
Subgroup C	*S. boydii* 1	170
	S. boydii 2	P288
Usually mannitol fermenters; each serologically distinct	*S. boydii* 3	D1
	S. boydii 4	P274
	S. boydii 5	P143
	S. boydii 6	D19
	S. boydii 7	Lavington 1; *S. etousae*
	S. boydii 8	Serotype 112
	S. boydii 9	Serotype 1296/7
	S. boydii 10	Serotype 430
	S. boydii 11	Serotype 34
	S. boydii 12	Serotype 123
	S. boydii 13	Serotype 425
	S. boydii 14	Serotype 2770-51
	S. boydii 15	Serotype 703
Subgroup D	*S. sonnei*	Duval's bacillus; *B. ceylanensis* A
Mannitol fermenter, late lactose and sucrose fermenter		

[a] Mannitol negative biotypes are sometimes known as *S. rabaulensis* or *S. rio*.
[b] See Table 6.
Modified from Anon. (1958*b*).

TABLE 5. Diagnostic biochemical reactions of *Shigella* subgroups

Test	*S. dysenteriae*	*S. flexneri*	*S. boydii*	*S. sonnei*
β-galactosidase	D/d	—	D/d	+/—
Ornithine decarboxylase[a]	—	—	D	+
Glucose[b]	A	A	A	A
Dulcitol[c]	—/+	—/+	D/d	—
Lactose	—	—	—	[+]/—
Mannitol	—	+	+	+
Raffinose	—	d	—	+/—
Sucrose	—	—	—	[+]/—
Xylose	—/+	—	d	—/+
Indole[d]	D	d	D	—

All subgroups are: non-motile and unable to grow on Simmons' citrate, do not produce H_2S, urease or phenylpyruvic acid, nor liquefy gelatin, nor ferment salicin, adonitol or inositol.

[a] Ornithine decarboxylase: *S. boydii* 13 and *S. sonnei* give positive results.
[b] Gas production from glucose: only certain biotypes of *S. flexneri* 6 (Table 6), and of *S. boydii* 13 (Rowe *et al.* 1975) and *S. boydii* 14 (Carpenter 1961) are aerogenic.
[c] Dulcitol: *S. dysenteriae* 5 and *S. flexneri* 6 may ferment dulcitol.
[d] Indole: *S. dysenteriae* 1, *S. flexneri* 6 and *S. sonnei* never produce indole while strains of *S. dysenteriae* 2 always produce indole.

TABLE 6. Biotypes of *Shigella flexneri* 6[a]

Old name	Glucose	Dulcitol[b]	Mannitol	Indole
Boyd type 88	A	—	A	—
Boyd type 88	A	A	A	—
Sh. newcastle	AG/A	AG/A/—	—	—
Manchester bacillus	AG	AG/—	AG	—

[a] All biotypes are serologically identical.
[b] Fermentation may be delayed.

TABLE 7. Differentiation of *Escherichia coli* and *Shigella*

Test	*E. coli*	*Shigella*
Sodium acetate[a]	+/—	—
β-galactosidase[b]	+/—	D
Christensen's citrate	+/—	—
Lysine decarboxylase	+/—	—
Glucose[c]	AG/A	A
Lactose[d]	+/—	—
Salicin	d	—
Indole	+/—	d
Sodium mucate[e]	+/—	—
Motility	d	—

[a] Sodium acetate: Some strains of *S. flexneri* 4a may grow on sodium acetate after more than 24 h incubation.
[b] β-galactosidase: *S. sonnei* and *S. dysenteriae* 1 usually give a positive result; other serotypes occasionally produce β-galactosidase.
[c] Gas from glucose: Some biotypes of *S. flexneri* 6 are aerogenic (Table 6).
[d] Lactose: *S. sonnei* frequently ferments lactose after incubation for more than 24 h.
[e] Sodium mucate: Some strains of *S. sonnei* may give positive results.

TABLE 8. Diagnostic biochemical reactions of *Salmonella* subgenera

Test	Subgenus I	II	III = Arizona	IV
β-galactosidase	—	—/+	+/—	—
Dulcitol	+	+	—	—
Lactose	—	—	+ or [+] or —	—
Salicin	—	—	—	+
Gelatin	—	[+]	[+]	[+]
KCN	—	—	—	+
Malonate	—	+	+	—
d-tartrate	+	— or [+]	— or [+]	—/[+]
i-tartrate	d	—	—	—
l-tartrate	d	—	—	—
Mucate	+	+	d	—
Sodium citrate	+	+	+	+

Modified from Kauffman (1966).

TABLE 9. Differentiation of some aberrant biotypes of *Salmonella*

Test	Typical *Salmonella* sp.	*S. choleraesuis*	*S. gallinarum*[a]	*S. paratyphi* A	*S. paratyphi* C	*S. pullorum*[a]	*S. typhi*
Decarboxylases:							
Arginine	[+]	[+]	—	+	+	d	[+] or —
Lysine	+	+	+	—	+	+	+
Ornithine	+	+	—	+	+	+	—
Gas from glucose	+	+	—	+	+	d	—
Arabinose	+	—	+	+	d	+	d
Dulcitol	+	—	+/[+]	+/[+]	d	—	—/[+]
Maltose	+	+	[+]/+	+	+	—	+
Rhamnose	+	+	+/[+]	+	+	+/—	—
Trehalose	+	—	+/[+]	+	[+][b]	+/—	+
Xylose	+	+	+/[+]	—	+	d	d
H_2S	+	D[c]	+/—	—/+	+	+	+[d]
Motility	+	+	—	+	+	—	+/—[e]

[a] Predominantly isolated from poultry.
[b] Acid and gas production after 3 days in trehalose is typical of *S. paratyphi* C.
[c] *S. choleraesuis* var. *kunzendorf* does not produce H_2S while the American variety does so.
[d] The majority of strains received at the Salmonella Reference Laboratory produce H_2S in one day though the reaction may be weak.
[e] Motility may be difficult to demonstrate in semi-solid media, although agglutination with specific antiflagellar antiserum is often present in such cases.

TABLE 10. Diagnostic biochemical reactions of the genus *Citrobacter*

Test	*C. freundii*	*C. koseri*[a]	*C.* (*Levinea*) *amalonaticus*[b]
Ornithine decarboxylase	d	+/—	+
Adonitol	—	+	—
Dulcitol	d	d	—
Lactose	d	+ or [+] or —	+ or [+] or —
Raffinose	d	—	—
Salicin	d	d	+ or [+] or —
Sucrose	d	d	—
H_2S	+/—	—	—
Indole	—/+	+	+
KCN	+	—	+
Malonate	—/+	+/—	—

[a] *C. koseri*, Frederiksen (1970): This organism is also referred to as *C. diversus* (Ewing & Davis 1972; Werkman & Gillen 1932) and as *Levinea malonatica* (Young *et al.* 1971).
[b] *C. amalonaticus*: Although *Levinea amalonatica* is considered by Ewing & Davis (1972) to correspond to an H_2S-negative, indole-positive biotype of *C. freundii*, Holmes *et al.* (1974) and Sakazaki *et al.* (1976) have suggested that *L. amalonatica* may be a species of *Citrobacter* separate from *C. freundii*.

TABLE 11. Diagnostic biochemical reactions of the genera *Enterobacter*, *Hafnia*, *Klebsiella* and *Serratia*

Test	*E. aerogenes*	*E. cloacae*	*Hafnia*	*Klebsiella*[a]	*Serratia*
Decarboxylases					
Arginine	—	+	—	—	—
Lysine	+	—	+	+	+/—
Ornithine	+	+	+	—	D
Adonitol	+	d	—	+/—	D
Inositol	+	—/+	—	+/—	+/—
Rhamnose	+	+	+	+/—	—
Gelatin	d[b]	[+]/—	—	—/[+]	+/[+]
Motility	+	+	+/—	—	+

[a] Excluding *K. rhinoscleromatis* (see Table 12 for the biochemical reactions of this species); some strains of *K. ozaenae* give a delayed or negative reaction in the lysine test.
[b] If positive, liquefaction is delayed.

TABLE 12. Diagnostic biochemical reactions of the genus *Klebsiella*[a]

Test	*K. aerogenes*	*K. atlantae*[b]	*K. edwardsii*[b]	*K. ozaenae*	*K. pneumoniae*[c]	*K. rhinoscleromatis*
β-galactosidase	+	+	+	+	+	—
Simmons' citrate	+	+	d	d	+	—
Lysine decarboxylase	+	+	+	d	+	—
Gas from glucose	+	+	—	d	+	—
Dulcitol	d	—	—	—	+	—
Lactose	+	[+]	[+]	[+]	+	—
Gluconate	+	d	+	—	d	—
KCN	+	+	+	+/—	—	+
Malonate	+	—	d	—	+	+
MR	—	+	d	+	+	+
Urease	+	+	+	—/+	+	—
VP	+	d	+	—	—	—

[a] Some authors consider that the indole positive, gelatin liquefying strains of *Klebsiella* justify a separate species, *K. oxytoca*; however, the DNA hybridization studies of Jain *et al.* (1974) indicated that such strains should be excluded from the genus *Klebsiella*.
[b] *K. atlantae* and *K. edwardsii*: These organisms are considered by Cowan (1974) as separate species, and are therefore added to this table although their species status was not maintained in the computer study of Bascomb *et al.* (1971).
[c] *K. pneumoniae*: This name is used in two senses, in *sensu lato* by Edwards & Ewing (1972) and other authors in which the name circumscribes both *K. aerogenes* and *K. pneumoniae*. *K. pneumoniae* is used in *sensu stricto* in this table may not justify species status but in the computer study of Bascomb *et al.* (1971) justified biochemical separation.
Chiefly after Cowan 1974.

TABLE 13. Diagnostic biochemical reactions of the genus *Serratia*

Test	*S. liquefaciens*	*S. marcescens*	*S. marinorubra*[a, b]	*S. plymuthica*[a]
Lysine	+/—	+	d	—
Ornithine	+	+	—	—
Gas from glucose	d	d	—	d
Adonitol	—	d	+	—
Arabinose	+/—	—	+	+
Lactose	d	—	+	d
Raffinose	+/—	—	+	d
Sorbitol	+	+/—	—	d
Xylose	+	d	+	+
Red pigment	—	d	d	d

[a] See Grimont *et al.* (1977).
[b] *S. marinorubra:* The name *Serratia rubidaea* has been used for this organism (Ewing *et al.* 1973).

TABLE 14. Diagnostic biochemical reactions and G + C content of the DNA of the genera *Proteus* and *Providencia*[a]

Test	*Prot. mirabilis*	*Prot. morganii*[b]	*Prot. rettgeri*[c]	*Prot. vulgaris*	*Prov. alcalifaciens*[d]	*Prov. stuartii*[e]
PPA	+	+	+	+	+	+
Urease	+	+	+	+	—	—
Simmons' citrate	d	—	+/—	—/+	+	+
Ornithine decarboxylase	+	+	—	—	—	—
Glucose, acid and gas	AG	AG	A	AG	Ag	A
Adonitol	—	—	+	—	+	—
Inositol	—	—	+	—	—	+
Maltose	—	—	—	+	—	—/+
Mannitol	—	—	+	—	—	—/+
Salicin	—/[+]	—	d	d	—	—
Sucrose	[+]/—	—	d	+	[+]/—	[+]/—
Trehalose	+	d	—	[+]/—	—	+/[+]
Xylose	+	—	—/+	+	—	—
Gelatin	+	—	—	+	—	—
H_2S	+	—	—	+	—	—
Indole	—	+	+	+	+	+
% G + C[f]	39·3	50	39	39·3	41·5	41·5

[a] Anomalies: anaerogenic strains of *Prot. mirabilis*, *Prot. morganii* and *Prot. vulgaris* and aerogenic strains of *Prot. rettgeri* occur occasionally. Some *Prot. mirabilis* strains are indole positive and some *Prot. vulgaris* strains are indole negative. Urease producing strains resembling *Prot. rettgeri* but which ferment trehalose and do not ferment adonitol and mannitol should be classified as *Providencia stuartii* (Farmer III *et al.* 1977).

[b] *Prot. morganii*: This species is placed by some authors in a separate genus, *Morganella* (see also Anon. 1963).

[c] *Prot. rettgeri*: This species is placed by some authors in a separate genus, *Rettgerella* (see also Anon. 1963).

[d] *Providencia alcalifaciens*: Corresponds to Subgroup A of *Providencia*.

[e] *Providencia stuartii*: Corresponds to Subgroup B of *Providencia*.

[f] From Buchanan & Gibbons (1974).

Modified from Carpenter (1964).

TABLE 15. Diagnostic biochemical reactions of the genus *Yersinia*

Tests	*Y. enterocolitica* biotype 1[a]	*Y. enterocolitica* biotype 2[a]	*Y. enterocolitica* biotype 3	*Y. enterocolitica* biotype 4[b]	*Y. enterocolitica* biotype 5[c]	*Y. pestis*	*Y. pseudotuberculosis*
β-galactosidase	+	+	+/—	+/—	—/+	+	+
Ornithine	+	+	+	+	—/+	—	—
Cellobiose	+	+	+	+	+	—	—
Rhamnose	—/+	—	—	—	—	—	+
Salicin	+/—	—/+	—	—	—	d	[+]
Sorbitol	+	+	+/—	+	—/+	d	—
Sucrose	+	+/—	+	+	+/—	—	—
Trehalose	+	+	+	+	—	+	+
Xylose	+	+	+	—	—	+/—	+
Indole	+	[+]	—	—	—	—	—
Nitrate	+	+	+	+/—	—	+/—	+
Urease	+	+	+	+	+	—	+
Voges-Proskauer R.T.[d]	+/—	+/—	+/—	+/—	—/+	—	—
Motility R.T.[d]	+	+	+	+	+	—	+

[a] *Y. enterocolitica* biotype 1 produces lipase at room temperature when grown on nutrient agar incorporating Tween 80 (1% v/v); biotype 2 (as well as biotypes 3–5) does not produce lipase. This test is not included in the table or in the 'Methods' section as this test is unlikely to be performed by the routine laboratory.
[b] Most human and porcine isolates belong to biotype 4.
[c] Strains of biotype 5 are isolated predominantly from hares.
[d] R.T., room temperature (18–22°C).
Modified from Niléhn 1969.

References

ANON. 1958*a* Report of the *Enterobacteriaceae* Subcommittee of the Nomenclature Committee of the International Association of Microbiological Societies. *International Bulletin of Bacteriological Nomenclature and Taxonomy* **8,** 25–70.

ANON. 1958*b* Supplement to the third report on the *Shigella* group. *International Bulletin of Bacteriological Nomenclature and Taxonomy* **8,** 93–95.

ANON. 1963 Report of the Subcommittee on taxonomy of the *Enterobacteriaceae*. *International Bulletin of Bacteriological Nomenclature and Taxonomy* **13,** 69–93.

BASCOMB, S., LAPAGE, S. P., WILLCOX, W. R. & CURTIS, M. A. 1971 Numerical classification of the tribe Klebsielleae. *Journal of General Microbiology* **66,** 279–295.

BUCHANAN, R. E. & GIBBONS, N. E. (eds) 1974 *Bergey's Manual of Determinative Bacteriology*, 8th edn. Baltimore: Williams & Wilkins.

BURGESS, N. R. H., MCDERMOTT, S. N. & WHITING, J. 1973 Aerobic bacteria occurring in the hind-gut of the cockroach, *Blatta orientalis*. *Journal of Hygiene, Cambridge* **71,** 1–7.

CARPENTER, K. P. 1961 The relationship of the Enterobacterium A12 (Sachs) to *Shigella boydii* 14. *Journal of General Microbiology* **26,** 535–542.

CARPENTER, K. P. 1964 The *Proteus*-Providence Group. In *Recent Advances in Clinical Pathology*, Series IV, ed. Dyke, S. C. London: Churchill.

CHRISTENSEN, W. B. 1946 Urea decomposition as a means of differentiating *Proteus* and paracolon cultures from each other and from *Salmonella* and *Shigella* types. *Journal of Bacteriology* **52,** 461–466.

CHRISTENSEN, W. B. 1949 Hydrogen sulfide production and citrate utilization in the differentiation of the enteric pathogens and the coliform bacteria. *Research Bulletin, Weld County Health Department* **1,** 3–16.

COWAN, S. T. 1974 *Cowan and Steel's Manual for the Identification of Medical Bacteria*, 2nd edn. London: Cambridge University Press.

CROSBY, N. T. 1967 The determination of nitrite in water using Cleve's acid, 1-naphthylamine-7-sulphonic acid. *Proceedings of the Society for Water Treatment and Examination* **16,** 51–55.

EDWARDS, P. R. & EWING, W. H. 1972 *Identification of Enterobacteriaceae*, 3rd edn. Minneapolis: Burgess Publishing Co.

EWING, W. H. & DAVIS, B. R. 1972 Biochemical characterization of *Citrobacter diversus* (Burkey) Werkman and Gillen and designation of the neotype strain. *International Journal of Systematic Bacteriology* **22,** 12–18.

EWING, W. H. & FIFE, M. A. 1972 *Enterobacter agglomerans* (Beijerinck) comb. nov. (The Herbicola-Lathyri bacteria). *International Journal of Systematic Bacteriology* **22,** 4–11.

EWING, W. H., DAVIS, B. R., FIFE, M. A. & LESSEL, E. F. 1973 Biochemical characterization of *Serratia liquefaciens* (Grimes and Hennerty) Bascomb *et al.* (formerly *Enterobacter liquefaciens*) and *Serratia rubidaea* (Stapp) comb. nov. and designation of type and neotype strains. *International Journal of Systematic Bacteriology* **23,** 217–225.

FARMER III, J. J., HICKMAN, F. W., BRENNER, D. J., SCHREIBER, M. & RICKENBACH, D. G. 1977 Unusual *Enterobacteriaceae: 'Proteus rettgeri'* that 'change' into *Providencia stuartii*. *Journal of Clinical Microbiology* **6,** 373–378.

FRAZIER, W. C. 1926 A method for the detection of changes in gelatin due to bacteria. *Journal of Infectious Diseases* **39,** 302–309.

FREDERIKSEN, W. 1970 *Citrobacter koseri* (n.sp.) a new species within the genus *Citrobacter*, with a comment on the taxonomic position of *Citrobacter intermedium* (Werkman and Gillen) *Spisy Přírodovědecké Fakulty University J.E. Purkyně Brně* **47,** *Series K,* 89–94.

GRIMONT, P. A. D., GRIMONT, F., DULONG DE ROSNAY, H. L. C. & SNEATH, P. H. A. 1977 Taxonomy of the genus *Serratia. Journal of General Microbiology* **98,** 39–66.

HOLMES, B., KING, A., PHILLIPS, I. & LAPAGE, S. P. 1974 Sensitivity of *Citrobacter freundii* and *Citrobacter koseri* to cephalosporins and penicillins. *Journal of Clinical Pathology* **27,** 729–733.

HUGH, R. & LEIFSON, E. 1953 The taxonomic significance of fermentative versus oxidative metabolism of carbohydrates by various Gram negative bacteria. *Journal of Bacteriology* **66,** 24–26.

JAIN, K., RADSAK, K. & MANNHEIM, W. 1974 Differentiation of the *Oxytocum* group from *Klebsiella* by deoxyribonucleic acid-deoxyribonucleic acid hybridization. *International Journal of Systematic Bacteriology* **24,** 402–407.

KAUFFMANN, F. 1966 *The Bacteriology of Enterobacteriaceae.* Copenhagen: Munksgaard.

KOVÁCS, N. 1928 Eine vereinfachte Methode zum Nachweis der Indolbildung durch Bakterien. *Zeitschrift für Immunitätsforschung und Experimentelle Therapie* **55,** 311–315.

KOVÁCS, N. 1956 Identification of *Pseudomonas pyocyanea* by the oxidase reaction. *Nature, London* **178,** 703.

LECLERC, H. 1962 Étude biochimique d'*Enterobacteriaceae* pigmentées. *Annales de l'Institut Pasteur, Paris* **102,** 726–741.

LOWE, G. H. 1962 The rapid detection of lactose fermentation in paracolon organisms by the demonstration of β-D-galactosidase. *Journal of Medical Laboratory Technology* **19,** 21–25.

MARMUR, J. & DOTY, P. 1962 Determination of the base composition of deoxyribonucleic acid from its thermal denaturation temperature. *Journal of Molecular Biology* **5,** 109–118.

MØLLER, V. 1954 Diagnostic use of the Braun KCN test within the *Enterobacteriaceae. Acta pathologica et microbiologica scandinavica* **34,** 115–126.

MØLLER, V. 1955 Simplified tests for some amino acid decarboxylases and for the arginine dihydrolase system. *Acta pathologica et microbiologica scandinavica* **36,** 158–172.

NILÉHN, B. 1969 Studies on *Yersinia enterocolitica* with special reference to bacterial diagnosis and occurrence in human acute enteric disease. *Acta pathologica et microbiologica scandinavica,* Supplement 206.

NORMORE, W. M. 1973 Guanine-plus-cytosine (GC) composition of the DNA of bacteria, fungi, algae and protozoa. In *CRC Handbook of Microbiology, Vol. II. Microbial Composition,* eds Laskin, A. I. & Lechevalier, H. A. Cleveland, Ohio: Chemical Rubber Company Press.

O'MEARA, R. A. Q. 1931 A simple delicate and rapid method of detecting the formation of acetylmethylcarbinol by bacteria fermenting carbohydrate. *Journal of Pathology and Bacteriology* **34,** 401–406.

OWEN, R. J. & LAPAGE, S. P. 1976 The thermal denaturation of partly purified bacterial deoxyribonucleic acid and its taxonomic applications. *Journal of Applied Bacteriology* **41,** 335–340.

ROWE, B., GROSS, R. J. & VAN OYE, E. 1975 An organism differing from *Shi-*

gella boydii 13 only in its ability to produce gas from glucose. *International Journal of Systematic Bacteriology* **25,** 301–303.

SAKAZAKI, R., TAMURA, K., JOHNSON, R. & COLWELL, R. R. 1976 Taxonomy of some recently described species in the family *Enterobacteriaceae*. *International Journal of Systematic Bacteriology* **26,** 158–179.

SHAW, C. & CLARKE, P. H. 1955 Biochemical classification of *Proteus* and Providence cultures. *Journal of General Microbiology* **13,** 155–161.

SIMMONS, J. S. 1926 A culture medium for differentiating organisms of typhoid-colon aerogenes groups and for isolation of certain fungi. *Journal of Infectious Diseases* **39,** 209–214.

SULKIN, S. E. & WILLETT, J. C. 1940 A triple sugar-ferrous sulfate medium for use in identification of enteric organisms. *Journal of Laboratory and Clinical Medicine* **25,** 649–653.

WERKMAN, C. H. & GILLEN, G. F. 1932 Bacteria producing trimethylene glycol. *Journal of Bacteriology* **23,** 167–182.

WILSON, G. S. & MILES, A. A. (eds) 1964 *Topley & Wilson's Principles of Bacteriology and Immunity*, 5th edn. London: Arnold.

YOUNG, V. M., KENTON, D. M., HOBBS, B. J. & MOODY, M. R. 1971 *Levinea*, a new genus of the family *Enterobacteriaceae*. *International Journal of Systematic Bacteriology* **21,** 58–63.

Identification of Human Vibrios

A. L. Furniss

Public Health Laboratory, Preston Hall Hospital, Maidstone, Kent, UK

During the past few years there has been a greater understanding of those species of vibrios that are pathogenic or potentially pathogenic for man. With hindsight it now seems remarkable that so much misunderstanding about the vibrios occurred at the start of the present cholera pandemic. It was not at first recognized as an outbreak of cholera for the simple reason that the causative vibrios were haemolytic. They were eltor vibrios; cholera vibrios had been defined as being non-haemolytic. Cholera itself had to be redefined as a disease that could be caused by the eltor vibrio as well as by the classical cholera vibrio. The subsequent discussion about whether classical and eltor vibrios were separate species or merely biotypes of a single species produced numerous publications and not a little controversy but it is now only of historical interest. *Vibrio cholerae* clearly includes classical and eltor types (biovars).

Controversy has now shifted to the so-called non-cholera vibrios (NCVs) and to whether they should be included within the species *V. cholerae*. A characteristic feature of NCVs is that, although closely resembling the cholera vibrios, they are not agglutinated by cholera antiserum. For this reason they have been referred to as NAGs or non-agglutinable vibrios. This is an obvious misnomer because they are agglutinated by homologous antisera. Non-cholera vibrios are potentially pathogenic, causing diarrhoea which is usually mild, but may be so severe as to resemble cholera itself. Outbreaks have been reported (McIntyre *et al.* 1965; Aldova *et al.* 1968; Kamal & Zinnaka 1971; Zafari *et al.* 1973). However, symptomless excreters of NCVs are common.

Vibrio parahaemolyticus is another pathogenic vibrio. It is one of the halophilic vibrios, which will not grow in the absence of salt in the medium. It has come to be recognized as one of the most common causes of food-poisoning in Japan where it was first described (Sakazaki *et al.* 1963). Outbreaks have occurred in Britain associated with the consumption of crustacean shellfish (Barrow 1974). It is widely distributed in the marine environment, but there is no evidence that all strains are pathogenic.

Cholera vibrios, NCVs and *V. parahaemolyticus* can be described as primary human pathogens, but other vibrios—such as *V. alginolyticus*—may become opportunistic pathogens. Ears may be infected; so also may open wounds. Such patients are most likely to be swimmers or those otherwise in contact with seawater. (Rubin & Tilton 1975; Ryan 1976).

Methods

Media for identification tests need to have a sodium chloride content that is adequate for the growth of *V. parahaemolyticus* as well as the non-halophilic cholera vibrios. Some non-pathogenic halophilic vibrios, however, may require much higher concentrations of sodium chloride.

Oxidase test. The technique of Kovács is used (Cowan 1974). The test should be made on a culture on nutrient agar and not on a medium containing a fermentable carbohydrate.

Oxidation-fermentation tests. Hugh & Leifson medium (Cowan 1974) is used. For some vibrios sodium chloride must be added to a final concentration of 3%.

Decarboxylases. The method of Møller (1955) is used, but is modified to contain 1% sodium chloride. Results are read after 48 h incubation.

Fermentation tests. Peptone water sugars are used with 1% sodium chloride. Results are read at 24 h and negative cultures are re-incubated for a further 24 h. Some sugars need particular care. For example, with mannose, fermentation may be weak thus leading to confusion; and arabinose is very liable to hydrolyse and give false positive results.

VP test. Semi-solid nutrient agar (0·3% agar) containing 1% glucose and 1% sodium chloride, which has been distributed in 3 ml amounts, is inoculated by stabbing and incubated at 30°C overnight. Then for the test add 0·2 ml of 5% α-naphthol in ethanol, and 0·1 ml of 40% KOH, containing 0·3% of creatine. A positive reaction is shown by a red colour which appears at the top of the agar in a few minutes.

It is recommended that the reagents are neutralized with acid before autoclaving and disposal, to prevent etching of the glassware.

Salt tolerance. This is tested by examination for turbidity in peptone water with 0, 3, 7, and 10% sodium chloride. Tubes are incubated with agitation at 30°C and read at 24 and 48 h.

Growth on CLED. As a short cut to distinguishing halophilic and non-halophilic vibrios their ability to grow on electrolyte-deficient medium

(CLED) (Bevis 1968) may be used. Halophilic vibrios will not grow on CLED.

Sensitivity to vibriostatic agent 0/129. Discs are prepared to contain 150 and 10 μg of the vibriostatic agent (2,4-diamino-6,7-di*iso*propyl-pteridine phosphate). The phosphate derivative is obtainable from British Drug Houses (Poole, UK) and is soluble in water. The test is performed as for a disc diffusion sensitivity test on nutrient agar. Special sensitivity testing agars should not be used.

Sensitivity to polymyxin. Discs are prepared to contain 50 iu of poly-myxin B and used as for a disc diffusion sensitivity test.

Haemolysis. Most human vibrios will form large colonies with large zones of haemolysis on horse blood agar. Haemolysis can be used as a means of differentiating vibrios but results vary according to the tech-nique used. For consistent results the technique of Sakazaki *et al.* (1971) is recommended.

Kanagawa test. This is a test for β-haemolysis of human blood by means of particular techniques as described by Miyamoto *et al.* (1969).

Chick cell haemagglutination. Growth from nutrient agar is emulsified on a slide in a drop of 2·5% suspension in saline of washed chick red cells. The slide is rocked for about 1 min. Clumping of the red cells indicates a positive result.

Phage sensitivity. Phages may be used for typing of cholera vibrios; two phages in particular may be used at routine test dilution to distinguish classical and eltor biotypes of *Vibrio cholerae.* These are Group IV classical phage and the Group 5 eltor phage of Mukerjee.

Identification

Microscopy

Vibrios are Gram negative, motile rods which may show curvature of their long axis. Curvature may not be obvious in a stained film; even if seen it is not a diagnostic characteristic of vibrios. Non-motile variants do very occasionally occur, particularly amongst old laboratory strains.

Appearance on TCBS

Thiosulphate-citrate-bile salt agar (TCBS) is commonly used for the isolation of vibrios, and the appearance of the colonies provides useful information for the purposes of identification (Furniss & Donovan 1974), (Table 1).

TABLE 1. Colony characteristics of human vibrios

Vibrio type	Colonies on TCBS
Vibrio choleral	
cholera vibrio	Yellow, 2–3 mm diam.
NCV	Yellow or green, 2–5 mm diam.
V. parahaemolyticus	Blue-green, 2–5 mm diam.
V. alginolyticus	Yellow, 2–5 mm diam.
Other vibrios	Yellow or green, variable size
Plesiomonas	Pale green, no growth or 1 mm diam.
Aeromonas	Yellow, variable size

Vibrio cholerae

Cholera vibrios

Serology. Cholera vibrios are agglutinated by O1 antiserum. This is commercially available and is commonly referred to as 'polyvalent' antiserum; it will agglutinate both Ogawa and Inaba types. These subserotypes can only be recognized by the use of absorbed sera. These are not immutable forms and variation does occur, so that the recognition of Ogawa and Inaba types is of limited epidemiological value and of no clinical importance.

Biotypes. No longer are eltor and classical cholera vibrios considered as separate species, but as biotypes or biovars. The production of haemolysin was once considered the main criterion for distinguishing eltor vibrios from classical vibrios. The modern strains of eltor vibrios are distinctly less haemolytic than the original strains, and other tests are used in addition to distinguish the different cholera vibrios (Table 3). Most strains, however, are intermediate between the two polar forms: the classical vibrio and the eltor vibrio, and proposals have been made to recognize more than the two forms (Feeley 1965). Such biotyping is not in fact very valuable; no isolate anywhere in the present cholera pandemic has resembled the classical cholera vibrio. Classical vibrios have remained confined to their original 'home' in Bengal and are now almost extinct elsewhere.

Phage-typing. Phage-typing of cholera vibrios is undertaken by the reference laboratory with an extended and modified schema based on the work of Mukerjee (Mukerjee & Takeya 1974).

Non-cholera vibrios (NCVs)

Serology. Non-cholera vibrios are not agglutinated by the O1 'polyvalent' cholera antiserum, but they are agglutinated by their specific O

TABLE 2. Biochemical and physiological characters

	V. cholerae		*V. parahaemolyticus*	*V. alginolyticus*	Other vibrios	*Plesiomonas*	*Aeromonas*
	Cholera	NCV					
Oxidase	+	+	+	+	d	+	+
O/F	F	F	F	F	F	F	F
Arginine dihydrolase	—	—	—	—	d	+	+
Lysine decarboxylase	+	+	+	+	d	+	d
Ornithine decarboxylase	+	+	+	+	d	+	—
Glucose fermentation	+	+	+	+	+	+	+
Gas from glucose	—	—	—	—	—	—	d
Sucrose	+	d	—	+	d	—	+
Mannose	+	d	+	+	d	+	+
Arabinose	—	—	d	+	d	—	+
Inositol	—	—	—	—	—	+	—
VP	d	d	—	+	d	—	—
CLED	+	+	—	—	d	+	+
0% NaCl	+	+	—	—	d	+	+
3% NACl	+	+	+	+	+	+	+
7% NaCl	—	+	+	+	d	—	—
10% NaCl	—	—	—	+	d	—	—
0/129 10 μg	S	S	R	R	d	S	R
150 μg	S	S	S	S	S	S	R
Cholera O1 serum	+	—	—	—	—	—	—

d: different strains show different reactions; +: 95–100% strains positive; —: 95–100% strains negative.

TABLE 3. Reactions of *Vibrio cholerae* biotypes

	Classical	Eltor
Haemolysis	—	+
VP	—	+
Chick cell haemagglutination	—	+
Polymyxin 50 iu	sensitive	resistant
Mukerjee classical Phage IV	sensitive	resistant
Mukerjee eltor Phage 5	resistant	sensitive

type antiserum (Sakazaki *et al.* 1970). More than 70 types (serovars) are recognized on the basis of specific O antigens, although some sharing of antigens does occur. Appropriate antisera are not commercially available, but serotyping is undertaken at the reference laboratory.

The 'H' antigen is common to all NCVs and to cholera vibrios.

Biotypes. A biotyping scheme for NCVs was proposed by Heiberg (1936). On the basis of fermentation reactions in the three sugars: sucrose, mannose and arabinose he described six groups of vibrios. It is now clear that true NCVs belong to one of only three groups (Table 4). All true cholera vibrios belong to Group I.

Group V strains appear to be distinct in other ways. The fact that they do not ferment sucrose means that they appear as green colonies on TCBS. Heiberg's biotyping scheme, however, is of little use in the identification of vibrios.

Roughness. Both cholera vibrios and NCVs are very liable to show rough colonies; some of these are so rugose and therefore so different from smooth colonies that the culture may be thought to be mixed. Rugose colonies are more obvious on a rich medium such as blood agar; they are usually auto-agglutinable. Colonies that are not obviously rough may still have lost the specific O antigen. They may be agglutinated with an antiserum against the R antigen (Shimada & Sakazaki 1973). Such an organism, agglutinated only by R antiserum, may have been derived from a cholera vibrio or from any other serotype.

Vibrio parahaemolyticus

Vibrio parahaemolyticus can be serotyped on the basis of O and K antigens. Such typing is undertaken at the appropriate reference laboratory.

The significance of the Kanagawa test is uncertain, but there is some correlation with pathogenicity. Strains isolated from human cases of *V. parahaemolyticus* food-poisoning are Kanagawa positive. Almost all

TABLE 4. A biotyping scheme for NCVs

	Fermentation of		
	Sucrose	Mannose	Arabinose
Heiberg Group I	+	+	—
Heiberg Group II	+	—	—
Heiberg Group V	—	+	—

From Heiberg 1936.

isolates from the environment—seawater and shellfish—are Kanagawa negative.

Plesiomonas shigelloides

Although *P. shigelloides* has been placed in a separate genus (Habs & Schubert 1962), it has also been classified as a vibrio (Shewan & Véron 1974). Its taxonomic position is uncertain, and so also is its pathogenicity. It is infrequently isolated, but when it has been isolated it has often come from diarrhoeic stools.

Like the vibrios it is oxidase-positive and ferments sugars without the formation of gas. It is sensitive to 0/129. It grows poorly on TCBS and is more likely to be isolated on DCA (desoxycholate-citrate agar) on which it grows well. It is non-halophilic, but unlike the non-halophilic vibrios its growth is not enhanced by sodium chloride.

The appearance of colonies on DCA is similar to *Shigella* colonies and many strains are strongly agglutinated by *Sh. sonnei* antiserum, a finding which may lead to confusion in identification.

References

ALDOVA, E., LAZNOCKOVA, K., STEPANKOVA, E. & LIETAVA, J. 1968 Isolation of non-agglutinable vibrios from an enteric outbreak in Czechoslovakia. *Journal of Infectious Diseases* **118,** 25–31.

BARROW, G. I. 1974 Microbiological and other hazards from seafoods with special reference to *Vibrio parahaemolyticus. Post-Graduate Medical Journal* **50,** 612–619.

BEVIS, T. D. 1968 A modified electrolyte deficient culture medium. *Journal of Medical Laboratory Technology* **25,** 38–41.

COWAN, S. T. 1974 *Cowan & Steel's Manual for the Identification of Medical Bacteria,* 2nd edn. Cambridge: University Press.

FEELEY, J. C. 1965 Classification of *V. cholerae* (*V. comma*), including El Tor vibrios, by infrasubspecific characteristics. *Journal of Bacteriology* **89,** 665–670.

FURNISS, A. L. & DONOVAN, T. J. 1974 The isolation and identification of *Vibrio cholerae. Journal of Clinical Pathology* **27,** 764–766.

HABS, H. & SCHUBERT, R. H. W. 1962 Ueber die biochemischen Mukmale und die taxonomische Stellung von *Pseudomonas shigelloides* (Bader). *Zentralblatt für*

Bakteriologie, Parasitenkunde, Infektionskrankheiten und Hygiene, Abt I Orig. **186,** 316–327.

HEIBERG, B. 1936 The biochemical reactions of vibrios. *Journal of Hygiene, Cambridge* **36,** 114–117.

KAMAL, A. M. & ZINNAKA, Y. 1971 Outbreak of gastro-enteritis by non-agglutinable (NAG) vibrios in the Republic of the Sudan. *Journal of the Egyptian Public Health Association* **46,** 125–174.

McINTYRE, O. R., FEELEY, J. C., GREENHOUGH, W. B. III, BENENSON, A. S., HASSAN, S. I. & SAAD, A. 1965 Diarrhoea caused by non-cholera vibrios. *American Journal of Tropical Medicine and Hygiene* **14,** 412–418.

MIYAMOTO, Y., KATO, T., OBARA, Y., AKIYAMA, S., TAKIZAWA, K. & YAMAI, S. 1969 *In vitro* haemolytic characteristic of *Vibrio parahaemolyticus*: its close correlation with human pathogenicity. *Journal of Bacteriology* **100,** 1147–1149.

MØLLER, V. 1955 Simplified tests for some amino acid decarboxylases and for the arginine dihydrolase system. *Acta pathologica et microbiologica scandinavica* **36,** 158–172.

MUKERJEE, S. & TAKEYA, K. 1974 Vibrio-phages and vibriocins. In *Cholera*, eds Barua, D. & Burrows, W. Philadelphia: W. B. Saunders.

RUBIN, S. J. & TILTON, R. C. 1975 Isolation of *V. alginolyticus* from wound infections. *Journal of Clinical Microbiology* **2,** 556–558.

RYAN, W. J. 1976 Marine vibrios associated with superficial septic lesions. *Journal of Clinical Pathology* **29,** 1014–1015.

SAKAZAKI, R., IWANAMI, S. & FUKUMI, H. 1963 Studies on the enteropathogenic facultatively halophilic bacteria *Vibrio parahaemolyticus*. I. Morphological, cultural and biochemical properties and its taxonomical position. *Japanese Journal of Medical Science and Biology* **16,** 161–188.

SAKAZAKI, R., TAMURA, K., GOMEZ, C. Z. & SEN, R. 1970 Serological studies on the cholera group of vibrios. *Japanese Journal of Medical Science and Biology* **23,** 13–20.

SAKAZAKI, R., TAMURA, K. & MURASE, M. 1971 Determination of the hemolytic activity of *Vibrio cholerae*. *Japanese Journal of Medical Science and Biology* **24,** 83–91.

SHEWAN, J. M. & VÉRON, M. 1974 Genus *Vibrio*. In *Bergey's Manual of Determinative Bacteriology*, 8th edn, eds Buchanan, R. E. & Gibbons, N. E. Baltimore: Williams & Wilkins.

SHIMADA, T. & SAKAZAKI, R. 1973 R antigen of *Vibrio cholerae*. Japanese *Journal of Medical Science and Biology* **26,** 155–160.

ZAFARI, Y., ZARIFI, A. Z., RAHMANZADEH, S. & FAKHAR, N. 1973 Diarrhoea caused by non-agglutinable *Vibrio cholerae* (non-cholera vibrio) *Lancet* **2,** 429–430.

Identification of *Aeromonas, Vibrio* and Related Organisms

J. V. Lee
Public Health Laboratory, Preston Hall, Maidstone, Kent, UK

Margaret S. Hendrie and J. M. Shewan
Torry Research Station, Aberdeen, UK

Since the publication of *Identification Methods for Microbiologists* Part B (Gibbs & Shapton 1968) considerable progress has been made in the differentiation and identification of the genera and species of the heterotrophic Gram negative fermentative bacteria particularly those within the Enterobacteriaceae and Vibrionaceae. Experience over the past decade has confirmed and extended the suggestions and ideas put forward in 1968. This has in a large measure been due to the intelligent use of three simple tests capable of being carried out in any microbiological laboratory, and without recourse to elaborate or sophisticated apparatus.

By the use of these three tests; Kovács oxidase (Kovács 1956); the Hugh & Leifson test for the mode of metabolism of glucose; and the sensitivity to the vibriostatic agent 0/129 (2,4-diamino-6,7-di-*iso*propylpteridine), together with presence or absence of motility, it is possible to differentiate fairly accurately the Enterobacteriaceae from the Vibrionaceae on the one hand (and the genera within the Vibrionaceae) and from the Pseudomonadaceae on the other. As Table 1 indicates, the Kovács oxidase test separates the Enterobacteriaceae from the Pseudomonadaceae and the Vibrionaceae, and the latter can be differentiated from the Pseudomonadaceae by the Hugh & Leifson test. It is recognized that the Hugh & Leifson test can give rise to difficulties in interpretation in inexperienced hands and several modifications have been proposed from time to time particularly for use with marine isolates. It would be most appropriate if some simple tests could be devised for differentiating the two main pathways involved in the respiratory and fermentative

TABLE 1. Some characters distinguishing the *Vibrionaceae* from the *Pseudomonadaceae* and *Enterobacteriaceae* and the genera within the *Vibrionaceae*

	Vibrionaceae			*Pseudomonadaceae*	*Enterobacteriaceae*
	Vibrio/Beneckea	*Aeromonas*	*Photobacterium*	*Pseudomonas/Alteromonas*	
Flagella	Polar or mixed	Polar or none	Polar	Polar or mixed	Peritrichous or none
Oxidase	+[b]	+	d	+[b]	—
Glucose metabolism					
O/F medium	F	F	F	O/NC	F
Gas from glucose	—	d	d	—	+
Sensitivity to 0/129					
10 μg	d	—	d	—	—
150 μg	+	—	+	—	—
Luminescence	d	—	d	—	—
Mol. % G+C	44–49	57–63	39–42	*Pseudomonas* 55–70 *Alteromonas* 42–54	39–59

F: fermentative; O: oxidative; NC: no change or alkaline; d: variable.
[a] All strains are Gram negative asporogenous rods.
[b] *V. metschnikovii* and *Ps. maltophilia* are oxidase negative.

attacks on a carbohydrate such as glucose by, for example, identifying the key enzymes unique to these pathways. Until such tests become a reality the Hugh & Leifson test, intelligently used, is the most useful test to employ for identification purposes. If it is accepted therefore that the Pseudomonadaceae and the Enterobacteriaceae can be separated from the Vibrionaceae as indicated above, the remaining problems are the differentiation of the genera and species within the Vibrionaceae, i.e. the species of *Aeromonas*, *Beneckea*, *Photobacterium*, *Plesiomonas* and *Vibrio*. *Plesiomonas* is discussed elsewhere in this book in the chapter 'Identification of Human Vibrios'.

Beneckea has been proposed as a separate genus for a group of Gram negative, facultatively anaerobic rods of marine origin which require Na^+ for growth and which are all oxidase positive, ferment glucose with the production of acid but no gas and are motile with sheathed polar flagella in liquid medium but frequently peritrichous on solid media (Baumann *et al.* 1971*a*). It is not certain, however, particularly for determinative purposes, that this genus can be sufficiently differentiated from that of *Vibrio* to stand on its own and accordingly the two genera are treated together. The main features differentiating the genera within the Vibrionaceae are also listed in Table 1. It will be noted that the main differentiating features are resistance to 0/129 and the production of gas from glucose by the *Aeromonas* group; together with the much higher guanine + cytosine content of the DNA in *Aeromonas* compared with the other genera of the Vibrionaceae.

The taxonomy of the genera within the Vibrionaceae is the subject of much research at the moment. Consequently the nomenclature of the species within the group is in a state of flux and is particularly confusing. For example the genus *Beneckea* as described above would include *Lucibacterium harveyi* and the marine vibrios such as *V. parahaemolyticus* and *V. alginolyticus* as well as the newly described *Beneckea* sp. (Baumann *et al.* 1971*a*, *b*, 1973; Reichelt *et al.* 1976). This is not the place for a discussion of the present taxonomic and nomenclatural arguments concerning the Vibrionaceae but it is necessary to make some comments on the species we have listed in Tables 2 and 3. Where possible we have used the species names as they appear in the 8th Edition of Bergey's Manual (Buchanan & Gibbons 1974) for the genera *Vibrio*, *Lucibacterium* and *Plesiomonas*.

Vibrio cholerae biotype *proteus* (Shewan & Véron 1974) we now consider a separate species which is correctly named *V. metschnikovii* (Lee *et al.* 1978). Group F has recently been described and contains organisms intermediate to *Vibrio* and *Aeromonas* (Furniss *et al.* 1977; Lee *et al.* 1978). We have not divided *Beneckea pelagia* into biotypes because many

TABLE 2. Synonyms

Species	Synonyms (including biotypes)
Vibrio cholerae serovar 1	*V. cholerae* biotypes *eltor* and *cholerae*
V. cholerae serovars 2–75	Non-agglutinable vibrios (NAGs) *V. cholerae* NAG Non-cholera vibrios (NCVs)
V. parahaemolyticus	*Beneckea parahaemolytica* *V. parahaemolyticus*, biotype 1
V. alginolyticus	*B. alginolytica* *V. parahaemolyticus*, biotype 2
Lucibacterium harveyi	*B. harveyi* *B. neptuna* (non-luminescent)
V. fischeri	*Photobacterium fischeri*
V. natriegens	*B. natriegens*, *Pseudomonas natriegens*
V. anguillarum	*B. anguillara* (*sic*)
V. metschnikovii	*V. proteus* *V. cholerae*, biotype *proteus*
Plesiomonas shigelloides	*Aeromonas shigelloides* *Pseudomonas shigelloides*, C27
Photobacterium leiognathi	*Photobacterium mandapamensis*

of the isolates from around Britain are intermediate to the biotypes proposed by Reichelt *et al.* (1976).

Recently strains of an organism which usually ferments lactose but which otherwise is very similar to *V. parahaemolyticus* have been isolated from cases of septicaemia in U.S.A. (Hollis *et al.* 1976). This group has been named *B. vulnifica* (Reichelt *et al.* 1976). So far we have had no experience of these organisms and can make no comments as to the validity of placing them in a separate species. They can be distinguished from *V. parahaemolyticus* by their ability to grow on butyrate, heptanoate, ethanol, L-serine, L-leucine and putrescine (Reichelt *et al.* 1976).

In the most recent edition of Bergey's Manual, Schubert (1974) differentiated three *Aeromonas* spp., *A. hydrophila*, *A. punctata* and *A. salmonicida*, with three subspecies (*hydrophila*, *anaerogenes* and *proteolytica*) under *A. hydrophila*, two (*punctata* and *caviae*) under *A. punctata* and three (*salmonicida*, *achromogenes* and *masoucida*) under *A. salmonicida*. Recent work, however (McCarthy 1975; Popoff & Véron 1976; Boulanger *et al.* 1977; Gibson *et al.* 1977; Kleeberger 1977), has indicated that: *A. hydrophila* subsp. *proteolytica* should be removed from the genus *Aeromonas*; that *A. hydrophila* and *A. punctata* be united as one species, *A. hydrophila* and that a new species *A. sobria* (Popoff & Véron 1976) be recognized. The general features distinguishing *A. salmonicida* from the other *Aeromonas* spp. are listed in Table 3. *Aeromonas sobria* does not appear on Table 3 because we have had little experience with the strains of this species but the general features distinguishing *A. hydrophila* from *A. sobria* are given in Table 4 (after Popoff & Véron 1976).

To help the reader of the current literature Table 2 lists the various species synonyms that are likely to be encountered. Table 3 lists the species that can be recognized at the present time and the characters most useful for their identification. It is hoped that the taxonomic arguments revolving around the Vibrionaceae will soon be resolved so that the nomenclature will be easier to follow for the bacteriologist interested in identifying these organisms.

Methods

Many of these organisms are marine and therefore require higher concentrations of certain cations notably Na^+, K^+ and Mg^{++} than is usual for terrestrial bacteria. We have found that increasing the sodium chloride concentration of all media to 1% gives satisfactory results for nearly all of these organisms. The media containing 1% sodium chloride are suitable for testing both freshwater and marine strains. Occasionally, strains still show inadequate growth or negative reactions even in the presence of 1% sodium chloride. This is particularly noticeable with Møller's test for decarboxylases and Thornley's test for the metabolism of arginine. These difficulties may be overcome by the addition to the medium of the following electrolyte supplement solution. The solution contains (g l^{-1}): NaCl, 100, $MgCl_2.6H_2O$, 40, KCl, 10. The solution is sterilized by autoclaving and added to the medium as required in the proportions of 0·1 ml of supplement to 1·0 ml of medium.

In all the methods described below the media contain 1% of sodium chloride unless otherwise stated.

Incubation. The usual temperature of incubation is 25°C, although the optimum for many species is higher than this, and for these 30°C or even 37°C may be used. On the other hand it may sometimes be necessary to use 20°C particularly for some strains of *Photobacterium* and *A. salmonicida*.

Gram reaction and morphology. Cultures are examined microscopically after three days incubation on Oxoid Blood Agar Base (CM 55). Some of the marine strains may be found to grow better on seawater agar (Evans Peptone, 1%; Oxoid Lab Lemco, 1%; made up in a 3:1 mixture of filtered aged sea water and distilled water) or Difco marine agar.

Motility. This is determined by phase contrast microscopic examination

TABLE 3. Characters useful for the identification of the species within the Vibrionaceae

Species	sensitivity to 0/129: 10 µg	sensitivity to 0/129: 150 µg	Luminescence	Pigment	Swarming	Growth on CLED
Vibrio cholerae (all serovars)	+	+	—	—	—	+
V. albensis	+	+	+	—	—	+
V. parahaemolyticus	—	+	—	—	d	—
V. alginolyticus	—	+	—	—	+	—
Lucibacterium harveyi	—	+	d	—	—	—
V. fischeri	+	+	+	+[1]	—	—
Beneckea campbellii	—	+	—	—	—	—
V. natriegens	—	+	—	—	—	—
B. pelagia (biotypes I and II)	+	+	—	—	—	—
B. nigrapulchrituda	—	+	—	+[2]	—	—
B. splendida I	·	·	+	·	—	—
B. splendida II	·	·	—	·	—	—
B. nereida	—	+	—	—	—	—
V. anguillarum	+	+	—	—	—	—/+
V. metschnikovii	+	+	—	—	—	+
Plesiomonas shigelloides	+	+	—	—	—	+
Group F	—	+	—	—	—	+
Aeromonas salmonicida	—	—	—	+[3]	—	+
Aeromonas (other spp.)	—	—	—	—	—	+
Photobacterium phosphoreum	·	+	+	—	—	—
Phot. leiognathi	·	+	+	—	—	—
Phot. angustum	·	+	—	—	—	—
V. costicola	+	+	—	—	—	—

+: character present for at least 90% of strains; —: character absent for at least 90% of strains; d: variable reaction from strain to strain; +/— or —/+: first reaction present in 80–90% of strains; 1: non-diffusible orange-yellow pigment; 2: non-diffusible blue-black pigment; 3: brown diffusible pigment.

Growth in % NaCl					Growth at °C	
0	3	6	8	10	37°	42°
+	+	d	—	—	+	+
+	—	—	—	—	+	—
—	+	+	+	—	+	+
—	+	+	+	+/—	+	+
—	+	+	+	d	+	d
—	+	+	+	—	d	—
—	+	+	d	—	+	—
—	+	+	d	—	+	d
—	+	+	d	—	d	—
—	+	—	—	—	—	—
·	·	·	·	·	·	—
·	·	·	·	·	·	—
—	+	+	+	+	+	d
—	+	+	d	—	d	—
—	+	+	d	—	+	d
+	+	d	—	—	+	·
—	+	+	+	d	+	d
+	+	d	—	—	—	—
+	+	d	—	—	+	d
—	+	d	d	—	—	—
—	+	d	d	—	d	—
—	+	d	d	—	+	—
—	d	+	+	+	d	—

TABLE 3—*cont.*

Species	Oxidase	Nitrate	VP	Arginine (Thornley)	Møller's decarboxylases: Lysine	Møller's decarboxylases: Ornithine	Gas from glucose
Vibrio cholerae (all serovars)	+	+	d	—	+	+	—
V. albensis	+	+	—	—	+	+	—
V. parahaemolyticus	+	+	—	—	+	+	—
V. alginolyticus	+	+	+	—	+	+	—
Lucibacterium harveyi	+	+	—	—	+	+	—
V. fischeri	+	+	—	—	+	—	—
Beneckea campbellii	+	+	—	—	+	—	—
V. natriegens	+	+	—	—	—	—	—
B. pelagia (biotypes I and II)	+	+	—	—	—	—	—
B. nigrapulchrituda	+	+	—	—	—	—	—
B. splendida I	+	+	—	+	·	·	—
B. splendida II	+	+	—	—	·	·	—
B. nereida	+	+	—	+	—	—	—
V. anguillarum	+	+	+	+	—	—	—
V. metschnikovii	—	—	+	+	d	—	—
Plesiomonas shigelloides	+	+	—	+	+	+	—
Group F	+	+	—	+	—	—	d
Aeromonas salmonicida	+	+	—	+	—	—	d
Aeromonas (other spp.)	+	+	d	+	d	—	d
Photobacterium phosphoreum	—	+	+	+	+	—	+
Phot. leiognathi	+	+	+	+	+	—	d
Phot. angustum	—/+	+	—	+	—	—	—
V. costicola	+	+	d	+	—	—	—

Acid from				
Arabinose	Mannose	Raffinose	Rhamnose	Sucrose
—	d	—	—	d
—	—	·	·	+
d	+	—	—	—
—	+	—	—	+
—	+	—	—	d
—	+	—	—	—
—	+	—	—	—
+	—	+	+	+
—	d	—	—	+
—	—	—	—	—
—	·	—	—	d
—	·	—	—	—
—	—	—	—	+
d	+	—	—	+
—	+/—	—	—	+
—	—	—	—	—
+	+/—	—	d	+
—	+	—	—	+
d	+	—	—	+
—	+	—	—	—
—	+	—	—	—
—	+	—	—	d
—	+	·	·	+

TABLE 3—*cont.*

Species	Growth on				
	Cellobiose	Xylose	Ethanol	Leucine	Putrescine
Vibrio cholerae (all serovars)	—	—	—	—	—
V. albensis	.	.	.	.	.
V. parahaemolyticus	—	—	+	+	+
V. alginolyticus	—	—	d	+	d
Lucibacterium harveyi	+	—	—	—	—
V. fischeri	+	—	—	—	—
Beneckea campbellii	d	—	—	—	—
V. natriegens	d	—	+	+	+
B. pelagia (biotypes I and II)	d	—	+/—	—	+
B. nigrapulchrituda	+	—	d	—	—
B. splendida I	+	—	—	—	—
B. splendida II	d	—	—	—	—
B. nereida	—	—	+	+	+
V. anguillarum	+/—	—	—	—	—/+
V. metschnikovii	—	—	—	—	—
Plesiomonas shigelloides	.	.	.	.	.
Group F	—	—	+	—	+
Aeromonas salmonicida	.	.	.	.	.
Aeromonas (other spp.)	d	—	—	—	—
Photobacterium phosphoreum	—	—	—	—	—
Phot. leiognathi	—	—	—	—	—
Phot. angustum	—	+	—	—	—
V. costicola	.	.	.	.	.

Alginase	Amylase	Gelatinase	Lecithinase	Tween 80
−	+	+	+	+
−	+	+	+	+
−	+	+	+	+
−	+	+	+	+
d	+	+	+	+
−	−	−	+	+
−	+	+	+	+
−	d	+	−	+
+	d	d	−	+
−	+	+	+	+
d	+	+	·	+
−	+	+	·	+
−	−	−	−	−
−	+	+	+	+/−
−	+	+	+	+
−	−	−	−	−
−	+	+	+	+
−	+	+	+	+
−	+	+	+	+
−	−	−	+	−
−	−	−	·	+
−	−	+	·	d
−	−	d	d	d

TABLE 4. Characters differentiating *Aeromonas hydrophila* from *A. sobria*

Character	*A. hydrophila*	*A. sobria*
Aesculin hydrolysis	+	—
Salicin fermentation	+	—
H_2S production	d	+
Growth on KCN medium	+	—
Growth on L-arabinose	+	—
salicin	+	—
L-arginine	+	—
L-histidine	+	—

After Popoff & Véron 1976.

of the culture after two days incubation in tryptone water (tryptone, 1%; NaCl, 1%).

Flagella. The arrangement of flagella is examined by electron microscopy or by the methods of Rhodes (1958) as modified by Gauthier *et al.* (1975).

Oxidase. The method of Kovács (1956) using a platinum wire, sterile toothpick, swab stick or glass rod, is employed.

Dissimilation of glucose. The medium of Hugh & Leifson (1953) is used, or marine O-F medium (Leifson 1963) (modified; see p. 4, this volume).

Sensitivity to the vibriostatic agent 0/129. A plate of CM 55 agar (for all strains where possible) is surface-seeded with the culture under test and the 0/129 applied in the form of discs containing 10 μg and 150 μg of the agent. The plates are incubated for 24 h before reading sensitivities. If commercial sensitivity agar is used instead of CM 55 many of the marine strains will not grow but in addition many enterobacteria will show a degree of sensitivity to 0/129.

The phosphate derivative of 0/129 (2,4-diamino-6,7-di-*iso*propyl-pteridine phosphate) is obtainable from BDH and is more soluble in water than the original pteridine derivative used by Shewan *et al.* (1954). Solutions containing 7500 μg ml^{-1} and 500 μg ml^{-1} are prepared and used to make 150 μg and 10 μg discs, respectively. Each disc has 20 μl of the appropriate solution spotted on it. The discs are then dried in a desiccating jar, freeze-dryer or simply by placing them at 37°C until dry.

Luminescence. Seawater agar plates are inoculated and examined for up to three days in a totally blacked-out room. The examiner should remain 10 min in the dark before reading the plates (Hendrie *et al.* 1970). Luminescence is most marked in a young culture, i.e. usually after 18–24 h incubation.

If seawater is not available the following medium has been found to give excellent results. It contains (g l^{-1}): Oxoid Nutrient Broth No. 2 (CM 67) dried base, 25; NaCl, 17·5; KCl, 1; $MgCl_2.6H_2O$, 4; Davis agar, 15.

Vibrio albensis is a non-halophilic vibrio that luminesces but only on CM 55 and not on the above medium.

Pigmentation. Production of a brown diffusible pigment is noted on Oxoid CM 55 after five days. Formation of a non-diffusible orange-yellow pigment or blue-black pigment is noted after five days on sea-water agar or marine agar.

Swarming. Marine agar is prepared and the sterile molten agar cooled in a water bath at 45–50°C for 30 min. Plates are then poured, allowed to set but not dried and 10 mm diam. gutters are cut in them dividing the plates into quadrants. Each quadrant is inoculated with a 10 mm streak of an organism. Plates are examined for swarming after incubation for 1 day. A known positive organism should be tested at the same time as a control.

DNA base composition. DNA may be extracted and purified by the method of Marmur (1961) and the mol. % G + C determined by thermal denaturation using the method of Marmur & Doty (1962).

Growth on CLED. One of us (JVL) has found it very convenient to use an electrolyte-deficient nutrient agar in the initial screening of isolates. Cysteine-lactose-electrolyte deficient medium (CLED) (Bevis 1968) is used in many clinical bacteriology laboratories. The non-halophilic strains will grow on CLED whereas most marine strains will not. The plate is lightly inoculated with the culture and incubated overnight.

Salt tolerance. Tryptone water (1% Oxoid Tryptone) containing 0, 3, 6, 8 and 10% of NaCl is used. The media are distributed in 4 ml amounts in bijou bottles or test tubes. Ten μl of a 3–5 h culture of the organisms in tryptone broth with 2% NaCl is inoculated into the various media. These are incubated and the growth recorded after 24 and 48 h.

Growth at 37° and 42°C. Tryptone broth with 2% NaCl is inoculated as for salt tolerance and growth recorded at 24 h after incubation at the appropriate temperatures.

Nitrate reduction. Nitrate broth (Nutrient broth + 0·1% KNO_3) is inoculated with the culture, incubated for one day and then tested for the presence of nitrite using the reagents A and B of Crosby (1967). Solution A contains: sulphuric acid, 0·5 g; glacial acetic acid, 30 ml; de-ionized water, 120 ml. Solution B contains: Cleeves acid (1-naphthyl-amine-7 sulphonic acid), 0·2 g; glacial acetic acid, 30 ml; de-ionized water, 120 ml.

For Solution B the water is added to the Cleeves acid and warmed on a water bath to dissolve it. The solution is filtered, cooled and the acetic acid is added. To test for the presence of nitrite, 1 ml of Solution A and 1 ml of Solution B are added to 1 ml of broth. Nitrite is indicated by the presence of a red colour. If nitrite is not present a small amount of zinc dust is added to the tube and the development of a red colour after a few minutes indicates the presence of nitrite. If nitrite has not been produced after one day the broth is reincubated for a further four days and then re-tested.

Voges-Proskauer (VP) test. A semi-solid medium (Barrow, pers. comm.) is used containing: yeast extract (Oxoid), 1 g; Trypticase Peptone (BBL), 7 g; Phyto Peptone (BBL), 5 g; glucose, 10 g; NaCl, 10 g; agar (Davis), 3 g; distilled water, 1 l. The pH is adjusted to 7·0 and the medium is distributed in 3 ml amounts and sterilized by autoclaving. The medium is inoculated by stabbing and incubated overnight at 30°C or two days at 25°C.

The presence of acetylmethylcarbinol is detected using the following two reagents: A. naphthol, 6 g; alcohol, 100 ml. B. potassium hydroxide, 40 g; creatine, 0·3 g; distilled water, 100 ml. Both reagents are kept in a refrigerator and may be used for two months. Solution A (0·2 ml) and Solution B (0·1 ml) are added to the culture which is then left at room temperature. A positive result is indicated by a deep red ring which develops in the top of the agar usually within a few minutes. Final readings are made after 30 min incubation on the bench.

Methods 1 and 2 of Cowan (1974) may also be used but the NaCl concentration of the media must be increased to 1%.

Arginine metabolism. The method of Thornley (1960) is used.

Lysine and ornithine decarboxylases. These are detected by the method of Møller (1955) using medium made with Difco decarboxylase base.

Action on carbohydrates. Tubes of 1% (w/v) carbohydrates in peptone water with Andrade's indicator, and containing Durham tubes, are used and examined after incubation periods of 1, 2, 4 and 7 days.

Growth on compounds as the sole organic source of carbon. The basal medium of Baumann *et al.* (1971*a*) is used for those strains requiring NaCl for growth and the medium of Clowes & Hayes (1968) for other strains. The carbon sources are used at a final concentration of 0·1%. Growth is recorded over a period of two weeks in comparison with a negative control (basal medium alone).

Alginase. Plates of Difco marine agar containing 2% of sodium alginate are spot-inoculated with the organism under test. The plates are incubated over a period of 7 days. Alginolytic activity is detected by marked pitting around the growth. *Vibrio alginolyticus* and certain strains of other species may swarm or spread markedly on this medium.

Amylase. CM 55 containing 0·1% of starch is spot-inoculated with a culture and incubated for two days. Lugol's or Gram's iodine solution is flooded over the plate. A positive result is indicated by a clear unstained zone around the growth.

Gelatinase. Gelatin hydrolysis is detected using Method 3 of Cowan (1974).

Lecithinase. Nutrient agar containing 1% of egg yolk emulsion (Oxoid) is inoculated with a spot or streak of culture and incubated and observed for seven days. A positive result shows as an opaque zone around the growth.

Tween 80 hydrolysis. Hydrolysis of Tween 80 is determined by the method of Sierra (1957).

References

BAUMANN, P., BAUMANN, L. & MANDEL, M. 1971*a* Taxonomy of marine bacteria: the genus *Beneckea*. *Journal of Bacteriology* **107,** 268–294.

BAUMANN, P., BAUMANN, L., MANDEL, M. & ALLEN, R. D. 1971*b* Taxonomy of marine bacteria: *Beneckea nigrapulchrituda* sp.n. *Journal of Bacteriology* **108,** 1380–1383.

BAUMANN, P., BAUMANN, L. & REICHELT, J. L. 1973 Taxonomy of marine bacteria: *Beneckea parahaemolytica* and *Beneckea alginolytica*. *Journal of Bacteriology* **113,** 1144–1155.

BEVIS, T. D. 1968 A modified electrolyte deficient culture medium. *Journal of Medical Laboratory Technology* **25,** 38–41.

BOULANGER, Y., LALLIER, R. & COUSINEAU, G. 1977 Isolation of enterotoxigenic *Aeromonas* from fish. *Canadian Journal of Microbiology* **23,** 1161–1164.

BUCHANAN, R. E. & GIBBONS, N. E. (eds) 1974 *Bergey's Manual of Determinative Bacteriology*, 8th edn. Baltimore: Williams & Wilkins.

CLOWES, R. C. & HAYES, W. 1968 *Experiments in Microbial Genetics*. Oxford & Edinburgh: Blackwell Scientific Publications.

COWAN, S. T. 1974 *Cowan & Steel's Manual for the Identification of Medical Bacteria*, 2nd edn. Cambridge: Cambridge University Press.

CROSBY, N. T. 1967 The determination of nitrate in water using Cleve's acid, 1-naphthylamine-7-sulphonic acid. *Proceedings of the Society for Water Treatment and Examination* **16,** 51–55.

FURNISS, A. L., LEE, J. V. & DONOVAN, T. J. 1977 Group F, a new vibrio? *Lancet* ii No. 8037, 565–566.

GAUTHIER, M. J., SHEWAN, J. M., GIBSON, D. M. & LEE, J. V. 1975 Taxonomic position and seasonal variations in marine neritic environment of some gram-negative antibiotic-producing bacteria. *Journal of General Microbiology*, **87,** 211–218.

GIBBS, B. M. & SHAPTON, D. A. (eds) 1968 *Identification Methods for Microbiologists* Part B. Society for Applied Bacteriology Technical Series No. 2. London & New York: Academic Press.

GIBSON, D. M., HENDRIE, M. S., HOUSTON, N. C. & HOBBS, G. 1977 The identification of some Gram negative heterotrophic aquatic bacteria. In *Aquatic Microbiology*. Society for Applied Bacteriology Symposium Series No. 6, eds Skinner, F. A. & Shewan, J. M. London & New York: Academic Press.

HENDRIE, M. S., HODGKISS, W. & SHEWAN, J. M. 1970 The identification, taxonomy and classification of luminous bacteria. *Journal of General Microbiology* **64,** 151–169.

Hollis, D. G., Weaver, R. E., Baker, C. N. & Thornsberry, C. 1976 Halophilic *Vibrio* species isolated from blood cultures *Journal of Clinical Microbiology* **3,** 425–431.

Hugh, R. & Leifson, E. 1953 The taxonomic significance of fermentative versus oxidative metabolism of carbohydrates by various Gram-negative bacteria. *Journal of Bacteriology* **66,** 24–26.

Kleeberger, A. 1977 Taxonomische Untersuchungen an Aeromonaden aus Milch, Wasser und Hackfleisch. *Zeitschrift für Lebensmittel-Untersuchung und-Forschung* **163,** 44–47.

Kovács, N. 1956 Identification of *Pseudomonas pyocyanea* by the oxidase reaction *Nature, London* **178,** 703.

Lee, J. V., Donovan, T. J. & Furniss, A. L. 1978 The characterization, taxonomy and emended description of *Vibrio metschnikovii. International Journal of Systematic Bacteriology* **28,** 99–111.

Leifson, E. 1963 Determination of carbohydrate metabolism of marine bacteria. *Journal of Bacteriology* **85,** 1183–1184.

Marmur, J. 1961 A procedure for the isolation of deoxyribonucleic acid from microorganisms. *Journal of Molecular Biology* **3,** 208–218.

Marmur, J. & Doty, P. 1962 Determination of the base composition of deoxyribonucleic acid from its thermal denaturation temperature. *Journal of Molecular Biology* **5,** 109–118.

McCarthy, D. H. 1975 The bacteriology and taxonomy of *Aeromonas liquefaciens.* Technical Reports Series No. 2. Ministry of Agriculture, Fisheries and Food, Fish Diseases Laboratories, Weymouth, Dorset.

Møller, V. 1955 Simplified tests for some amino acid decarboxylases and for the arginine dihydrolase system. *Acta pathologica et microbiologica scandinavica* **36,** 158–172.

Popoff, M. & Véron, M. 1976 A taxonomic study of the *Aeromonas hydrophila-Aeromonas punctata* Group. *Journal of General Microbiology* **94,** 11–22.

Reichelt, J. L., Baumann, P. & Baumann, L. 1976 Study of genetic relationships among marine species of the genera *Beneckea* and *Photobacterium* by means of *in vitro* DNA/DNA hybridization. *Archiv für Mikrobiologie* **110,** 101–120.

Rhodes, M. E. 1958 The cytology of *Pseudomonas* spp. as revealed by silver-plating staining method. *Journal of General Microbiology* **18,** 639–648.

Schubert, R. H. W. 1974 Genus *Aeromonas.* In *Bergey's Manual of Determinative Bacteriology,* 8th edn, eds Buchanan, R. E. & Gibbons, N. E. Baltimore: Williams & Wilkins.

Shewan, J. M. & Véron, M. 1974 Genus *Vibrio.* In *Bergey's Manual of Determinative Bacteriology,* 8th edn, eds Buchanan, R. E. & Gibbons, N. E. Baltimore: Williams & Wilkins.

Shewan, J. M., Hodgkiss, W. & Liston, J. 1954 A method for the rapid differentiation of certain non-pathogenic asporogenous bacilli. *Nature, London* **173,** 208–209.

Sierra, G. 1957 A simple method for the detection of lipolytic activity of micro-organisms and some observations on the influence of the contact between cells and fatty substrates. *Antonie van Leeuwenhoek* **23,** 15–22.

Thornley, M. J. 1960 The differentiation of *Pseudomonas* from other Gram-negative bacteria on the basis of arginine metabolism. *Journal of Applied Bacteriology* **23,** 23–52.

Identification Methods Applied to *Chromobacterium*

P. H. A. SNEATH

Department of Microbiology, Leicester University, Leicester, UK

The genus *Chromobacterium* contains two well-defined species, *C. violaceum* and *C. lividum*. They are motile Gram negative rods. For a general study see Sneath (1960); many details are also in Sneath (1956*a*, *b*, 1974), Leifson (1956), Moffett & Colwell (1968) and Bascomb *et al.* (1973).

The genus is probably not a natural one. A separate genus *Janthinobacterium* has been suggested for *C. lividum* (De Ley *et al.* 1978). Some unpigmented bacteria resembling aeromonads or vibrios, and possibly some resembling agrobacteria or pseudomonads, may prove on closer study to be related to the two species mentioned above. Nevertheless, the two species do share some unusual features:

(1) Both produce the same violet pigment, violacein.
(2) Both have an unusual flagellar morphology.
(3) Both are highly sensitive to peroxides.

Further study may justify separation of *C. lividum* into two or more groups, and there is a distinctive group occurring in river water (Moss *et al.* 1978) which has been named *C. fluviatile*. This group shares the first two properties mentioned above, but the third has not been studied. In addition a number of Gram negative violet-pigmented marine bacteria have recently been described, *C. marinum* by Hamilton & Austin (1967) and *Alteromonas luteoviolaceus* by Gauthier (1976). They are still not very well known, as they are difficult to maintain in culture, but they share the first and last of the properties listed above. The arrangement of flagella is predominantly if not entirely monopolar. Whatever their correct systematic position, the marine forms seem similar to one another, and likely to be regarded on first isolation as strains of *Chromobacterium*, so for this reason their properties are also considered here, under the designation of the 'marine group'.

In practice strains are recognized initially by their pigment; if this is too faint to be noticed, the strains may be thought to belong to one of the other groups mentioned above. Non-pigmented strains of *C. violaceum* are not uncommon, and Sivendra *et al.* (1975) discuss ways of recognizing them. A few gas-producing strains of this species have also been reported (Sivendra 1976). These non-pigmented variants should be suspected if Gram negative, motile rods that are catalase positive and oxidase positive are also found to produce hydrogen cyanide, as well as being fermentative and positive for arginine decarboxylase but negative for ornithine and lysine decarboxylases. The other best-known organism that produces HCN on occasion is *Pseudomonas aeruginosa*, but this can be distinguished by being oxidative, urease positive, by not reducing gluconate or utilizing malonate, and by its monopolar flagellum. For such difficult problems, and also the recognition of non-pigmented strains of *C. lividum*, it may be necessary to use the computer techniques described by Bascomb *et al.* (1973).

Methods Useful in Recognizing Members of the Genus

The pigment violacein

Violacein is a complex indole-pyrrole pigment, found within the bacterial cells. Not all media give abundant pigment; glycerol, peptone and suboptimal growth temperatures favour its production.

Violacein is insoluble in water and in chloroform, but is soluble in ethyl alcohol, giving a violet solution. To obtain this, shake pigmented culture with 96% ethyl alcohol and filter off the cells through filter paper. The solution shows an absorption maximum at 580 nm and a minimum at 430 nm. On adding 10% (v/v) of sulphuric acid (care!) the solution becomes green (absorption maximum at 700 nm, see Fig. 1). If caustic soda is added it also becomes green, but rapidly turns red-brown and decomposes.

Spot-testing cultures for violacein is simple:

(i) First check that the pigment does not diffuse away from the violet colonies on agar plates (slight diffusion is seen in old cultures). (ii) Stir a loopful of pigmented growth in chloroform and confirm that the pigment is insoluble. (iii) Stir a loopful of growth in 96% ethyl alcohol in a hollowed white tile, and check that it gives a violet solution. (iv) Transfer a loopful of solution to a drop of dilute (25% v/v) sulphuric acid (turns green). (v) Add 10% (w/v) NaOH to the rest (turns green and then reddish). So far as is known, these reactions are not given by any other bacterial pigment.

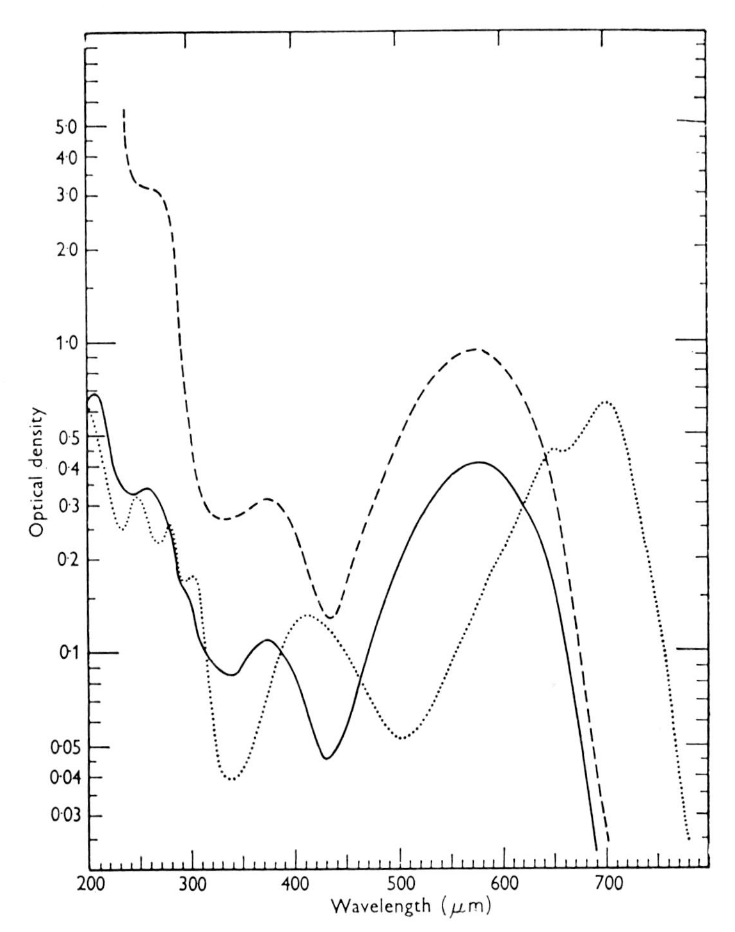

FIG. 1. Absorption of violacein. (—) Ethanolic solution of crystalline violacein; (......) crystalline violacein in 10% (v/v) sulphuric acid in 96% ethanol; (- - - -) crude violacein in ethanol. (Reproduced from Sneath 1956*a*, by permission of the *Journal of General Microbiology*.)

Flagella

Most strains of *Chromobacterium* in young agar cultures have an unusual form of flagella arrangement: many cells have both polar and lateral (peritrichous) flagella. These are distinguishable in several ways in good flagellar-stained preparations:

Site of insertion. While the single polar flagellum is always inserted at the tip, the peritrichate flagella (often as many as 3 to 6) may be inserted anywhere on the cell—most often as sub-polar flagella (on the 'shoulder').

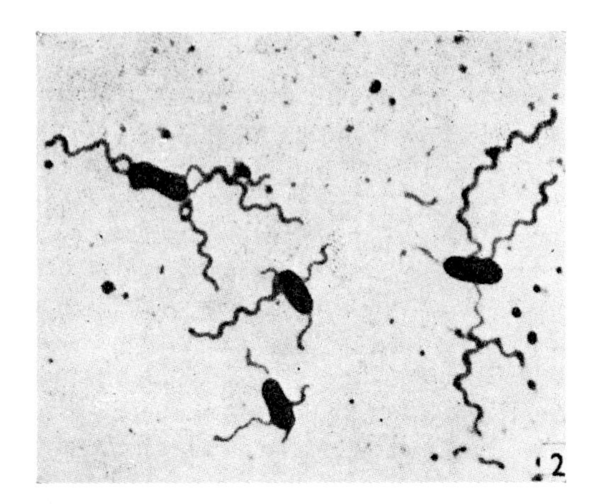

FIG. 2. Flagella-stained preparation of *C. violaceum* showing organisms with both polar and lateral flagella. × 2250. (Reproduced from Sneath 1956*b*, by permission of the *Journal of General Microbiology*.)

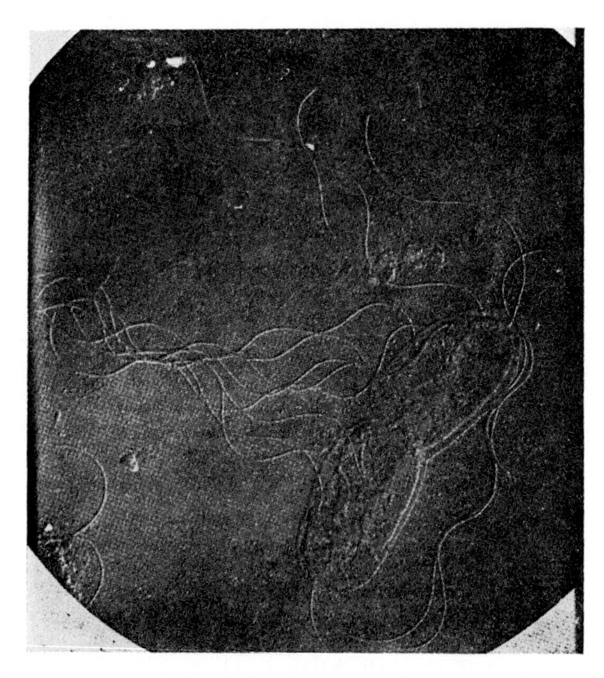

FIG. 3. Electron micrograph of *C. violaceum*, showing both polar and lateral flagella × 12 000. (Reproduced from Sneath 1956*b*, by permission of the *Journal of General Microbiology*.)

Shape and staining. The polar flagellum usually shows long, shallow waves, and often stains faintly. The peritrichate (lateral) flagella usually show deep, short waves, stain well and are often very long.

The two forms of flagella are also antigenically different. Old cultures and cultures in liquid media show few lateral flagella. Occasional strains never show lateral flagella.

Peroxide sensitivity

Not enough is yet known of the distribution of peroxide sensitivity nor of the best techniques for its study to make this a useful diagnostic test.

Tests for Distinguishing between *Chromobacterium violaceum, C. lividum, C. fluviatile* and the Marine Group

Chromobacterium violaceum is mesophilic and strongly proteolytic. It is usually fermentative on a few carbohydrates, and is a facultative anaerobe.

Chromobacterium lividum is psychrophilic and weakly proteolytic. It is a strict aerobe and attacks many carbohydrates oxidatively.

In the earlier literature these two species are much confused, and both have been commonly called *C. violaceum.* This name is now kept for the mesophilic group (Anon. 1958).

C. fluviatile is psychrophilic but fermentative on several carbohydrates, facultatively anaerobic and proteolytic. Colonies are flat and thin, unlike the convex colonies seen in the other groups.

The marine group has a restricted range of temperature for growth, from about 10–30°C, is strictly aerobic and oxidative on a few carbohydrates, and also proteolytic. It has an obligate requirement for at least 1% NaCl.

The most convenient tests for distinguishing these four groups are given below. It should be emphasized that occasional exceptions are found to all tests. For the marine group 3% (w/v) NaCl should be added to all media.

Growth temperature

Growth temperature is tested on nutrient agar slants incubated at 37°C and at 4°C for 7 days after inoculation with one drop of a broth culture.

Hydrogen cyanide production

Production of HCN is conveniently tested for in stab cultures of semi-solid medium (nutrient agar diluted with an equal volume of water). An indicator paper placed between tube and plug turns from yellow to brick red in 1–2 days at 25°C if the cultures produce hydrogen cyanide (some pseudomonads are also positive in this test). The papers are made as follows: filter paper is dipped in saturated aqueous picric acid and dried, and then dipped in 10% aqueous sodium carbonate and again dried. They keep well.

Acid from glucose and trehalose in Hugh and Leifson medium

Strains of *C. violaceum* are not invariably fermentative with glucose, though *C. lividum* is never strongly fermentative. Trehalose is also a useful carbohydrate, though a few exceptional strains of *C. lividum* are known to ferment it (Moss, pers. comm.). The preferred medium is that of Hugh & Leifson (1953), i.e. agar, 0·3 g; peptone, 0·2 g; NaCl, 0·5 g; K_2HPO_4, 0·3 g; bromothymol blue, 0·003 g; either glucose or trehalose, 1 g; distilled water, 100 ml. Adjust to pH 7·1, fill into tubes to a depth of 4 cm and autoclave. Inoculate duplicate tubes by stabbing and incubate at 25°C for 4 days with and without a vaseline seal.

Acid from arabinose or from xylose in Hugh and Leifson medium

Use medium (see above) with 1% of L-arabinose or D-xylose instead of the glucose or trehalose; incubate for 7 days, as acid production may be slow.

Aesculin hydrolysis

The medium contains: peptone, 1 g; sodium citrate, 0·1 g; aesculin, 0·1 g; ferric citrate, 0·005 g; water, 100 ml; pH 7·0. Tube in 5 ml quantities and autoclave. After inoculation incubate at 25°C for 3–4 days. This medium is less toxic than the formulation used earlier (Sneath 1960, p. 376).

Casein hydrolysis

Plates of 50% skim milk with 1·5% agar added are streak inoculated and incubated for 4 days at 25°C.

TABLE 1. Differential features of species and groups of *Chromobacterium*

Test	*C. violaceum*	*C. lividum*	*C. fluviatile*	Marine group
Growth at 37°C in 7 days	Abundant	None or very slight	None	None
Growth at 4°C in 7 days	None	Moderate	Slight or moderate	None
Hydrogen cyanide production	Positive	Negative	Negative	Negative
Attack on glucose	Usually fermentative	Oxidative	Fermentative	Oxidative
Trehalose	Acid (prompt)	No change or alkali	Acid	Usually acid
Arabinose	No change or alkali	Acid (often slow)	Usually no change	Not recorded
Xylose	No change or alkali	Acid (often slow)	No change	Usually no change
Aesculin hydrolysis	Negative	Positive (brown)	Negative	Usually negative
Casein hydrolysis	Positive (clear zone)	Negative or slight	Positive	Positive
Egg-yolk reaction	Positive (turbid zone)	Negative	Weak positive (under colony)	Positive
Arginine hydrolysis	Positive (red)	Negative	Negative	Negative
Growth on media without added NaCl	Present	Present	Present	Absent

Turbidity from egg-yolk

Plates of nutrient agar containing one part of egg-yolk emulsion (egg-yolk removed aseptically and suspended in 5% (w/v) saline to give a concentration of 20% v/v) to nine parts of nutrient agar after melting and cooling to 55°C. Streak inoculate and incubate for 4 days at 25°C.

Hydrolysis of arginine

The medium contains: peptone, 1 g; NaCl, 5 g; K_2HPO_4, 0·3 g; agar, 3 g; phenol red, 0·01 g; L(+) arginine HCl, 10 g; distilled water, 1 l. Adjust to pH 7·2, dispense in tubes to a depth of 2 cm and sterilize by autoclaving. Inoculate by stabbing, cover with sterile melted petroleum jelly and incubate for 4 days at 25°C.

Results of the differential tests are shown in Table 1.

References

ANON. 1958 Judicial Commission. Opinion 16. *International Bulletin of Bacteriological Nomenclature and Taxonomy* **8,** 151–152.

BASCOMB, S., LAPAGE, S. P., CURTIS, M. A. & WILLCOX, W. R. 1973 Identification of bacteria by computer: identification of reference strains. *Journal of General Microbiology* **77,** 291–315.

DE LEY, J., SEGERS, P. & GILLIS, M. 1978 Intra- and intergeneric similarities of *Chromobacterium* and *Janthinobacterium* ribosomal ribonucleic acid cistrons. *International Journal of Systematic Bacteriology* **28,** 154–168.

GAUTHIER, M. J. 1976 Morphological, physiological, and biochemical characteristics of some violet-pigmented bacteria isolated from seawater. *Canadian Journal of Microbiology* **22,** 138–149.

HAMILTON, R. D. & AUSTIN, K. E. 1967 Physiological and cultural characteristics of *Chromobacterium marinum* sp. n. *Antonie van Leeuwenhoek* **33,** 257–264.

HUGH, R. & LEIFSON, E. 1953 The taxonomic significance of fermentative versus oxidative metabolism of carbohydrates by various gram negative bacteria. *Journal of Bacteriology* **66,** 24–26.

LEIFSON, E. 1956 Morphological and physiological characteristics of the genus *Chromobacterium*. *Journal of Bacteriology* **71,** 393–400.

MOFFETT, M. L. & COLWELL, R. R. 1968 Adansonian analysis of the *Rhizobiaceae*. *Journal of General Microbiology* **51,** 245–266.

MOSS, M. O., RYALL, C. & LOGAN, N. A. 1978 The classification and characterization of chromobacteria from a lowland river. *Journal of General Microbiology* **105,** 11–21.

SIVENDRA, R. 1976 Unusual *Chromobacterium violaceum*: aerogenic strains. *Journal of Clinical Microbiology* **3,** 70–71.

SIVENDRA, R., LO, H. S. & LIM, K. T. 1975 Identification of *Chromobacterium violaceum*: pigmented and non-pigmented strains. *Journal of General Microbiology* **90,** 21–31.

SNEATH, P. H. A. 1956*a* Cultural and biochemical characteristics of the genus *Chromobacterium*. *Journal of General Microbiology* **15,** 70–98.

SNEATH, P. H. A. 1956*b* The change from polar to peritrichous flagellation in *Chromobacterium* spp. *Journal of General Microbiology* **15,** 99–105.
SNEATH, P. H. A. 1960 A study of the bacterial genus *Chromobacterium*. *Iowa State Journal of Science* **34,** 243–500.
SNEATH, P. H. A. 1974 Genus *Chromobacterium* Bergonzini 1881. In *Bergey's Manual of Determinative Bacteriology*, 8th edn, eds Buchanan, R. E. & Gibbons, N. E. Baltimore: Williams & Wilkins.

The Identification of Gram Negative, Yellow Pigmented Rods

P. R. HAYES

Department of Microbiology, Leeds University, Leeds, UK

T. A. McMEEKIN

Department of Agricultural Science, University of Tasmania, Hobart, Australia

AND

J. M. SHEWAN

Torry Research Station, Aberdeen, UK

The classification and identification of yellow pigmented, Gram negative, aerobic, chemoorganotrophic rods, the 'flavobacteria', remains a complex problem. In the 8th edition of *Bergey's Manual of Determinative Bacteriology* (Buchanan & Gibbons 1974), Weeks (1974) lists only 12 species in the genus *Flavobacterium* as compared with 26 species in the 7th edition (Breed *et al.* 1957) but it is still heterogeneous and has been arranged in two sections. The first contains the non-motile strains with low guanine + cytosine mol % (30–42%) in their DNA—usually stated as G + C ratios; the second contains motile and non-motile strains with G + C ratios in the range of 60–70%.

A major difficulty in the present situation is that the original type strain of *Flavobacterium aquatile* no longer exists and the proposed neotype has a number of features, including a low G + C ratio, in common with the Cytophagaceae (Leadbetter 1974), members of which exhibit gliding motility on solid media. In fact, it has been claimed (Perry 1973) that *F. aquatile* shows gliding motility, but others have suggested that the movement is similar to twitching rather than gliding (Henrichsen, pers. comm).

Since the 8th edition of *Bergey's Manual* was published in 1974 at least six taxonomic studies on a total of almost 600 strains of presumptive flavobacteria have been undertaken. These strains have been obtained from a wide variety of ecological situations, such as soils, foods, water, human and hospital environments (Byrom 1971; McMeekin *et al.* 1971, 1972; Gavini & Leclerc 1975; Owen & Lapage 1974; Owen & Snell 1973, 1976; Geringer & Kielwein 1976; Hayes 1977). These studies have clarified the situation considerably and as a result the following identification scheme has been suggested. It should, however, be emphasized that the most that can be proposed at present is a differentiation into groups which may, or may not, represent genera and no attempt is made to identify to the species level. It is unfortunate that none of the species listed by Weeks (1974) can be fitted satisfactorily into any of our proposed groups and it is believed that in some measure this is due to the fact that few of the species listed are adequately described or have been examined by the tests used by the workers already mentioned. Moreover, in our opinion, most of these species are infrequently encountered. Detailed descriptions are clearly required as has, for example, been provided for *F. odoratum* and *F. breve* by Holmes *et al.* (1977 and pers. comm.); it might also be added here that on present evidence *F. aquatile* seems not to be a good choice as the type species for *Flavobacterium.*

Test Methods

The following test methods have been found most useful in the group differentiation proposed below.

Pigmentation

Initial selection from isolation media depends on the possession of a yellow pigment. The hue and intensity of pigmentation varies considerably and may be markedly affected by, among other things, the composition of the growth medium, incubation temperature and time. Organisms with colonies described as cream, pale yellow, yellowish green, orange, pinkish and brown, have been allocated to the genus *Flavobacterium.* In our experience, pigmentation on nutrient agar is usually conspicuous and very poorly pigmented isolates (off-white or cream) should be viewed with suspicion particularly if pigmentation remains poor under a variety of cultural conditions, e.g. using casein, starch or milk agars or by variation of the incubation temperatures. It is usually assumed that the pigments of both flavobacteria and Cytophagaceae (i.e. *Cytophaga* and *Flexibacter*) are carotenoids (Weeks 1974; Leadbetter 1974) but non-carotenoid pigments have been described (McMeekin *et al.* 1971). Reichenbach *et al.* (1974) reported that the main pigments in the *Cytophaga-Flexibacter* group were flexirubins (phenate esters of polyene acids) the pigment chemistry of which has been elucidated by Achenbach *et al.* (1976). However, the pigments of many flavobacteria appear to be flexirubins (Hayes, unpublished results) on preliminary examination (addition of N NaOH to the culture resulting in a red coloration, Reichenbach *et al.* 1974).

Gram stain and morphology

Two- or three-day cultures on nutrient agar should be used; in addition, Gram staining or phase contrast microscopy on one- or two-day nutrient broth cultures is also recommended. Although all Gram positive species were excluded from the genus in the 7th edition of *Bergey's Manual* (Breed *et al.* 1957), many such strains have subsequently been isolated and described as *Flavobacterium* spp. (Ferrari & Zannini 1958). Often these isolates are easily decolourized during Gram staining (e.g. by the use of acetone) and may appear Gram negative. Great care must be paid, therefore, in the staining of the prepared films and Weeks (1974) recommended the use of the Kopeloff-Beerman modification of the Gram stain (Conn 1957), confirmed by the ability of the organism to grow on crystal violet agar ($2 \cdot 5 \times 10^{-5}$ w/v and 1×10^{-6} w/v) or on sodium dodecyl sulphate (1% w/v) agar. Treatment of a thick cell suspension in a 10% (w/v) solution of sodium dodecyl sulphate may also prove useful since Gram negative cell walls are lysed immediately by the detergent resulting in a marked increase in the viscosity of the suspension.

Motility and flagellar arrangement

Electron microscopy is preferred in order to differentiate polar from peritrichous flagella but a staining method may be successfully employed, such as a modification of the Casares-Gil technique (Gemmell & Hodgkiss 1964) or that recently described by Mayfield & Inniss (1977). Motility examinations should be made from several media before an organism is recorded as non-motile and, where initial examinations have been made on cultures incubated at 30–37°C, a re-examination should be made at 20°C.

Oxidative-fermentative attack on carbohydrates

The method of Hugh & Leifson (1953) should be used; for marine isolates the Leifson (1963) modification is recommended. Demonstration of the fermentative action of certain species of *Cytophaga* requires special media such as those of Bachmann (1955) and Anderson & Ordal (1961).

Gliding motility

The demonstration of gliding motility amongst flexibacters and cytophagas sometimes presents difficulties. It is essential to culture the organism on low nutrient concentration agar such as that of Anacker & Ordal (1959), which contains (% w/v): yeast extract, 0·05; Lab-Lemco, 0·02; tryptone (Oxoid), 0·05; sodium acetate, 0·02; agar (Oxoid No. 28), 0·80. Gliding is also affected by the amount of surface moisture and best results are obtained with freshly poured plates and incubation in a humid atmosphere (Henrichsen 1972). Gliding is determined by direct microscopic observation of the swarming edge on a thinly poured plate following overnight incubation. Three methods are currently in use. Henrichsen (1972) uses a high power dry lens arguing that the addition of a coverslip to allow observation under oil introduces additional surface forces which may lead to the observation of artefacts. Others such as McMeekin *et al.* (1971), Perry (1973) and Hayes (1977) observe under oil. Perry (1973) additionally recommends the use of small glass beads to provide pools of liquid of varying depths.

The definition and hence the interpretation of gliding varies among different workers. Thus, Henrichsen (1972) defines gliding as 'movement which is continuous and regularly follows the long axis of the cells which are predominately organized in bundles during the movement'. Perry's definition (pers. comm.) is somewhat wider and includes strains showing 'unrestricted movement' on a solid substratum without necessarily continuous directional movement. Thus, Perry (1973) recorded *F. aquatile*, the type species of the genus *Flavobacterium*, as a glider, a view not shared by some other workers such as Henrichsen (pers. comm.) and Hayes (1977). Clearly, further work on the mechanism of gliding motility is required but, meanwhile, it is probably better to accept Henrichsen's (1972) definition which, in our opinion, is the least controversial and includes only those isolates which show unequivocal gliding motility.

Spreading growth

Isolates exhibiting gliding motility should also produce spreading growth on low nutrient concentration media, since spreading colonies are theoretically a macroscopic manifestation of gliding motility. However, it has been found that spreading growth and gliding motility do not always correlate. Thus, Henrichsen (1972) suggested that spreading may result from other forms of surface translocation mechanisms, whilst several workers have recorded an apparently anomalous result for *F. pectinovorum*, NCIB 9059, which shows true gliding but fails to spread (Lund 1969; McMeekin *et al.* 1971; Perry 1973; Hayes 1977). This organism is now considered to be *Cytophaga johnsonae* (Christensen 1973). However, many of these anomalous results could be explained by the use of different media when comparing gliding and spreading and observations on both these characteristics should be performed using the same medium.

Oxidase test

The method of Kovács (1956) is used. Alternatively, Kovács's reagent may be added directly to the culture plates (Hayes 1977).

Phosphatase test

The method of Barber & Kuper (1951) is recommended. Cultures should be incubated for 3–5 days to allow good growth before exposure to the ammonia vapour. Occasionally colonies of some positive strains fail to change colour, possibly due to masking by the normal yellow colour of the colony; a halo of mauve pink can be seen in the medium surrounding the colonies in such cases.

Growth in inorganic salts + glucose medium

Isolates should be examined for their ability to grow in a simple inorganic salts + glucose medium: KH_2PO_4, 1·5 g; $(NH_4)_2HPO_4$, 7·0 g; $MgSO_4.7H_2O$, 0·5 g; $CaCl_2.2H_2O$, 0·3 g; $MnSO_4.4H_2O$, 4·0 mg; $FeSO_4.7H_2O$, 2·5 mg; ammonium molybdate, 2·0 mg; distilled water 1 l; pH 7·0. The solution is boiled, filtered, dispensed in 5 ml amounts and sterilized for 20 min at 115°C (Knight & Proom 1950); 0·5 ml of a sterile stock solution of glucose (30% w/v) is added to each 5 ml of basal medium. It is stressed that growth should be evident after a minimum of three serial subcultures.

Ultraviolet sensitivity

Samples (5 μl) of a dilution series prepared from 24–48 h nutrient broth cultures are applied in duplicate on nutrient agar, using a semi-automatic pipette, and one of each pair is exposed to u.v. radiation. McMeekin

(1977) used an exposure time of 90 s, 55 cm from a 15 watt germicidal lamp in a laminar flow cabinet. This resulted in at least a 10^5-fold decrease in survivors of low % G + C cultures and a < 10^3-fold decrease in survivors of high % G + C cultures. As different u.v. outputs may be emitted from different lamps it will be necessary to determine the optimum exposure time for each system using organisms of known G + C ratios.

G + C ratios

There are several methods available for the extraction and analysis of the base compositions of the DNA (see Skidmore & Duggan 1971). Accurate determinations of the G + C contents usually employ measurements of the thermal denaturation temperature (Marmur & Doty 1962) or buoyant densities in CsCl (Schildkraut *et al.* 1962). Simpler, though somewhat less accurate methods, including that of Hayes *et al.* (1977), are particularly useful where there are gross differences in the G + C contents such as are found with the organisms discussed here. However, the most rapid screening can be obtained by using the u.v. light sensitivity method described above.

Identification

Isolates should be examined initially for Gram reaction, morphology, motility (with the position of the flagella), the method of attack on glucose (Hugh & Leifson test), and the Kovács's oxidase reaction. This should result in the exclusion of strains which can be allocated to already well defined genera (Table 1). All Gram positive isolates should be excluded since most are likely to belong to the coryneform group.

The motile strains

The great majority of motile strains may be allocated to the *Pseudomonas*, *Xanthomonas*, *Alteromonas*, *Aeromonas*, *Vibrio* or *Erwinia* genera, depending on the position of the flagella and the initial test reactions given in Table 1. Accordingly, reference should be made to the appropriate chapters in this volume or to the 8th edition of *Bergey's Manual of Determinative Bacteriology*. Undoubtedly, difficulties will arise when attempts are made to identify to the species level, and identification to the genus level may have to suffice.

In the six most recent taxonomic studies already mentioned (page 178), about 35% (190 strains) of the isolates were motile. All but three of these

TABLE 1. Initial screening of yellow pigmented rods

Test(s)	Suspected group
Gram positive oxidase — ve	Coryneforms
Gram negative Motile (polar) non-fermentative oxidase + ve	*Pseudomonas*, *Alteromonas*
As above but oxidase — ve or sometimes weakly + ve	*Xanthomonas*
Motile (polar) fermentative oxidase + ve	*Aeromonas*, *Vibrio*
Motile, (peritrichous) fermentative oxidase — ve	Enterobacteriaceae (particularly *Erwinia*)
Non-motile or motile (peritrichous) Non-fermentative oxidase + ve	*Flavobacterium* *Flexibacter*/*Cytophaga*

were allocated to genera other than *Flavobacterium*; two were unnamed (McMeekin *et al.* 1971) and one was presumptively designated a *Flavobacterium* (Hayes 1977). Thus, it is apparent that only a very few yellow pigmented, peritrichous, oxidase positive, non-fermentative rods are likely to be encountered in the wide variety of habitats already mentioned. The taxonomic position of these strains remains in doubt but as they have G + C contents of more than 50% they could conveniently be grouped with the three *Flavobacterium* species listed by Weeks (1974). Accordingly, they are classified as members of the genus *Flavobacterium* Section 2, Table 2.

The non-motile strains

The gliding isolates

The non-motile non-flagellate strains, which composed more than 65% of the isolates in the above mentioned studies, also present problems to the taxonomist. Strains which show unequivocal gliding motility according to Henrichsen's definition should be assigned to the Cytophagaceae. These organisms often form long slender flexuous rods on nutrient agar, whilst in nutrient broth they are almost invariably markedly elongated (> 8 μm). In addition, gliding strains are relatively inexacting nutritionally (McMeekin *et al.* 1971, Christensen 1973,

TABLE 2. Scheme for the identification of *Flavobacterium* and related genera

	Flavobacterium Section 1*	*Flavobacterium* Section 2*	*Flexibacter Cytophaga*
Motility	—	—/+	—
Flagellation	—	— Peritrichous	—
Gliding	—	—	+
u.v. sensitivity	+	—	+
Morphology	Short to medium rods	Short rods	Long, flexuous rods
Growth in NH_4/glucose	—	—	+
Oxidase	+	±	+
Phosphatase	+	—	+
% G + C	30–45	55–70	30–45

*Bergey's Manual

Hayes 1977) and they generally grow in the inorganic salts and glucose medium. They are also sensitive to u.v. radiation indicating a low G + C ratio (30–45%, Table 2), and are oxidase and phosphatase positive. Occasionally, however, morphologically identical gliding strains are isolated which have high G + C contents (Shilo 1970; Daft & Stewart 1971; Stewart & Brown 1971) and these are thought to be non-fruiting variants of the higher Myxobacterales as described by Lewin (1969). The majority of the non-flagellate isolates (85%) exhibit neither gliding nor spreading growth; surface translocation movements such as twitching or the motions ascribed for *F. aquatile* by Perry (1973) may occasionally be seen in non-spreading types. In addition, restricted spreading on agar of low nutrient status may occasionally occur. However, in the absence of unequivocal gliding none of these types may be assigned to the Cytophagaceae.

Non-gliding non-flagellate strains

Two groups of non-flagellate non-gliding isolates may be distinguished on the basis of their G + C contents. A very large majority (over 90%) of the isolates examined in the six studies already mentioned were found to have low G + C values and would thus be sensitive to u.v. light. The non-motile, non-gliding types with low G + C contents (*Flavobacterium*, Section 1, Table 2) have certain characteristics in common with the Cytophagaceae and are both oxidase and phosphatase positive. However, flavobacteria exhibit great variation in their nutritional requirements and strains growing without supplementary amino acids and/or growth factors are only rarely isolated; thus flavobacteria generally fail to grow in the inorganic salts + glucose medium. Furthermore, flavobacteria can be distinguished morphologically by being much shorter in length (Table 2), rarely exceeding 5 μm even in nutrient broth.

The other group of non-gliding non-flagellated strains has a high G + C content (55–70%) (*Flavobacterium* Section 2, Table 2). Unfortunately only a few strains have been isolated in which the G + C values have been determined including the two species described by Weeks (1974). Additional strains have been listed by McMeekin & Shewan (1978). These are distinguished by being both oxidase and phosphatase negative (Table 2). Some of these are now regarded as being Gram positive strains which are easily decolourized during Gram staining (Hendrie *et al.* 1968, Bousfield 1972).

It will be evident from what has been said above that the taxonomy of the yellow pigmented Gram negative rods is still in a state of flux. However, it has been suggested (Hayes 1977; McMeekin & Shewan 1978) that the generic title *Flavobacterium* be retained for the non-flagellate non-gliding low G + C species, whilst the high G + C content strains be assigned to the genus *Empedobacter* (Brisou 1957) as already suggested by Shewan (1973).

References

ACHENBACH, H., KOHL, W. & REICHENBACH, H. 1976 Flexirubin, ein neuartiges Pigment aus *Flexibacter elegans*. *Chemische Berichte* **109,** 2490–2502.

ANACKER, R. L. & ORDAL, E. J. 1959 Studies on the myxobacterium *Chondrococcus columnaris*. I. Serological typing. *Journal of Bacteriology* **78,** 25–32.

ANDERSON, R. T. & ORDAL, E. J. 1961 *Cytophaga succinicans* n.sp., a facultatively anaerobic aquatic myxobacterium. *Journal of Bacteriology* **81,** 130–138.

BACHMANN, B. J. 1955 Studies on *Cytophaga fermentans* n.sp., a facultatively anaerobic lower myxobacterium. *Journal of General Microbiology* **13,** 541–551.

BARBER, M. & KUPER, S. W. A. 1951 Identification of *Staphylococcus pyogenes* by the phosphatase reaction. *Journal of Pathology and Bacteriology* **63,** 65–68.

BOUSFIELD, I. J. 1972 A taxonomic study of some coryneform bacteria. *Journal of General Microbiology* **71,** 441–455.

BREED, R. S., MURRAY, E. G. D. & SMITH, N. R. (eds) 1957 *Bergey's Manual of Determinative Bacteriology* 7th edn, London: Baillière, Tindall & Cox.

BRISOU, J. 1957 Contribution à l'étude des Pseudomonadaceae. D.Sc. Thesis, Université de Paris.

BYROM, N. A. 1971 The Adansonian taxonomy of some cannery flavobacteria. *Journal of Applied Bacteriology* **34,** 339–346.

BUCHANAN, R. E. & GIBBONS, N. E. (eds) 1974 *Bergey's Manual of Determinative Bacteriology*, 8th edn. Baltimore: Williams & Wilkins.

CHRISTENSEN, P. J. 1973 *Studies on Soil and Freshwater Cytophagas*. Ph.D. Thesis, University of Alberta, Edmonton, Canada.

CONN, H. J. 1957 Staining methods. In *Manual of Microbiological Methods*. Society of American Bacteriologists. London: McGraw-Hill.

DAFT, M. J. & STEWART, W. D. P. 1971 Bacterial pathogens of freshwater blue-green algae. *New Phytologist* **70,** 819–829.

Ferrari, A. & Zannini, E. 1958 Richerche sulle specie del genere *Flavobacterium*. *Annali di Microbiologia ed Enzimologia* **8,** 138–204.

Gavini, F. & Leclerc, H. 1975 Etude de bacilles Gram negatif pigmentés en jaune isolés de l'eau. *Revue internationale d'oceanographie medicale* **37/38,** 17–68.

Gemmell, M. & Hodgkiss, W. 1964 The physiological characters and flagellar arrangement of motile homofermentative lactobacilli. *Journal of General Microbiology* **35,** 519–526.

Geringer, M. & Kielwein, G. 1976 Untersuchungen zur Einteilung der in Milch vorkommenden Flavobakterien. *Archiv fur Lebensmittelhygiene* **27,** 11–17.

Hayes, P. R. 1977 A taxonomic study of flavobacteria and related Gram negative yellow pigmented rods. *Journal of Applied Bacteriology* **43,** 345–367.

Hayes, P. R., Wilcock, A. P. D. & Parish, J. H. 1977 Deoxyribonucleic acid base composition of flavobacteria and related Gram negative yellow pigmented rods. *Journal of Applied Bacteriology* **43,** 111–115.

Hendrie, M. S., Mitchell, T. G. & Shewan, J. M. 1968 The identification of yellow pigmented rods. In *Identification Methods for Microbiologists*, Part B, eds Gibbs, B. M. & Shapton, D. A. London: Academic Press.

Henrichsen, J. 1972 Bacterial surface translocation: a survey and classification. *Bacteriological Reviews* **36,** 478–503.

Holmes, B., Snell, J. J. S. & Lapage, S. P. 1977 Revised description, from clinical isolates, of *Flavobacterium odoratum* Stutzer and Kwaschnina 1929 and description of the neotype strain. *International Journal of Systematic Bacteriology* **27,** 330–336.

Hugh, R. & Leifson, E. 1953 The taxonomic significance of fermentative versus oxidative metabolism of carbohydrates by various Gram negative bacteria. *Journal of Bacteriology* **66,** 24–26.

Knight, B. C. J. G. & Proom, H. 1950 A comparative study of the nutrition and physiology of mesophilic species in the genus *Bacillus*. *Journal of General Microbiology* **4,** 508–538.

Kovács, N. 1956 Identification of *Pseudomonas pyocyanea* by the oxidase reaction. *Nature, London* **178,** 703.

Leadbetter, E. R. 1974 Cytophagales. In *Bergey's Manual of Determinative Bacteriology* 8th edn, eds Buchanan, R. E. & Gibbons, N. E. Baltimore: Williams & Wilkins.

Leifson, E. 1963 Determination of carbohydrate metabolism of marine bacteria. *Journal of Bacteriology* **85,** 1183–1184.

Lewin, R. A. 1969 A classification of flexibacteria. *Journal of General Microbiology* **58,** 189–206.

Lund, B. M. 1969 Properties of some pectolytic, yellow pigmented, Gram negative bacteria isolated from fresh cauliflowers. *Journal of Applied Bacteriology* **32,** 60–67.

McMeekin, T. A. 1977 Ultraviolet light sensitivity as an aid for the identification Gram negative yellow pigmented rods. *Journal of General Microbiology* **103,** 149–151.

McMeekin, T. A., Patterson, J. T. & Murray, J. G. 1971 An initial approach to the taxonomy of some Gram negative, yellow pigmented rods. *Journal of Applied Bacteriology* **34,** 699–716.

McMeekin, T. A. & Shewan, J. M. 1978 Taxonomic strategies for *Flavobacterium* and related genera. *Journal of Applied Bacteriology* **45,** 321–332.

McMeekin, T. A., Stewart, D. B. & Murray, J. G. 1972 The Adansonian taxonomy and the deoxyribonucleic acid base composition of some Gram negative yellow pigmented rods. *Journal of Applied Bacteriology* **35,** 129–137.

MARMUR, J. & DOTY, P. 1962 Determination of the base composition of deoxyribonucleic acid from its thermal denaturation temperature. *Journal of Molecular Biology* **5,** 109–118.

MAYFIELD, C. I. & INNISS, W. E. 1977 A rapid, simple method for staining bacterial flagella. *Canadian Journal of Microbiology* **23,** 1311–1313.

OWEN, R. J. & LAPAGE, S. P. 1974 A comparison of strains of King's Group IIb of *Flavobacterium* with *Flavobacterium meningosepticum. Antonie van Leeuwenhoek* **40,** 255–264.

OWEN, R. J. & SNELL, J. J. S. 1973 Comparison of group IIf with *Flavobacterium* and *Moraxella. Antonie van Leeuwenhoek* **39,** 473–480.

OWEN, R. J. & SNELL, J. J. S. 1976 Deoxyribonucleic acid reassociation in the classification of flavobacteria. *Journal of General Microbiology* **93,** 89–102.

PERRY, L. B. 1973 Gliding motility in some non-spreading flexibacteria. *Journal of Applied Bacteriology* **36,** 227–232.

REICHENBACH, H., KLEINIG, H. & ACHENBACH, H. 1974 The pigments of *Flexibacter elegans*: novel and chemosystematically useful compounds. *Archives of Microbiology* **101,** 131–144.

SCHILDKRAUT, C. I., MARMUR, J. & DOTY, P. 1962 Determination of the base composition of deoxyribonucleic acid from its buoyant density in CsCl. *Journal of Molecular Biology* **4,** 430–443.

SHEWAN, J. M. 1973 Recent progress in the taxonomy and identification of some genera of marine bacteria. In *Proceedings of the 5th International Colloquium of Medical Oceanography*, Messina 1971, pp. 57–71.

SHILO, M. 1970 Lysis of blue-green algae by myxobacter. *Journal of Bacteriology* **104,** 453–461.

SKIDMORE, W. D. & DUGGAN, E. L. 1971 Base composition of nucleic acids. In *Methods of Microbiology* Vol. 5B, eds Norris, J. R. & Ribbons, D. W. London & New York: Academic Press.

STEWART, J. R. & BROWN, R. M. 1971 Algicidal non-fruiting myxobacteria with high G + C ratios. *Archiv für Mikrobiologie* **80,** 176–190.

WEEKS, O. B. 1974 Genus *Flavobacterium* Bergey *et al.* 1923. In *Bergey's Manual of Determinative Bacteriology* 8th edn, eds Buchanan, R. E. & Gibbons, N. E. Baltimore: Williams & Wilkins.

Methods for the Characterization of the Bacteroidaceae

ELLA M. BARNES

A.R.C. Food Research Institute, Colney Lane, Norwich NR4 7UA, UK

In the present classification the family *Bacteroidaceae* consists of anaerobic non-sporing Gram negative rods which are separated into three genera, *Fusobacterium*, *Leptotrichia* and *Bacteroides* (Holdeman & Moore 1974). It has been proposed by Moore *et al.* (1976) that the family should be emended to include bacteria that are monotrichous or lophotrichous and should therefore include a number of other genera. These will not be discussed here.

The anaerobic Gram negative bacteria form a major part of the normal flora of the alimentary tract of man and other animals and some species are involved in human and animal infections. These organisms differ considerably in their sensitivity to oxygen and in practical terms there is a whole range of strains from those which show little sensitivity to oxygen to others which can only be grown when all traces of oxygen are removed from the environment. Thus, one of the major problems has been the difficulty of growing all of the organisms under optimal conditions coupled with the problem of keeping them alive so that type cultures and reference strains are available. Evidence now suggests that the pathogenic anaerobes are amongst the more oxygen tolerant strains (Rosenblatt *et al.* 1973). Thus, both because of their clinical importance and because it has been easier to maintain the reference strains, much of the present classification is based on these organisms. With the current interest in the normal flora of the alimentary tract and improvements in anaerobic techniques many organisms are now being isolated which cannot be related to known species.

Following the development of the Hungate (1950) technique for the isolation of the strict anaerobes from the rumen e.g. *Bacteroides succinogenes*, *B. amylophilus* and *B. ruminicola* this technique was then used by Holdeman & Moore (1974) to make a comparative study of all the type and reference strains of anaerobes from many sources. Further tests have now shown that a number of the strains can be grown using simpler anaerobic techniques. This is particularly important with clinical isolates where rapid identification is needed. The problem arises when unknown organisms are isolated by one of the simpler methods (see below) and a comparison is required with an organism such as the type strain of *B. ruminicola* which can only be grown using the Hungate or similar technique. The sensitivity to oxygen may vary within a species so that it cannot be assumed that an organism isolated using one method is not identical with an organism which requires a much stricter anaerobic technique for growth. This situation is at present handled in many different ways and will depend on the type of study being undertaken, but the best way of finally ensuring the identity of an organism is to relate it to the reference strain grown under optimal conditions.

There are now a number of detailed reference books and papers which can be consulted for the methods to use in identifying these anaerobes, so this chapter will be confined to a discussion of the more important tests and some of the problems which may arise.

Methods

Anaerobic techniques

For practical purposes growth conditions for the anaerobes can be divided into three categories.

Method 1

For strict anaerobes where the organisms are either grown in an anaerobic chamber (Drasar & Crowther 1971) or using the technique first described by Hungate (1950). Media preparation and all other operations are carried out under a continual flow of carbon dioxide or nitrogen from which the last traces of oxygen have been removed by passing over heated copper.

Method 2

Inoculation of pre-reduced agar plates or broths and incubation in an anaerobic jar containing 90% of hydrogen + 10% of carbon dioxide together with a catalyst to remove traces of oxygen.

Method 3

Growth in media containing reducing agents, either as deep agar in tubes, or broths in sealed or capped bottles. Any dissolved oxygen is removed before inoculation by holding the media in a boiling water bath for at least 20 min. They are then cooled and inoculated immediately.

Media and methods using the Hungate technique (Method 1)

The media and methods used by Holdeman & Moore (1974) for the description of all the species of Gram negative anaerobes whatever their origin have been described in detail by Holdeman *et al.* (1977) and are based on the use of a PY medium containing peptone, trypticase, yeast extract, salts, haemin and cysteine hydrochloride, to which carbohydrates and other substances are added. Resazurin is generally included in all media as a redox indicator, the media staying colourless if the E_h is below −100 mV.

The earlier methods were designed particularly for the isolation and characterization of the rumen anaerobes (Bryant 1963) and the media were supplemented with rumen fluid. Subsequently Caldwell & Bryant (1966) developed a medium without rumen fluid (Medium 10) for the non-selective isolation of rumen bacteria. When this medium was tested for the isolation and growth of anaerobic organisms from chicken caeca it was found necessary to supplement it with chicken faecal extract and liver extract for optimal recovery and growth of the organisms. Modifications of this supplemented M10 medium are therefore used for characterizing the chicken isolates and reference strains are grown under comparable conditions (Barnes & Impey 1974).

Traditional techniques (Methods 2 and 3)

Many of the organisms within the family *Bacteroidaceae* can be grown by the traditional and much simpler techniques. The media and methods described by Barnes & Impey (1968) are based on those of Beerens & Tahon-Castel (1965) and were used successfully by Barnes & Goldberg (1968) in their numerical taxonomy study of these organisms.

It was shown by Rosenblatt *et al.* (1973) that the anaerobes most frequently isolated from clinical specimens could be grown successfully without using the Hungate technique and various schemes have now been established for the rapid identification of the clinically important organisms (Moore *et al.* 1975; Sutter *et al.* 1975).

Biochemical and other tests

Fermentation products

Gas-liquid chromatography (g.l.c.) is now used in the analysis of various growth media (particularly glucose broths) to determine fermentation products. Methods for estimating (a) volatile fatty acids e.g. formic, acetic, propionic, butyric etc., (b) non-volatile acids such as pyruvic, lactic and succinic or (c) alcohols, have been described by Holdeman *et al.* (1977), Sutter *et al.* (1975) and Salanitro & Muirhead (1975).

Threonine deaminase test

Beerens *et al.* (1959) showed that certain Gram negative anaerobes deaminated L(−) threonine with the production of α-keto-butyrate and ammonia; the α-keto-butyrate could then be converted to propionic acid, hydrogen and carbon dioxide. Several methods for carrying out this test are described by Barnes (1969). However, if g.l.c. is available the method most frequently used is the detection of propionic acid after growth in a basal growth medium (without added carbohydrate) e.g. PY of Holdeman *et al.* (1977) containing 0·3% threonine and comparing with propionic acid production in the same medium without threonine.

Growth in the presence of bile

This test was originally developed by Beerens & Castel (1960). Holdeman *et al.* (1977) incorporate ox gall to give a final concentration of 20% whilst Shimada *et al.* (1970) prefer to use 20% bile together with 0·1% sodium desoxycholate which is a more inhibitory medium. For tests not requiring the use of the Hungate technique the author now uses VL + haemin (Barnes & Impey 1968) containing 20% bile + 0·1% sodium desoxycholate. Growth is compared with growth in the medium without bile and is recorded as inhibited, no effect, or stimulated.

Glutamic acid decarboxylation

The ability of some species to form γ-amino-*n*-butyric acid from glutamic acid has been studied by Suzuki *et al.* (1966) and Werner (1970).

Antibiotics and dyes

Differing susceptibility to a range of antibiotics has been used as part of rapid identification schemes by Sutter & Finegold (1971) whilst dyes, especially victoria blue 4R, brilliant green or gentian violet have been found useful for differentiating some of the *Fusobacterium* spp. from *Bacteroides* spp. (Suzuki *et al.* 1966). Further studies which include the use of antibiotics and dyes for differentiating the clinically important Gram negative anaerobes have been described by Duerden *et al.* (1976).

Serology

Reference should be made to the comprehensive reviews of Sonnenwirth (1973, 1975). Detailed studies have been published by Werner & Sebald (1968), Beerens *et al.* (1971) and Sharpe (1971).

Identification of Organisms

Before assigning an unknown isolate to the family *Bacteroidaceae* it must

TABLE 1. Family Bacteroidaceae[a]

Gram negative, uniform or pleomorphic, non-sporing rods.
Non-motile or motile with peritrichous flagella. Obligate anaerobes.
Fusobacterium
Organisms produce *n*-butyric acid as a major product (without *iso*butyric and *iso*valeric acids) from peptone or glucose. DNA Base ratios: G + C, 26–34 moles %.
Leptotrichia
(Only 1 species, *L. buccalis*).
Produces mainly lactic acid. DNA Base ratios: G + C, 32–34 moles %.
Bacteroides
Organisms produce (from peptone or glucose) mixtures of acids including succinic, acetic, formic, lactic, propionic; *n*-butyric acid usually not a major product. Some species produce a mixture of *n*-butyric, *iso*butyric and *iso*valeric acids along with major amounts of succinic acid. DNA Base ratios: G + C, 40–55 moles % with some exceptions.

[a] Data of Holdeman & Moore (1974).

first be confirmed that the organism is a Gram negative non-sporing rod. It can be seen from Table 1 that further differentiation into the genera *Fusobacterium*, *Leptotrichia* or *Bacteroides* depends on the fermentation products. Thus, in identifying these organisms, particular attention should be paid to the following tests.

Anaerobic growth

The organism must be an obligate anaerobe which can be defined in practical terms as failure to grow on an optimal agar growth medium incubated in air, or air + 10% carbon dioxide. There are a few exceptions in the present list of species assigned to this family one of them being *Bacteroides ochraceus* some strains of which can grow in air + 10% carbon dioxide.

Morphology

Young cultures (generally 24 h) should be examined in wet mounts by phase contrast microscopy for size, shape, motility, etc. Morphologically, it is often difficult to differentiate between rods and cocci and some of the *Bacteroides* spp. often appear as coccobacilli. Extreme pleomorphism is also encountered in many species.

Gram reaction

The present classification is based on a Gram reaction determined using Kopeloff's modification as recommended by Holdeman *et al.* (1977).

Any sign of a Gram positive reaction should be viewed with caution and the test repeated with a 6 h culture.

Spores

It is insufficient to record the absence of spores on direct microscopical examination. It is now known that a number of organisms which have been characterized as Gram negative non-sporing rods do in fact form spores. There are several ways of confirming this but in principle the organisms are preferably grown in one of the optimal buffered growth media for 7 and 21 days. The broths are then tested for the presence of surviving organisms after heating at both 70°C and 80°C for 10 min. In some instances the spores will not survive heating at 80°C. The Gram negative pointed rods described by Holdeman & Moore (1974) as *Bacteroides clostridiiformis* subspecies *clostridiiformis* (ATCC 25537) together with *B. clostridiiformis* subspecies *girans* and *B. biacutus* have now been shown to form spores and have been transferred to the genus *Clostridium* as *Clostridium clostridiiforme* (Kaneuchi *et al.* 1976). Similarly, the organism described as *Fusobacterium symbiosum* (ATCC 14940) is now classified as *Cl. symbiosum* (Kaneuchi *et al.* 1976).

Fermentation products

The determination of fermentation products using g.l.c. is one of the most reliable methods of identifying these organisms. It has been found that tested under standard conditions the range of volatile fatty acids, non-volatile acids and alcohol produced is very reproducible. The relative proportions will however vary with the medium and anaerobic conditions for growth, so in order to confirm identification of a difficult organism, it is best to compare the products produced with those of the reference or type strain grown under comparable conditions.

The production of *n*-butyric acid is particularly important because it assigns an organism to the genus *Fusobacterium* whilst the genus *Leptotrichia* is defined as containing organisms which produce lactic acid as the major product (Table 1). The genus *Bacteroides* contains a variety of species which vary considerably in their fermentation products, some producing *iso*-butyric and other branch chain acids but most species do not produce *n*-butyric acid. In many cases the fermentation products also help to differentiate species or groups of species.

Terminal pH in glucose broth

Whether tested using the Hungate technique or by traditional methods the terminal pH in glucose broth is a useful guide to the differentiation of some of the species. Generally the *Fusobacterium* strains have a terminal pH above 5·5 whilst many of the *Bacteroides* are below pH 5·5, unless they do not utilize glucose (Table 2). When assessing utilization of glucose by a change in pH it is necessary to compare the terminal pH obtained in the growth medium with and without the addition of glucose, as some organisms can cause a lowering of the pH from the utilization of other energy sources present.

Other tests

Amongst other tests which have proved particularly useful in differentiating some of the species are the production of propionic acid from threonine, decarboxylation of glutamic acid and growth in the presence of bile.

For a final characterization of the Gram negative anaerobes the properties listed by Holdeman & Moore (1974) as typical of the species, should be studied.

Properties of some of the more important organisms

The properties of some of the more commonly encountered anaerobes in human and animal infections are shown in Table 2. Much of the development of quick testing methods has been carried out in relation to these organisms as a rapid diagnosis is required and the numbers of different types of anaerobes generally involved in these infections is known to be limited.

In earlier classifications the genera *Fusobacterium* and *Sphaerophorus* were separated but all of the *n*-butyric acid producing *Sphaerophorus* species are now included in the genus *Fusobacterium*. Thus, the important human and animal pathogen *Sphaerophorus necrophorus* is now *Fusobacterium necrophorum*.

One of the difficult areas has been the differentiation of *Bacteroides fragilis* from *B. vulgatus*, *B. thetaiotamicron*, *B. distasonis* and *B. ovatus*. These organisms share many important properties including similar fermentation products, the ability to decarboxylate glutamic acid and growth stimulation by bile. Sugar fermentation tests were sufficiently variable for the organisms to be placed by Holdeman & Moore (1974) in a single species *B. fragilis* with the appropriate subspecies. Cato & Johnson (1976) have shown that the subspecies are genetically distinct and have proposed that they should be restored to species rank. Other differences have been discussed by Werner (1974). Evidence is accumulating that the pathogenic member of this group is *B. fragilis* subspecies *fragilis*. Werner (1974) states that the mechanisms associated with the virulence of *B. fragilis* subspecies *fragilis* are little understood but among

TABLE 2. Differential properties of some of the more commonly encountered species in human and animal infections

	Fermentation products[a]	Glucose fermented	Terminal pH in glucose broth	Propionic acid from threonine	Bile 20% + 0·1% desoxycholate
Fusobacterium nucleatum	Baps (Lf)	+	5·8–6·2	+	I[b]
F. necrophorum	Bap (Lsf)	+	5·8–6·3	+	I
F. varium	BLA (Sp)	+	5·6–6·0	+	NE or S
F. mortiferum	BAp (LF, iv, s4)	+		+	NE or S
Bacteroides fragilis subspecies *fragilis*	Sap (ib, iv, lf)	+	5·4	—	NE or S
B. melaninogenicus subspecies *intermedius*	SA, ib, iv, (lp)	+	4·9–5·4	—	I
subspecies *melaninogenicus*	SA (fpb, iv, l)	+	5·1	—	I
subspecies *asaccharolyticus* (*B. asaccharolyticus*)	ABp, iv, ib, (ls)	—	6·4	—	I
B. corrodens	sa (fpl)	—		—	I
B. nodosus[c]	asp (ib, b, iv)	—	7·2–7·4	— or +	I

[a] Data of Holdeman *et al.* (1977) from PYG broth: a: acetic; b: *n*-butyric; ib: *iso*butyric; f: formic; l: lactic; p: propionic; s: succinic; iv: *iso* valeric; 4: butanol: < 1 mEq 100 ml^{-1}. Caps e.g. A B etc. > 1 mEq 100 ml^{-1}. ( ): some strains.

[b] I: growth inhibited; NE: no effect; S: stimulated.

[c] Information mostly provided by Skerman (pers. comm.) and Skerman (1975).

Glutamic acid decarboxylase	Indole produced	Aesculin hydrolysed	Base ratios (mol % G + C)	Other properties
−	+	−	27–28	Long slender rods and filaments with pointed ends
−	+	−	31–34	Long slender rods and filaments with round ends. Copious gas production
	+ or −	−	27	
	−	+	26–28	
+	−	+	42–44	
−	+	−	41–45	Forms black colonies due to production of haematin
−	−	+	40–42	
−	+	−	50–54	
−	−	−	28–30	Colonies show 'pitting' of the agar
−	−	−	45	Requires arginine for growth

the factors with possible pathogenic significance are (a) endotoxic polysaccharides (b) the capsule (c) a neuraminidase (d) a fibrinolysin (e) a haemolysin'. The other species related to *B. fragilis* are commonly found as part of the normal flora of man and animals.

Other succinic acid-producing *Bacteroides* spp. such as *B. ruminicola*, saccharolytic strains of *B. melaninogenicus* and possibly *B. oralis*, differ from *B. fragilis* in that they are inhibited by bile salts and do not decarboxylate glutamic acid. Recent studies with *B. melaninogenicus* (Finegold & Barnes 1977) have shown that two distinct species are involved both of which produce a black pigment haematin (Table 2).

General Conclusions

With the current interest in ecology and the increasing use of techniques designed for the isolation and growth of strict anaerobes (Method 1) it is inevitable that many organisms will be isolated which cannot be identified. When Moore & Holdeman (1974) studied the human faecal flora, very many types of anaerobes were found which could not be assigned to known species. A similar situation occurs with the avian caecal flora (Barnes & Impey 1974). Except in the few cases of really well characterized clinically important organisms no one at present should expect to be able to identify all of their isolates to the species level. Before assigning an organism to an established species it is advisable to compare it with the reference strain.

References

BARNES, E. M. 1969 Methods for the Gram-negative non-sporing anaerobes. In *Methods in Microbiology* Vol. 3B, eds Norris, J. R. & Ribbons, D. W. London: Academic Press.

BARNES, E. M. & GOLDBERG, H. S. 1968 The relationship of bacteria within the family *Bacteroidaceae* as shown by numerical taxonomy. *Journal of General Microbiology* **51,** 313–324.

BARNES, E. M. & IMPEY, C. S. 1968 Anaerobic Gram-negative non-sporing bacteria from the caeca of poultry. *Journal of Applied Bacteriology* **31,** 530–541.

BARNES, E. M. & IMPEY, C. S. 1974 The occurrence and properties of uric acid decomposing anaerobic bacteria in the avian caecum. *Journal of Applied Bacteriology* **37,** 393–409.

BEERENS, H. & CASTEL, M. M. 1960 Action de la bile sur la croissance de certaines bactéries anaérobies à Gram-négatif. *Annales de l'Institut Pasteur* **99,** 454–456.

BEERENS, H., GUILLAUME, J. & PETIT, H. 1959 Etude de la fermentation propionique de la L(—) thréonine par 45 souches de bactéries anaérobies non sporulées à Gram négatif. *Annales de l'Institut Pasteur* **96,** 211–216.

BEERENS, H. & TAHON-CASTEL, M. M. 1965 *Infections humaines à bactéries anaérobies non toxigènes*. Bruxelles: Presses Academiques Européennes.

BEERENS, H., WATTRE, P., SHINJO, T. & ROMOND, C. 1971 Premier resultats d'un essai de classification sérologique de 131 souches de *Bacteroides* du groupe *fragilis* (*Eggerthella*). *Annales de L'Institut Pasteur* **121,** 187–198.

BRYANT, M. P. 1963 Symposium on microbial digestion in ruminants; identification of groups of anaerobic bacteria active in the rumen. *Journal of Animal Science* **22,** 801–813.

CALDWELL, D. R. & BRYANT, M. P. 1966 Medium without rumen fluid for non selective enumeration and isolation of rumen bacteria. *Applied Microbiology* **14,** 794–801.

CATO, E. P. & JOHNSON, J. L. 1976 Reinstatement of species rank for *Bacteroides fragilis, B. ovatus, B. distasonis, B. thetaiotamicron* and *B. vulgatus.* Designation of neotype strains *Bacteroides fragilis* (Veillon and Zuber) Castellani and Chalmers and *Bacteroides thetaiotamicron* (Distaso) Castellani and Chalmers. *International Journal of Systematic Bacterology* **26,** 230–237.

DRASAR, B. S. & CROWTHER, J. S. 1971 The cultivation of human intestinal bacteria. In *Isolation of Anaerobes,* eds Shapton, D. A. & Board, R. G. Society for Applied Bacteriology Technical Series No. 5 London: Academic Press.

DUERDEN, B. I., HOLBROOK, W. P., COLLEE, J. G. & WATT, B. 1976 The characterization of clinically important Gram negative anaerobic bacilli by conventional bacteriological tests. *Journal of Applied Bacteriology* **40,** 163–188.

FINEGOLD, S. M. & BARNES, E. M. 1977 Report of the ICSB Taxonomic Subcommittee on Gram-negative Anaerobic Rods. Proposal that the saccharolytic and asaccharolytic strains at present classified in the species *Bacteroides melaninogenicus* (Oliver and Wherry) be reclassified in two species as *Bacteroides melaninogenicus* and *Bacteroides asaccharolyticus. International Journal of Systematic Bacteriology* **27,** 388–391.

HOLDEMAN, L. V., CATO, E. P. & MOORE, W. E. C. 1977 *Anaerobe Laboratory Manual 4th edn.* VPI Anaerobe Laboratory, Virginia Polytechnic Institute and State University, P.O. Box 49, Blacksburg, Virginia 24060.

HOLDEMAN, L. V. & MOORE, W. E. C. 1974 Family I *Bacteroidaceae.* In *Bergey's Manual of Determinative Bacteriology* 8th edn, eds Buchanan, R. E. & Gibbons, N. E. Baltimore: Williams & Wilkins.

HUNGATE, R. E. 1950 The anaerobic mesophilic cellulolytic bacteria. *Bacteriological Reviews* **14,** 1–49.

KANEUCHI, C., WATANABE, K., TERADA, A., BENNO, Y. & MITSUOKA, T. 1976 Taxonomic study of *Bacteroides clostridiiformis* subsp. *clostridiiformis* (Burri and Ankersmit) Holdeman and Moore and of related organisms. Proposal of *Clostridium clostridiiforme* (Burri and Ankersmit) comb. nov. and *Clostridium symbiosum* (Stevens) comb. nov. *International Journal of Systematic Bacteriology* **26,** 195–204.

MOORE, W. E. C. & HOLDEMAN, L. V. 1974 Human fecal flora: the normal flora of 20 Japanese-Hawaiians. *Applied Microbiology* **27,** 961–979.

MOORE, W. E. C., JOHNSON, J. L. & HOLDEMAN, L. V. 1976 Emendation of *Bacteroidaceae* and *Butyrivibrio* and descriptions of *Desulfomonas,* gen. nov. and ten new species in the genera *Desulfomonas, Butyrivibrio, Eubacterium, Clostridium* and *Ruminococcus. International Journal of Systematic Bacteriology* **26,** 238–252.

MOORE, H. B., SUTTER, V. L. & FINEGOLD, S. M. 1975 Comparison of three procedures for biochemical testing of anaerobic bacteria. *Journal of Clinical Microbiology* **1,** 15–24.

ROSENBLATT, J. E., FALLON, A. & FINEGOLD, S. M. 1973 Comparison of methods for isolation of anaerobic bacteria from clinical specimens *Applied Microbiology* **25,** 77–85.

SALANITRO, J. P. & MUIRHEAD, P. A. 1975 Quantitative method for the gas chromatographic analysis of short-chain monocarboxylic and dicarboxylic acids in fermentation media. *Applied Microbiology* **29,** 374–381.

SHARPE, M. E. 1971 Serology of rumen bacteroides. *Journal of General Microbiology* **67,** 273–288.

SHIMADA, K.; SUTTER, V. L. & FINEGOLD, S. M. 1970 Effect of bile and desoxycholate on Gram-negative anaerobic bacteria. *Applied Microbiology* **20,** 737–741.

SKERMAN, T. M. 1975 Determination of some *in vitro* growth requirements of *Bacteroides nodosus*. *Journal of General Microbiology* **87,** 107–119.

SONNENWIRTH, A. C. 1973 Serology of *Bacteroidaceae*. Abstracts Vol. 1. 1st International Congress for Bacteriology, Jerusalem, Sept. 2–7th 1973.

SONNENWIRTH, A. C. 1975 Serology of *Bacteroidaceae*—A review. Symposium *Actual Data on the Biology and Pathology of Anaerobic Bacteria* Cantacuzino Institute, Bucharest, Romania, July 24–25.

SUTTER, V. L. & FINEGOLD, S. M. 1971 Antibiotics disc susceptibility tests for rapid presumptive identification of Gram-negative anaerobic bacilli. *Applied Microbiology* **21,** 13–20.

SUTTER, V. L., VARGO, V. L. & FINEGOLD, S. M. 1975 Wadsworth *Anaerobic Bacteriology Manual* 2nd edn. Anaerobic Bacteriology Laboratory, Wadsworth Hospital Center, Veterans Administration, Wilshire and Sawtelle Blvds, Los Angeles, Ca, 90073 and the Department of Medicine, UCLA School of Medicine, Los Angeles, Ca, 90024.

SUZUKI, S., USHIJIMA, T. & ICHINOSE, H. 1966 Differentiation of *Bacteroides* from *Sphaerophorus* and *Fusobacterium*. *Japanese Journal of Microbiology* **10,** 193–200.

WERNER, H. 1970 Glutaminsäuredecarboxylase-aktivität bei Bacteroides-Arten, *Zentralblatt fur Bakteriologie, Parasitenkunde, Infektionskrankheiten und Hygiene, Abt I Orig.* **215,** 320–326.

WERNER, H. 1974 Differentiation and medical importance of saccharolytic intestinal *Bacteroides*. *Arzneimittel-Forschung* **24,** 340–343.

WERNER, H. & SEBALD, M. 1968 Étude serologique d'anaérobies Gram négatifs asporulés, et particulièrement de *Bacteroides convexus* et *Bacteroides melaninogenicus*. *Annales de L'Institut Pasteur* **115,** 350–366.

Methods for Identifying Staphylococci and Micrococci

A. C. Baird-Parker

Unilever Research, Colworth House, Sharnbrook, Bedford, UK

Since the publication of the first edition of this manual there have been a number of significant advances in the classification and identification of staphylococci and micrococci. These have come from a better understanding of their physiology and biochemistry, the relatedness of their DNA and the structure and composition of their cell walls and membranes.

The purpose of this chapter is to outline those methods that can be used to identify isolates of staphylococci and micrococci in the laboratory; it is not intended to be a taxonomic treatise. The classification followed is based on that described in the 8th Edition of Bergey's manual (Baird-Parker 1974) updated to take account of the more recent taxonomic studies by Hájek, Kloos, Oeding, Schleifer and others.

Preliminary Screening and Maintenance of Isolates

Check by phase-contrast and light microscopy that the organism is a Gram positive, cluster-forming coccus and that it is capable of forming catalase when grown for 24 h at 30°C on a medium containing (% w/v): Difco Bacto peptone, 1·0; Difco yeast extract, 0·1; glucose, 1·0; sodium chloride, 0·5; agar, 1·5; pH 7·0–7·2 (Medium A). The medium is sterilized by autoclaving at 121°C for 15 min. Test for catalase by applying, using a Pasteur pipette, a drop of 3% (v/v) hydrogen peroxide to a colony and examining for the release of gas bubbles from the applied peroxide solution.

After purification, isolates should be maintained on slopes of Medium A (without glucose) and stored at 4°C. Before attempting to identify an isolate, subculture into Heart Infusion Broth and incubate overnight at 30°C (see section 'Identification Methods'). It is important that a vigorously growing culture is used for all tests but subculturing should be minimal as variants are always likely to arise in laboratory media.

Once purified, avoid picking single colonies when purity checks are made.

Separation of Members of the Genus *Staphylococccus* from Members of the Genus *Micrococcus*

Some of the key characters currently accepted as separating staphylococci from micrococci are listed in Table 1. The first of these characters, namely the ability of staphylococci to grow anaerobically and to ferment glucose was first proposed by Evans *et al.* (1955) and adopted by the ICSB Sub-Committee on the Taxonomy of Staphylococci and Micrococci (Anon. 1965) as the routine test for separating staphylococci from the aerobic glucose-oxidizing micrococci.

The use of this character has been justly criticized on the grounds that *S. saprophyticus* and related species (Kloos & Schleifer 1975*a*, *b*) grow poorly or not at all under anaerobic conditions and are thus misclassified as micrococci. Many alternatives to this test have been proposed, the most significant of which is the test system proposed by Schleifer & Kloos (1975*a*). This test system makes use of the greater resistance of staphylococci to erythromycin together with their sensitivity to lysostaphin endopeptidase and their resistance to lysozyme. Lysostaphin lyses cell walls containing glycyl-glycine links in their cross bridges. The lytic activity of lysozyme is inhibited by teichoic acids and other molecules in the intact staphylococcal cell wall (Table 1).

Preparation of media for use in the Schleifer and Kloos test

Growth in the presence of erythromycin

Prepare Purple Agar Base (Difco) according to the manufacturer's instructions (Medium B). Sterilize the basal medium in 90 ml amounts in screw-capped bottles by autoclaving at 121° for 15 min. Dissolve 4 mg of pure erythromycin in 0·5 ml of 95% ethanol, make up to 100 ml with distilled water in a volumetric flask and sterilize by filtration. For use, add 10 ml of sterile 10% (w/v) glycerol solution and 1 ml of the erythromycin solution to 90 ml of molten and cooled (46°–48°C) Medium B. Mix thoroughly and pour into Petri dishes (15 ml per 9 cm dish). After drying, radially streak 6 isolates per plate. Incubate plates for 2 days at 37°C and observe for growth and acid production.

Sensitivity to lysostaphin and lysozyme

To a bottle of molten and cooled medium A add a sterile filtered solution of lysostaphin (Schwartz/Mann, Orangeburg, New York) to give a final concentration of 200 μg ml^{-1}. To a further bottle, add egg

TABLE 1. Characters separating Staphylococci from Micrococci

		Staphylococcus	*Micrococcus*
Growth	Anaerobic and fermentation of glucose	+	—
	Erythromycin (0·4 μg ml^{-1})	R	S
Cell wall	Glycine containing penta- and hexapeptides crossbridge in the peptidoglycan	+	—
	Ribitol and/or glycerol teichoic acids	+	—
Cell membrane	aliphatic hydrocarbons	—	+
	hydrogenated menaquinones	+	—
Nucleic acid	G + C content (mol %)	30–40	66–75

+: > 80% +ve; R: resistant; —: > 80% —ve; S: sensitive.

white lysozyme (Sigma Chemical Co., St. Louis, Mo.) to give a final concentration of 25 μg ml^{-1}. Pour plates on a flat and horizontal surface (15 ml per 9 cm dish) and after setting and drying, radially streak with the test organisms and incubate as described above for growth in the presence of erythromycin.

Interpretation of Schleifer and Kloos Test

	Growth in the presence of erythromycin	Sensitivity to Lysostaphin	Sensitivity to Lysozyme
Staphylococcus	+	+	—
Micrococcus	—	—	Variable

Identification of Species

Staphylococci

The number of species recognized in this genus continues to rise as isolates from a wider range of habitats are studied in more detail. In the 8th Edition of *Bergey's Manual* (Baird-Parker 1974) 3 species 'groups' are recognized within the genus *Staphylococcus*: *S. aureus* (Baird-Parker's *Staphylococcus* subgroup I); *S. epidermidis* (Baird-Parker's *Staphylococcus* subgroups II–V); *S. saprophyticus* (Baird-Parker's *Micrococcus* subgroups 1–4). These species groups have since been subdivided into 10 additional species largely as a result of the detailed studies of

Kloos and Schleifer using a variety of modern microbiological and chemical techniques (Kloos & Schleifer 1975*a*, *b*; Schleifer & Kloos 1975*b*; Kloos *et al.* 1976). The probable relationship between the classification in the 8th Edition of *Bergey's Manual* and that of Kloos and Schleifer is outlined in Table 2. It is impossible to be certain of the

TABLE 2. Probable relationship between classification in 8th edition of Bergey's Manual and that of Kloos & Schleifer (1975a, b)

Bergey's Manual (8th Edition)	Kloos & Schleifer (1975a, b)
Staphylococcus aureus	*S. aureus*[a]
S. epidermidis biotype I	*S. epidermidis*
biotype II	*S. hyicus* subsp. *chromogenes* and subsp. *hyicus*[b]
biotypes III/IV	*S. capitis, S haemolyticus, S hominis, S. warneri*
Not recognized	*S. simulans*
S. saprophyticus biotype 3	*S. saprophyticus*
biotypes 1, 2 4 and *M. lactis*	*S. cohnii*? *S. xylosus*?
Not recognized	*S. sciuri*[c]

[a] *S. aureus* biotypes E and F (Hàjek & Maršàlek 1971) now recognized as *S. intermedius* (Hàjek 1976).
[b] Proposed by Devriese *et al.* (1978).
[c] Proposed by Kloos *et al.* (1976).

exact relationship of a number of Kloos and Schleifer 'species' to those in Bergey as substantially different characters were used to delineate the species. A number of Kloos and Schleifers' 'species' are very closely related and should probably be regarded as subspecies (Dr. K. Feltham, pers. comm. 1978).

The most commonly found species on man are *S. aureus*, *S. epidermidis* and *S. saprophyticus* with *S. haemolyticus* and *S. cohnii* occurring less frequently (Oeding & Digranes 1977). Occasionally *S. simulans*, *S. hominis* and *S. capitis* are isolated but by and large the other species are found only on specific animals. In this regard it is important to note the existence of *S. intermedius*, a species proposed by Hájek (1976) for *S. aureus* biotypes E and F of the now classical subdivision of *S. aureus* into biotypes proposed by Hájek & Maršálek (1971). Isolates of this species have been obtained from dogs, foxes, mink and pigeons. It is distinguished from *S. aureus* by a number of important biochemical and cell wall characters (Schleifer *et al.* 1976) but it may be confused with *S. aureus* as it possesses both coagulase and a thermo-stable nuclease; both enzymes are generally considered to be produced only by *S. aureus*. A further species that may be confused with *S. aureus* is *S. hyicus* which is often coagulase and thermo-stable nuclease positive and occurs quite commonly on pigs, poultry and occasionally cows (Devriese 1977). Also, strains weakly producing thermo-stable nuclease, belonging to further species of staphylococci, have been reported by a number of workers and are discussed in detail by Gramoli & Wilkinson (1978). The problem of distinguishing such organisms from *S. aureus* is not easy for a routine laboratory. Many can be eliminated by only recording strong thermo-stable nuclease and coagulase reactions for the presumptive identification of *S. aureus* (Rayman *et al.* 1975) or by the use of additional tests such as mannitol *fermentation* and production of acetoin. *Staphylococcus intermedius* will not be confused with *S. aureus* on a tellurite-containing selective medium such as Baird-Parker agar (Baird-Parker 1969) as these organisms fail to reduce tellurite. However, some strains of *S. hyicus* and other staphylococci may be indistinguishable from *S. aureus* on such media. Characters separating the main species of staphylococci are listed in Table 3. For characters separating the biotypes of *S. aureus* see papers by Hájek & Maršálek (1971) and Hájek (1976).

TABLE 3. Characters separating the main species of staphylococci

	S. aureus	*S. intermedius*	*S. hyicus*	*S. epidermidis*	*S. capitis*	*S. hominis*	*S. warneri*	*S. haemolyticus*	*S. cohnii*	*S. saprophyticus*
Coagulase[a]	+	+	±/V	—	—	—	—	—	—	—
Thermo-stable nuclease	+	+	V[c]	—	—	—	—	—	—	—
Haemolysis[b]	+	V	—	—	—	—	—	+	—	—
Acetoin	+	—	—	+	V	V	+	+	V	+
Acid, aerobically from										
Sucrose	+	+	+	+	+	+	+	+	—	+
Trehalose	+	+	+	—	—	+	+	+	+	+
Mannitol	+	±	—	—	+	—	V	V	+	+
Phosphatase	+	+	+	+	—	—	—	—	—	—
Novobiocin (1·6 μg ml^{-1})	S	S	S	S	S	S	S	S	R	R

—: > 80% —ve; +: > 80% +ve; ±: weak; V: variable; S: Sensitive; R: Resistant.
[a] Rabbit plasma.
[b] Bovine/human blood.
[c] *S. hyicus* subsp. *hyicus* +ve; *S. hyicus* subsp. *chromogenes* —ve (Devriese *et al.* 1978).

Micrococci

The 8th Edition of *Bergey's Manual* recognizes only three species of micrococci: *M. varians*, *M. luteus* and *M. roseus*. Most isolates of micrococci can be fitted into these species and indeed most will be found to belong either to *M. varians* or *M. luteus*. In addition to these species a number of further species have been described in recent years. These include *M. lylae*, *M. kristinae*, *M. sedentarius* and *M. nishinomiyaensis*. However, as these species are rather uncommon and their identification difficult they will not be considered further; for further information see Kloos *et al.* (1974). The characters separating the three species recognized in *Bergey's Manual* are listed in Table 4.

TABLE 4. Characters separating the main species of micrococci

	M. varians	*M. luteus*	*M. roseus*
Pigment	yellow	yellow	pink
Growth in 10% NaCl	+	+	—
Acid aerobically from glucose	+	—	—
Nitrate reduction	+	—	+
Oxidase activity	—	+	— or ±

—: > 80% strains —ve; +: > 80% strains +ve; ±: weak.

Identification Methods

For all tests use cultures grown overnight at 30°C in Heart Infusion Broth (HIB).

Coagulase production

Pipette 0·1 ml of the HIB culture into a sterile 10 × 75 mm tube and add 0·3 ml of rabbit plasma plus EDTA (Difco). After plugging the tube and mixing the contents incubate at 37°C. Examine for clotting of the plasma after 4 h. If negative, incubate at room temperature overnight and examine again. Traditionally any degree of coagulation of plasma is considered positive evidence of coagulase activity. However, when using the coagulase test as the sole means for the presumptive identification of *S. aureus* it is considered by many workers that only strong reactions should be recorded; for further information on this see Rayman *et al.* 1975; Sperber & Tatini 1975. Ideally, if weak coagulase activity is recorded, further tests, as listed in Table 3, should be applied.

Thermo-stable nuclease production

Prepare toluidine blue-DNA agar by dissolving tris (hydroxymethyl) amino-methane 6·1 g in 1 l of distilled water and adjust pH to 9·0. Add DNA (Calf thymus, Difco), 0·3 g; agar, 10 g; calcium chloride (anhydrous), 0·11 g; sodium chloride, 10 g; dissolve these ingredients by boiling. Add O-toluidine blue 0·083 g to the molten agar, mix thoroughly and dispense in 25 ml amounts into bottles. Store under refrigeration. Melt the DNA Agar and spread 3 ml over a microscope slide using a Pasteur pipette. When set, cut 2 mm diam. wells in the agar (up to 12 per slide) using a cannula needle attached to a vacuum source. By means of a suitable Pasteur pipette add to each well *ca.* 10 μl of HIB culture that has been heated for 15 min in a boiling water bath; heat no more than 5 ml in a narrow tube. Incubate slide in a moist chamber for 4 h at 37°C. The presence of a thermo-stable nuclease is indicated by the appearance of a bright pink halo extending at least 1 mm beyond the periphery of the well. *Staphylococcus aureus* strains generally give halos with diameters of at least 1 cm.

Acetoin production

Prepare medium containing % (w/v) tryptone, 1; Lab-lemco, 0·3; yeast extract, 0·1; glucose, 2·0; pH 7·2. Dispense in 5 ml amounts into appropriate tubes or bottles and sterilize by autoclaving at 115°C for 20 min. Inoculate medium from HIB culture and incubate for 14 days at 30°. Test for acetoin production using Barritt's modification of the VP test (Barritt 1936). It is necessary to shake the bottles vigorously for at least 30 sec after addition of the reagents in order to detect weak acetoin producers. A definite reddening of the culture supernatant indicates the production of acetoin.

Acid production from carbohydrates

To molten (46°–48°C) Purple Agar Base (Difco) add filter-sterilized carbohydrate to give a final concentration of 1% (w/v). Pour plates and after drying streak inoculate with the HIB culture. Up to 4 cultures can be radially streaked on each plate. Plates should be incubated for up to 5 days at 30°C but read plates daily if possible as pH inversions can occur.

Phosphatase production

Prepare a 0·005 mol l^{-1} solution of phenolphthalein monophosphate (sodium salt) in 0·01 mol l^{-1} citric/sodium citrate buffer (pH 5·8). Thoroughly rinse all glassware used in this test in order to remove any contaminating phosphates. To 10 × 75 mm tubes add 0·5 ml of the above solution followed by a loopful of a well grown HIB culture, i.e.

$10^8 - 10^9$ CFU ml^{-1}; if a strain does not grow to such a density in HIB make a suspension from cells grown on a Heart Infusion Agar slope. Incubate at 37°C in a water bath for 4 h and then add 0·5 ml of 0·5 N sodium hydroxide followed by 0·5 ml of 0·5 mol ml^{-1} sodium bicarbonate. Test for presence of phosphatase by adding 0·5 ml of 4-aminoantipyrine solution (0·6 g 100 ml^{-1}) and 0·5 ml of potassium ferricyanide solution (2·4 g 100 ml^{-1}). The development of a red colour indicates phosphatase production (Kloos & Schleifer 1975*a*).

Novobiocin susceptibility

Prepare a filter-sterilized solution of novobiocin (Upjohn) such that when added to molten Medium A a concentration of 1·6 μg ml^{-1} is obtained. Radially streak test strains and incubate at 37°C for 24 h. Observe for growth (resistant) or no growth (sensitive).

Salt susceptibility

Prepare Medium A containing 10% of sodium chloride. Radially streak isolates and incubate for 3 days at 30°C; observe for growth.

Nitrate reduction

Prepare nitrate broth (Difco). Inoculate and incubate for 5 days at 30°C. Test for nitrite presence by adding 0·5 ml of 0·8% sulphanilic acid in 5N acetic acid followed by 0·5 ml of 0·25% 5-amino-2-naphthylene sulphonic acid in a 15% aqueous solution of glacial acetic acid. Examine for the development of a pink or red colour after standing at room temperature for *ca.* 5 min. If no colour develops add a knife tip of zinc dust and observe again. If no colour develops the nitrate has been reduced beyond nitrite and the result should be recorded as showing nitrite reduction.

Oxidase production

On a piece of filter paper (7 cm diam.) in a Petri dish place 2 to 3 drops of a 1% (w/v) solution of tetramethyl-p-phenylene-diamine dihydrochloride (Kovács's reagent) in 1% (w/v) aqueous ascorbic acid; do not allow to dry. Remove a portion of the growth from an overnight culture (30°C) on Heart Infusion Agar with a glass rod and smear over the surface of the paper impregnated with Kovács's reagent. Observe for a dark purple colour appearing within 10 s.

Pigment production

Streak on Medium A and incubate for 5 days at 30°C; note pigmentation of colonies.

Haemolysis

Streak on to plates of Medium A containing 5% of bovine blood and incubate for 3 days at 30°C; observe for haemolysis of the blood.

References

ANON. 1965 Recommendations of International Subcommittee on the taxonomy of staphylococci and micrococci. *International Bulletin of Bacteriological Nomenclature and Taxonomy* **15,** 109–110.

BAIRD-PARKER, A. C. 1969 The use of Baird-Parker's medium for the isolation and enumeration of *Staphylococcus aureus*. In *Isolation Methods for Bacteriologists*. Society for Applied Bacteriology, Technical Series No. 3, eds Shapton, D. A. & Gould, G. W. London & New York: Academic Press.

BAIRD-PARKER, A. C. 1974 Micrococcaceae. In *Bergey's Manual of Determinative Bacteriology* 8th edn, eds Buchanan, R. E. & Gibbons, N. E. Baltimore: Williams & Wilkins.

BARRITT, M. M. 1936 The intensification of the Voges-Proskauer reaction by the addition of α-naphthol. *Journal of Pathology and Bacteriology* **42,** 441–446.

DEVRIESE, L. A. 1977 Isolation and identification of *Staphylococcus hyicus* *American Journal of Veterinary Research* **38,** 787–791.

DEVRIESE, L. A., HÁJEK, V., OEDING, P., MEYER, S. A. & SCHLEIFER, K. H. 1978 *Staphylococcus hyicus* (Sompolinsky 1953) comb. nov. and *Staphylococcus hyicus* subsp. *chromogenes* subsp. nov. *International Journal of Systematic Bacteriology* **28,** 482–490.

EVANS, J. B., BRADFORD, W. L. & NIVEN, C. F. JR. 1955 Comments concerning the taxonomy of the genera *Micrococcus* and *Staphylococcus*. *International Bulletin of Bacteriological Nomenclature and Taxonomy* **5,** 61–66.

GRAMOLI, J. L. & WILKINSON, B. J. 1978 Characterization of coagulase-negative, heat stable deoxyribonuclease-positive staphylococci. *Journal of General Microbiology* **105,** 275–285.

HÁJEK, V. 1976 *Staphylococcus intermedius*, a new species isolated from animals *International Journal of Systematic Bacteriology* **26,** 401–408.

HÁJEK, V. & MARŠÁLEK, E. 1971 The differentiation of pathogenic staphylococcus and a suggestion for their taxonomic classification. *Zentralblatt für Bakteriologie, Parasitenkunde, Infektionskrankheiten und Hygiene Abt. I Orig.* **A217,** 176–182.

KLOOS, W. E., TORNABENE, T. G. & SCHLEIFER, K. H. 1974 Isolation of micrococci from human skin including two new species: *Micrococcus lylae* and *Micrococcus kristinae*. *International Journal of Systematic Bacteriology* **24,** 79–101.

KLOOS, W. E. & SCHLEIFER, K. H. 1975*a* Simplified scheme for routine identification of human *Staphylococcus* species. *Journal of Clinical Microbiology* **1,** 82–88.

KLOOS, W. E. & SCHLEIFER, K. H. 1975*b* Isolation and characterization of staphylococcus from human skin. II. Descriptions of four new species: *Staphylococcus capitis*, *Staphylococcus hominis*, *Staphylococcus warneri* and *Staphylococcus simulans*. *International Journal of Systematic Bacteriology* **25,** 62–79.

KLOOS, W. E., SCHLEIFER, K. H. & SMITH, F. R. 1976 Characterization of *Staphylococcus sciuri* sp. nov. and its subspecies. *International Journal of Systematic Bacteriology* **26,** 22–37.

OEDING, P. & DIGRANES, A. 1977 Classification of coagulase-negative staphylococci in the diagnostic laboratory. *Acta pathologica et microbiologica scandinavica* **85,** 136–142.

RAYMAN, M. K., PARK, C. E., PHILPOTT, J. & TODD, E. C. D. 1975 Reassessment of the coagulase and thermostable nuclease tests as means of identifying *Staphylococcus aureus*. *Applied Microbiology* **29,** 451–454.

SCHLEIFER, K. H. & KLOOS, W. E. 1975*a* A simple test system for the separation of staphylococci from micrococci. *Journal of Clinical Microbiology* **1,** 337–338.

SCHLEIFER, K. H. & KLOOS, W. E. 1975*b* Isolation and characterization of staphylococci from human skin. 1. Amended descriptions of *Staphylococcus epidermidis* and *Staphylococcus saprophyticus* and descriptions of three new species: *Staphylococcus cohnii*, *Staphylococcus haemolyticus* and *Staphylococcus xylosus*. *International Journal of Systematic Bacteriology* **25,** 50–61.

SCHLEIFER, K. H., SCHUMACHER-PERDREAU, F., GÖTZ, F. & POPP, B. 1976 Chemical and biochemical studies for the differentiation of coagulase-positive staphylococci. *Archives of Microbiology* **110,** 263–270.

SPERBER, W. H. & TATINI, S. R. 1975 Interpretation of the tube coagulase test for identification of *Staphylococcus aureus*. *Applied Microbiology* **29,** 502–505.

Colonial Morphology and Fluorescent Labelled Antibody Staining in the Identification of Species of the Genus *Clostridium*

P. D. Walker and Irene Batty

The Wellcome Research Laboratories, Beckenham, Kent, UK

A recent modification of the McIntosh and Fildes jar (McIntosh & Fildes 1916) which utilizes a catalyst of palladianized alumina, active at room temperature (Heller 1954), provides the simplest and most convenient way of isolating and obtaining surface cultures of anaerobes. By the use of appropriate media in which suitable indicator systems can be incorporated (Hobbs *et al.* 1971) many species of clostridia can readily be identified by this method. More recently a modification of the BTL anaerobic jar manufactured by Baird and Tatlock (London) Ltd has been introduced by Don Whitley Scientific, using Schrader valve vented jars. Other recent improvements have concerned the hydrogen source, such as the GasPak anaerobic system introduced by Brewer & Allgeier (1966), and the Gaskit system introduced by Don Whitley Scientific (4 Wellington Crescent, Shipley, West Yorkshire BD18 3PH). The relative merits of these systems are discussed by Willis (1977). Nevertheless, there still exists opposition to use of the anaerobic jar, mainly because either, it is allegedly difficult to use and requires special apparatus, or the method only allows the growth of those anaerobes that are relatively easy to culture. The first view is taken by the French school (Prévot, pers. comm.), who prefer the use of Veillon tubes for isolation of anaerobic organisms in pure culture, and by a number of other workers, e.g. Shank (1963), who offer alternative techniques such as the use of plastic films to exclude oxygen. The second view is taken by some Canadian workers (Fredette 1956). Many of these techniques are open to the serious objection that many anaerobes have the property of forming a thin spreading layer between surfaces, e.g. glass/agar and plastic/agar, a property which makes the resolution of mixtures of anaerobes extremely difficult. It is frequently forgotten that many aerobes are difficult to cultivate on plates unless the ordinary medium is suitably supplemented, and that failure to grow is not necessarily a result of culture on a surface.

In the present paper the colonial appearances of a number of anaerobes which are examined routinely in our laboratories are described. Many of these organisms are usually regarded as difficult to grow, and it is hoped that some of the points which will emerge from these descriptions will be useful to those working in other fields.

The use of fluorescent labelled antisera for the identification of clostridia has previously only been investigated with *Clostridium botulinum* (Bulatova & Kabanova 1960; Kalitina 1960) and *Cl. welchii* (*perfringens*) (Geck & Szanto 1961). The method, where applicable, appears to have many advantages in the identification of clostridia, and some recent work will be reviewed.

Materials and Methods

Organisms

The following organisms were used: *Cl. welchii* (*perfringens*) (Wellcome Research Laboratories Culture Collection (CN) 5385, 5386); *Cl. tetani* (CN655); *Cl. sporogenes* (CN642); *Cl. bifermentans* (CN1617); *Cl. sordellii* (CN1620); *Cl. botulinum* type A (NCTC2916), type B (CN5009), type C (CN4946), type D (CN4947) and type E (E20 Dolman); *Cl. novyi* (*oedematiens*) type A (CN1496), type B (CN755) and type D (CN3629); *Cl. chauvoei* (CN5002); *Cl. septicum* (CN5293); *Cl. histolyticum* (CN647) and *Cl. tertium* (CN4603).

Media

Blood agar will support the growth of most species of pathogenic clostridia and is the routine medium of choice. All types of *Cl. welchii*, *Cl. sordellii*, *Cl. bifermentans*, *Cl. septicum*, *Cl. histolyticum*, *Cl. tertium* and *Cl. oedematiens* type A will grow readily on this medium. Only *Cl. chauvoei*, *Cl. botulinum* (types C and D) and *Cl. oedematiens* (types B and D) require special media.

Media (see Table 1 for the media used in our laboratories) are invariably based on meat extract with added peptone and should be used as soon as possible after preparation. Peptone may be prepared quite simply as a peptone digest of meat by the following method: one litre of tap water is added to 1 kg of fresh minced muscle and stirred. Any fat which rises to the surface is skimmed off and 3·5 g of papain powder is added to the mixture. The temperature is raised to 60°C for 2 h after

TABLE 1. Media for the isolation and identification of clostridia

Normal blood agar	nutrient broth containing 1·8% of agar (New Zealand agar = 1·6 times Oxoid no. 3 is used throughout), 75 parts; pancreatic autodigest, 25 parts; horse blood, 5 parts, by volume
Stiff blood agar	as above but containing 3% of agar
Cl. chauvoei medium	nutrient broth containing 1·8% of agar, 75 parts; 50% glucose, 2 parts; liver extract, 3 parts; sheep's blood, 5 parts, by volume
Cl. oedematiens type B medium (Moore 1968)	peptone, 10 g; yeast extract, 5 g; proteolysed liver, 5 g; glucose, 10 g; agar, 20 g; cysteine hydrochloride, 100 mg; glutamine, 50 mg; dithiothreitol, 100 mg; horse blood, 100 ml; salts solution, 5 ml; glass-distilled water to 1000 ml; pH, 7·6–7·8.

which the enzyme activity is destroyed by boiling for 20 min. The mixture is then filtered through paper until clear. Total nitrogen is measured by the Kjeldahl procedure and the appropriate amount is added to the meat extract; it is our experience that a 3% solution of 'commercial peptone' has a total nitrogen content of *ca.* 5 g l^{-1}. After addition of the peptone the pH is adjusted to 7·8 and the medium boiled to bring down phosphates. After removal of phosphates by filtration, agar can be added to give the complete medium.

In the case of *Cl. chauvoei* the nutrient agar is supplemented with liver extract and glucose (Batty & Walker 1964). The liver extract is prepared as follows: a sheep's liver is finely minced, covered with tap water and boiled for 30 min. After straining off the coarse particles through muslin, the supernatant liquid is Seitz filtered and stored at 4°C. It is advisable to make the extract fresh each week.

Plates of the above media were poured and inoculated as soon as dry with a loopful of an overnight culture in Robertson's meat broth. The inoculated plates were transferred immediately to an anaerobic jar and incubated for appropriate periods before examination. The colonies were photographed under a Vickers projection microscope (Vickers Ltd), the angle of incidence of the light being varied to reveal their shape and surface texture. The illumination was either by tungsten light or carbon arc, depending on the configuration of the colony.

Fluorescent staining. The preparation of antisera, conjugation procedures and staining techniques were as described in Batty & Walker (1963*a*, *b*).

Results

Surface culture

In the several plates, the magnification of the organisms and colonies has been standardized at × 2000 and × 6, respectively. The appearance of colonies at this magnification corresponds approximately to those seen with a hand lens. The photographs are designed to bring out surface details and no attempt has been made to show haemolysis, which in many cases is an important ancillary factor in identifying the organism.

Clostridium welchii (*perfringens*)

The morphological appearance is shown (Fig. 1*a*); spores are rarely seen and the typical appearance is of short Gram positive rods with square ends. The characteristic smooth, circular, convex colonies with an entire margin are shown (Fig. 1*b*). Occasionally rough colonies are found (Fig. 1*c*). There appears to be little correlation between colonial morphology and the various toxigenic types of this organism.

Clostridium tetani

The typical drumstick appearance is seen on smears (Fig. 2*a*). On ordinary agar plates, isolated surface colonies are difficult to obtain, because *Cl. tetani* commonly spreads as a fine rhizoidal film over the surface. Isolated colonies can be obtained, however, on stiff agar plates (Fig. 2*b*). The rhizoidal edges can be seen clearly, but the colony remains quite discrete.

Clostridium sporogenes

This organism is a common contaminant. It produces oval spores which swell the sporangium (Fig. 2*c*). The colonies are umbonate with an opaque greyish-white centre, and a flattened irregular periphery (Fig. 2*d*).

Clostridium bifermentans/Cl. sordellii

These two species are regarded as distinct by some workers and as varieties of the same species by others (Tataki & Huet 1953; Smith 1955; Ellner & Green 1963). A comprehensive treatment is to be found in Brooks & Epps (1959), who draw attention to the characters which can be used to separate *Cl. bifermentans* from *Cl. sordellii*. Morphologically the two organisms are indistinguishable; both are large rods with cylindrical spores causing little swelling of the sporangium (Figs 3*a*, *c*). There are, however, slight differences in appearance on surface culture, colonies of *Cl. bifermentans* being in general low, convex and with an entire margin (Fig. 3*b*), whilst those of *Cl. sordellii* usually have a markedly irregular margin and tend to follow the line of the streak (Fig. 3*d*).

Clostridium botulinum, types A, B, C, D and E

In the current edition of *Bergey's Manual* (Buchanan & Gibbons 1974), the name *Cl. botulinum* is reserved for the non-proteolytic types C, D and E, while the name *Cl. parabotulinum* is used to distinguish the proteolytic types A and B; the recently isolated type F is also included in this latter group. These differences in metabolic activity are reflected in the morphology and colonial appearance of the organisms. Both *Cl. botulinum* type A and type B are morphologically similar and produce oval spores swelling the sporangium (Figs 4*a*, *c*). Both have umbonate colonies with a granular periphery and irregular margin (Figs 4*b*, *d*).

Clostridium botulinum types C, D and E resemble each other both morphologically and culturally, though in this latter aspect types C and D are more fastidious but they grow on the medium prescribed for *Cl. novyii*. As Fig 5*a*, *c* and *e* show, all three are large bacilli giving rise to spores only infrequently. The colonies of the usual toxigenic wild types are large with a granular surface and a rhizoidal margin (Fig. 5*b*, *d* and *f*). Colonies of the various mutants described by Dolman (1957) are not shown.

Clostridium novyii (*oedematiens*)

This species is divisible into four types, A, B, C and D. *Clostridium novyii* type D is probably one of the most fastidious of the pathogenic clostridia. The organisms are large Gram positive rods with oval spores (Fig. 6*a*). Colonies of type A are flat with finely granular surface and an irregular edge (Fig. 6*b*), whilst those of type B, although similar, are smaller (Fig. 6*c*) and those of type D are more umbonate, with entire irregular margins (Fig. 6*d*).

Clostridium chauvoei/Cl. septicum

Clostridum chauvoei and *Cl. septicum* are two species which though very similar morphologically can be distinguished culturally. Both organisms are pleomorphic and this quality is well illustrated in the case of *Cl. chauvoei* (Fig. 7*a*) and of *Cl. septicum* (Fig. 7*c*). *Clostridium chauvoei* is a more fastidious organism than *Cl. septicum* and to ensure growth the special medium already described should be used. The majority of colonies of *Cl. chauvoei* are umbonate with a raised lip, but occasionally smooth colonies are seen (Fig. 7*b*) which resemble the smooth colonies of

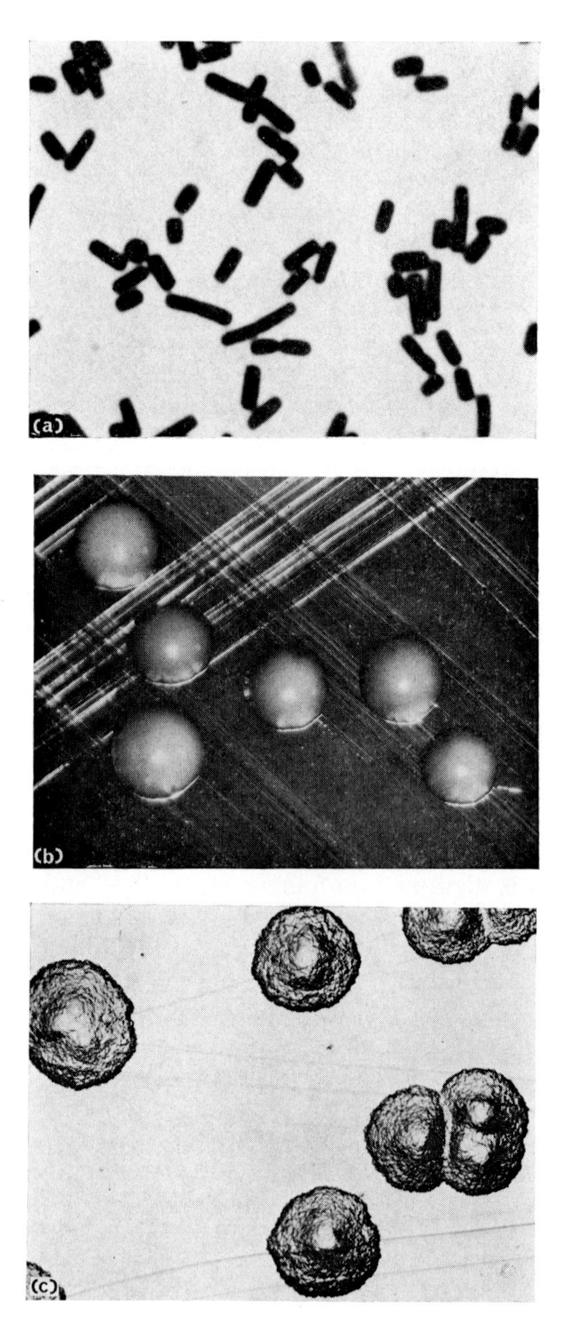

FIG. 1. Morphological appearances of *Cl. welchii* (a) from a surface colony 24 h old (× 2000); (b) smooth surface colonies 24 h old (× 6); (c) rough surface colonies 24 h old (× 6).

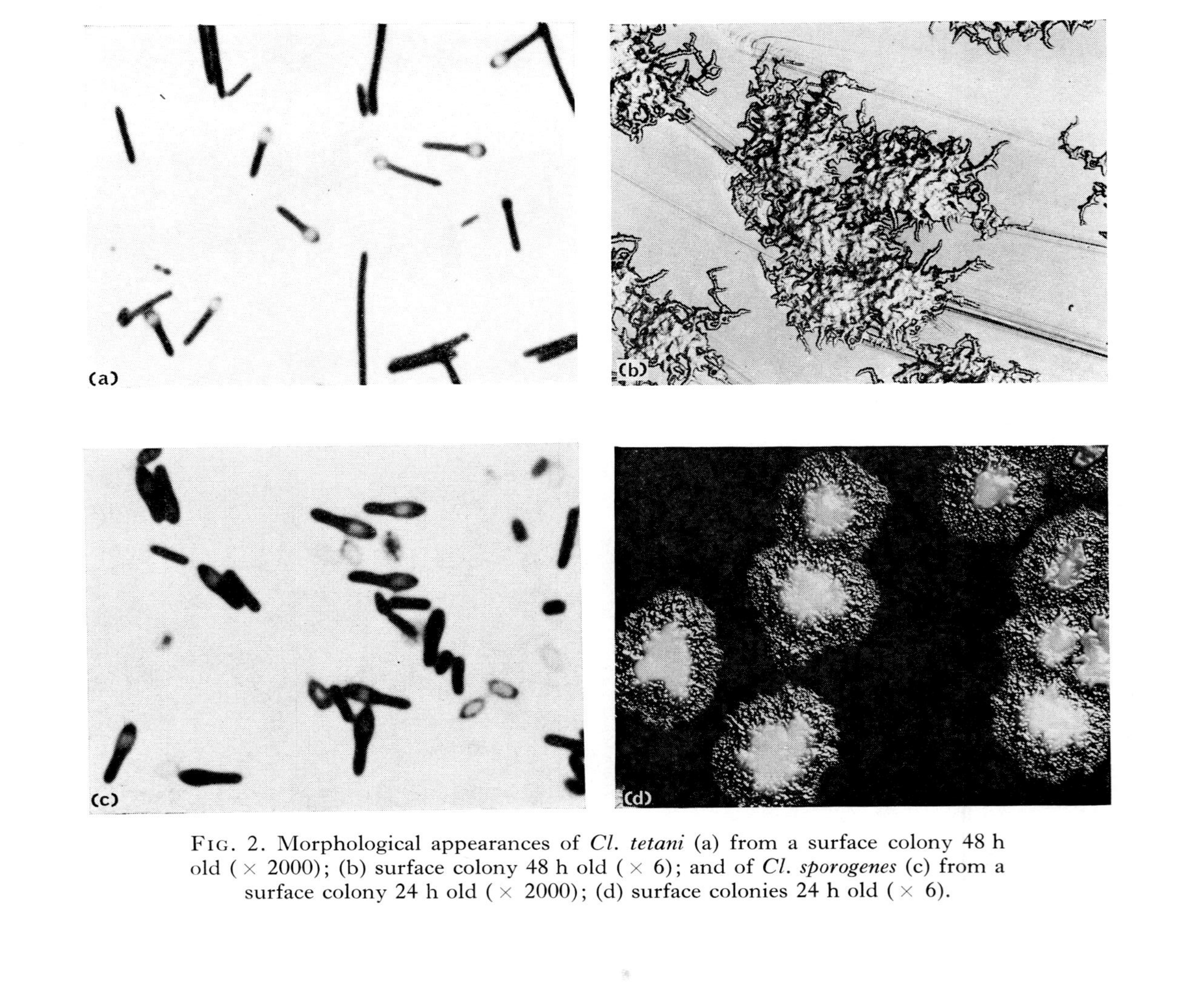

FIG. 2. Morphological appearances of *Cl. tetani* (a) from a surface colony 48 h old ($\times$ 2000); (b) surface colony 48 h old ($\times$ 6); and of *Cl. sporogenes* (c) from a surface colony 24 h old ($\times$ 2000); (d) surface colonies 24 h old ($\times$ 6).

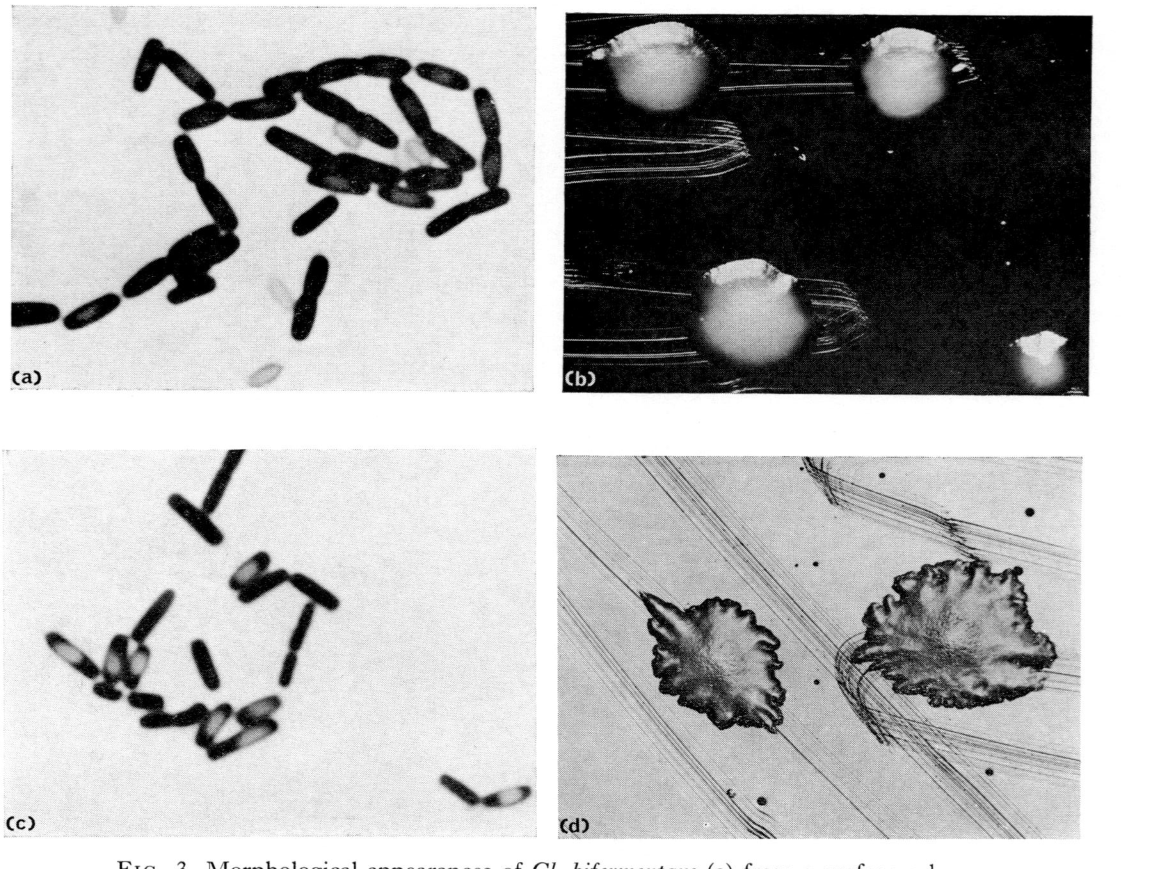

FIG. 3. Morphological appearances of *Cl. bifermentans* (a) from a surface colony 24 h old (× 2000); (b) surface colonies 24 h old (× 6); and of *Cl. sordellii* (c) from a surface colony 24 h old (× 2000); (d) surface colonies 24 h old (× 6).

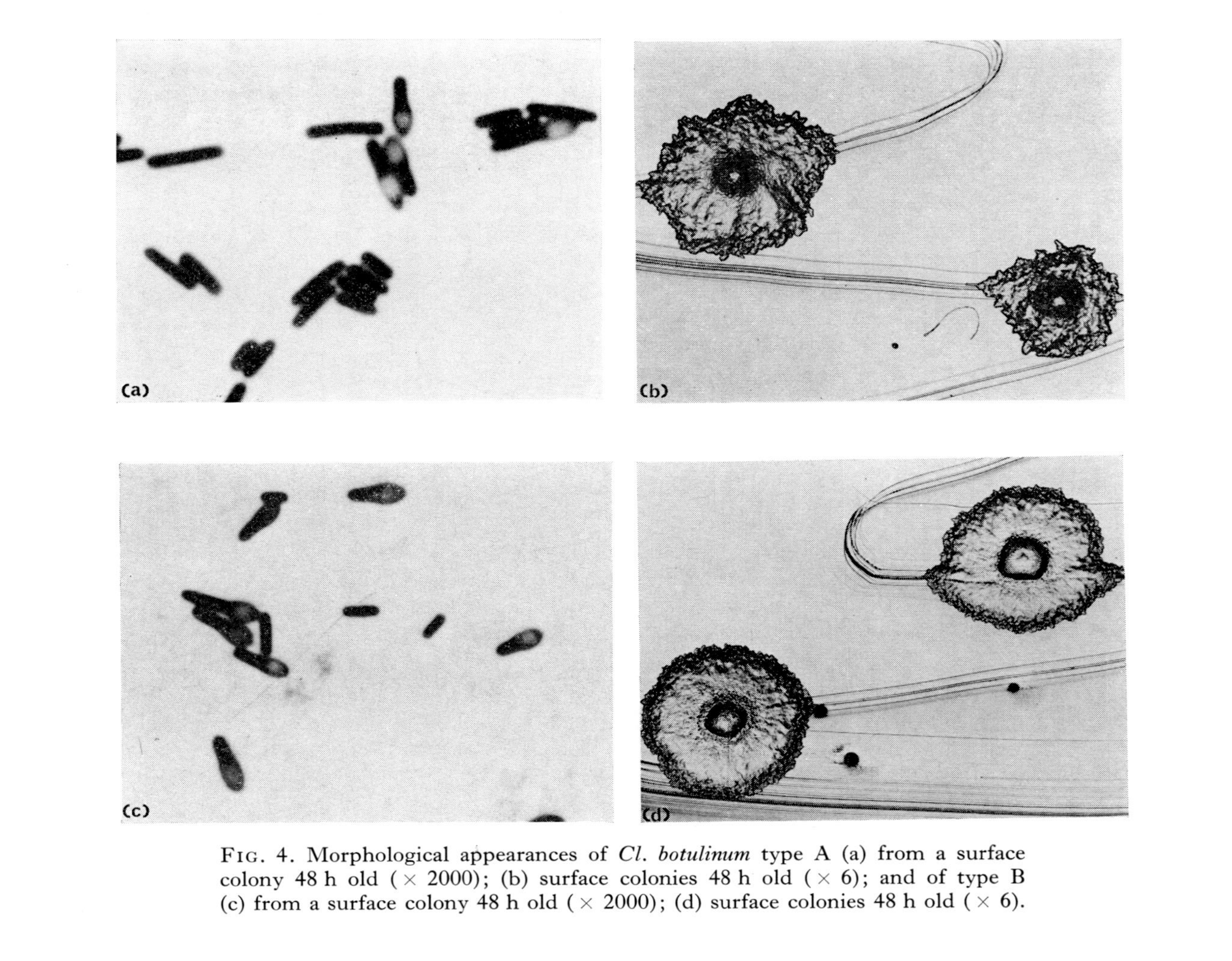

FIG. 4. Morphological appearances of *Cl. botulinum* type A (a) from a surface colony 48 h old ($\times$ 2000); (b) surface colonies 48 h old ($\times$ 6); and of type B (c) from a surface colony 48 h old ($\times$ 2000); (d) surface colonies 48 h old ($\times$ 6).

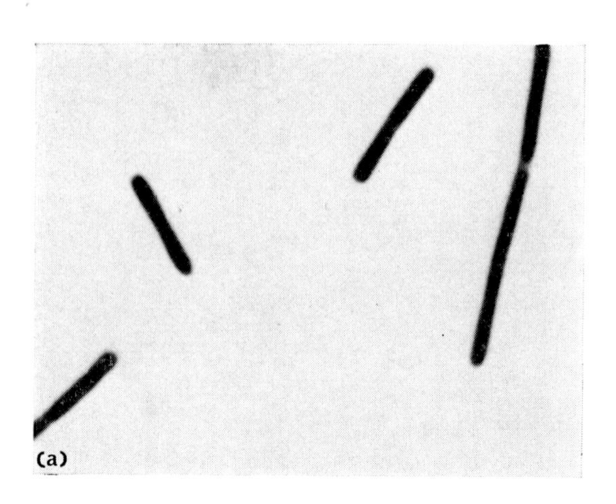

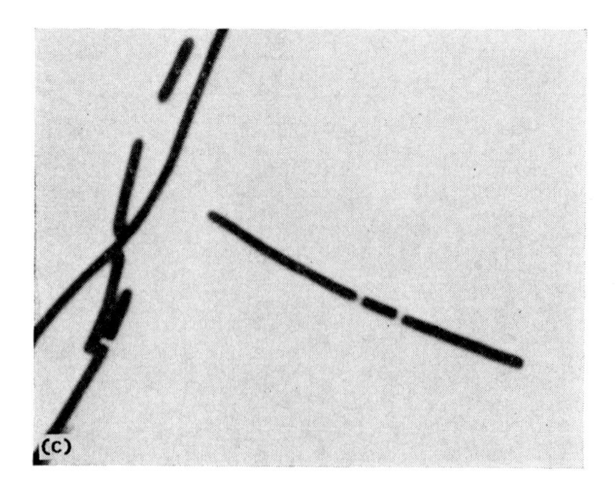

FIG. 5.

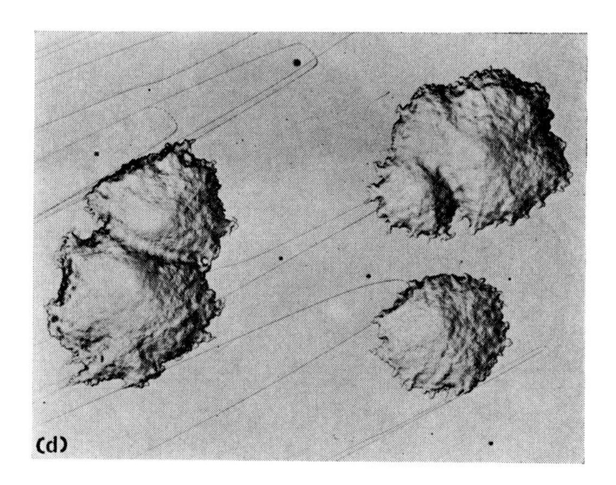

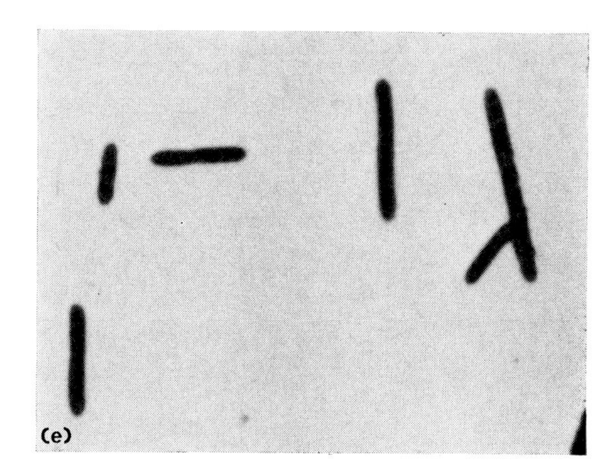

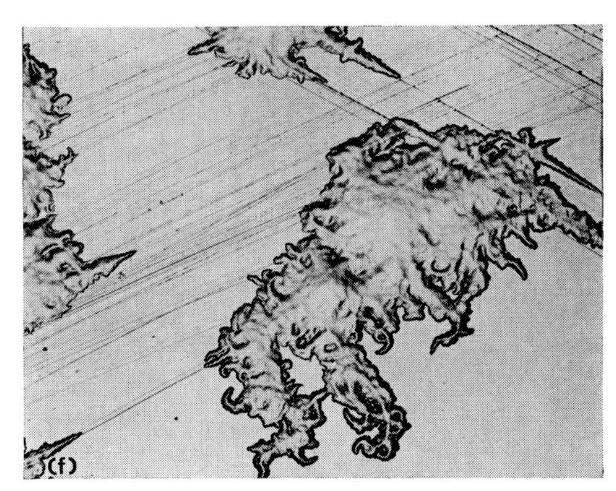

FIG. 5. Morphological appearances of *Cl. botulinum* type C (a) from a surface colony 48 h old (× 2000); (b) surface colonies 48 h old (× 6); of type D (c) from a surface colony 48 h old (× 2000); (d) surface colonies 48 h old (× 6); and of type E (e) from a surface colony 48 h old (× 2000); (f) surface colonies 48 h old (× 6).

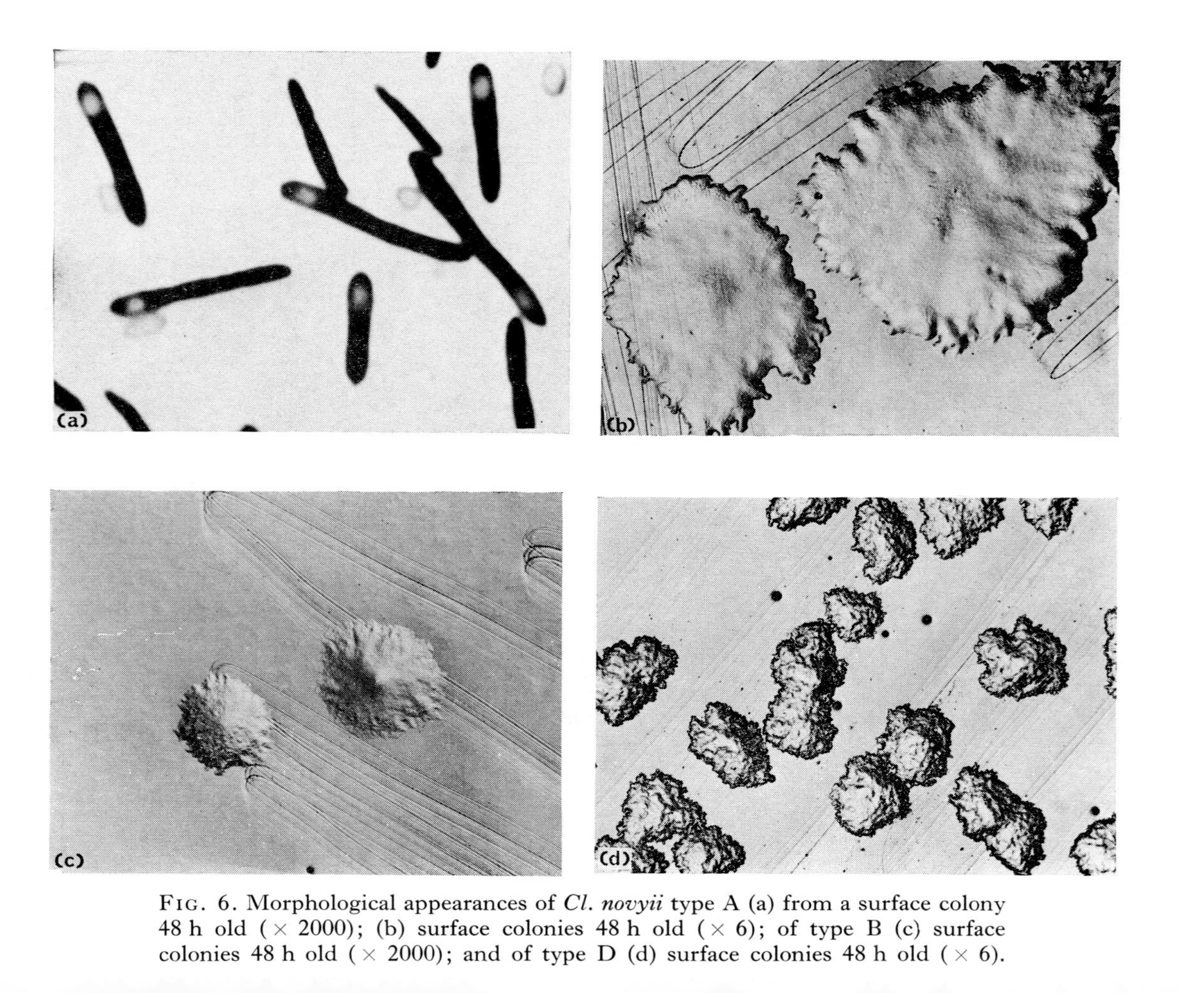

FIG. 6. Morphological appearances of *Cl. novyii* type A (a) from a surface colony 48 h old (× 2000); (b) surface colonies 48 h old (× 6); of type B (c) surface colonies 48 h old (× 2000); and of type D (d) surface colonies 48 h old (× 6).

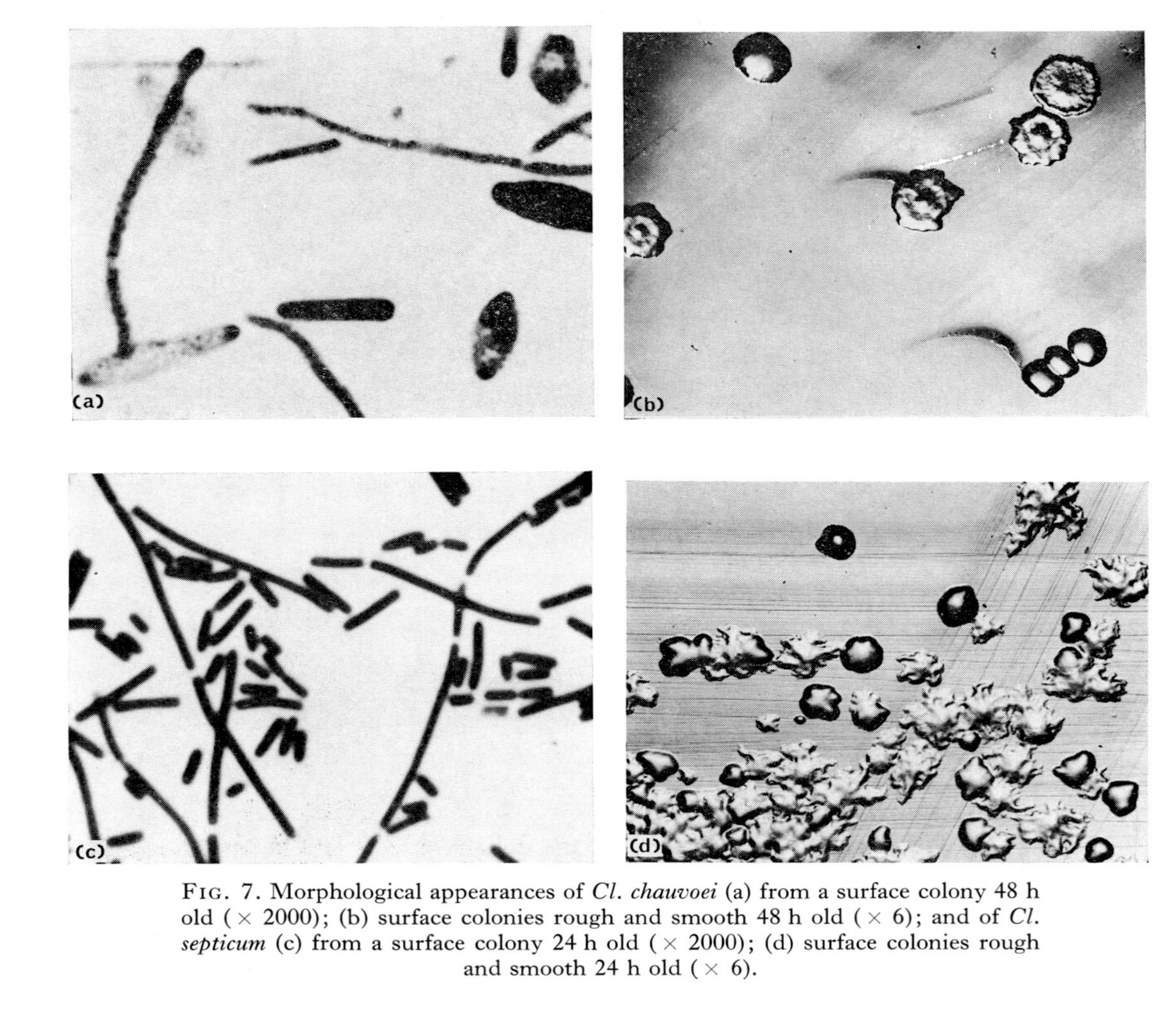

FIG. 7. Morphological appearances of *Cl. chauvoei* (a) from a surface colony 48 h old (× 2000); (b) surface colonies rough and smooth 48 h old (× 6); and of *Cl. septicum* (c) from a surface colony 24 h old (× 2000); (d) surface colonies rough and smooth 24 h old (× 6).

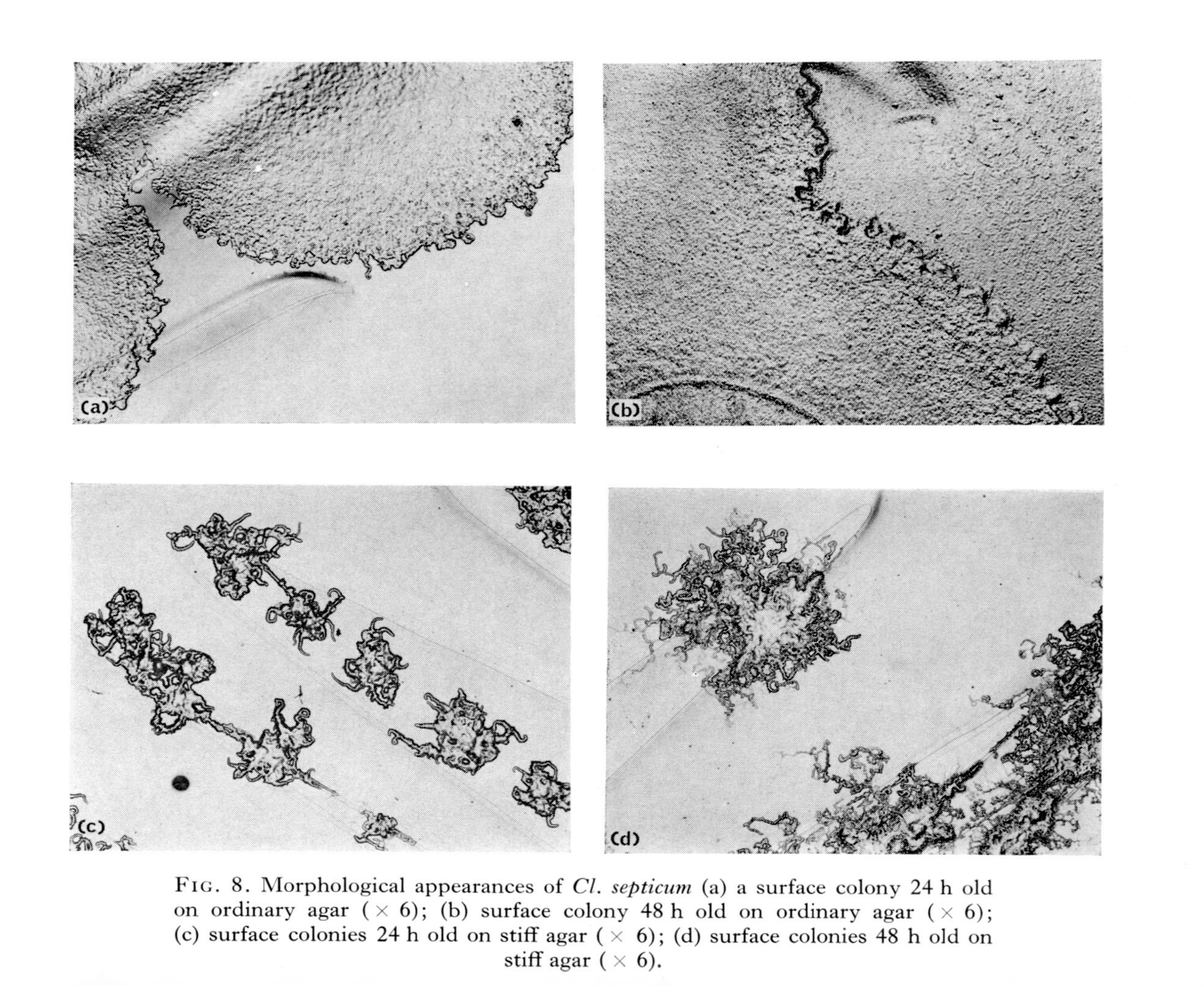

FIG. 8. Morphological appearances of *Cl. septicum* (a) a surface colony 24 h old on ordinary agar ($\times$ 6); (b) surface colony 48 h old on ordinary agar ($\times$ 6); (c) surface colonies 24 h old on stiff agar ($\times$ 6); (d) surface colonies 48 h old on stiff agar ($\times$ 6).

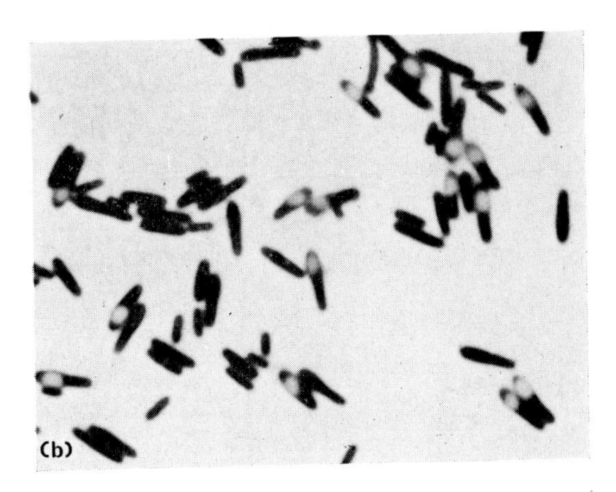

FIG. 9.

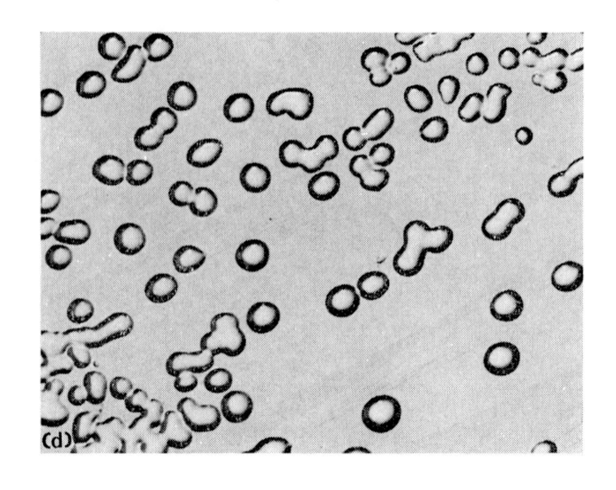

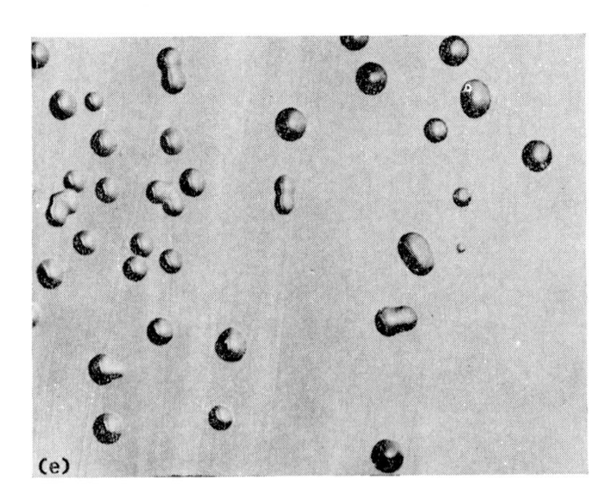

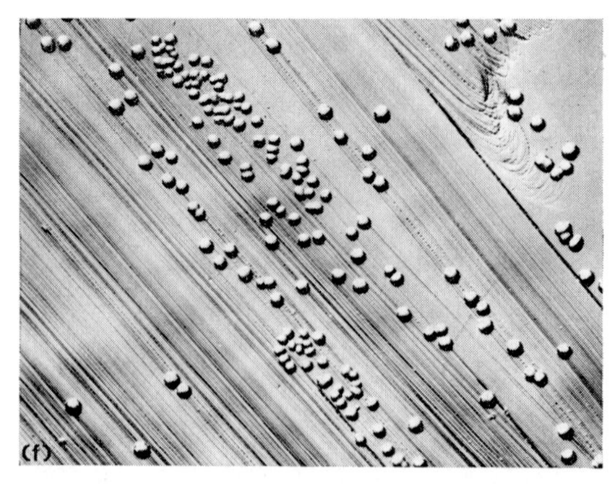

FIG. 9. Morphological appearances of *Cl. tertium* (a) from a surface colony 24 h old (× 2000); (c) surface colonies anaerobic 24 h old (× 6); (e) surface colonies aerobic 24 h old (× 6); and of *Cl. histolyticum* (b) from a surface colony 24 h old (× 2000); (d) surface colonies anaerobic 24 h old (× 6); (f) surface colonies aerobic 24 h old (× 6).

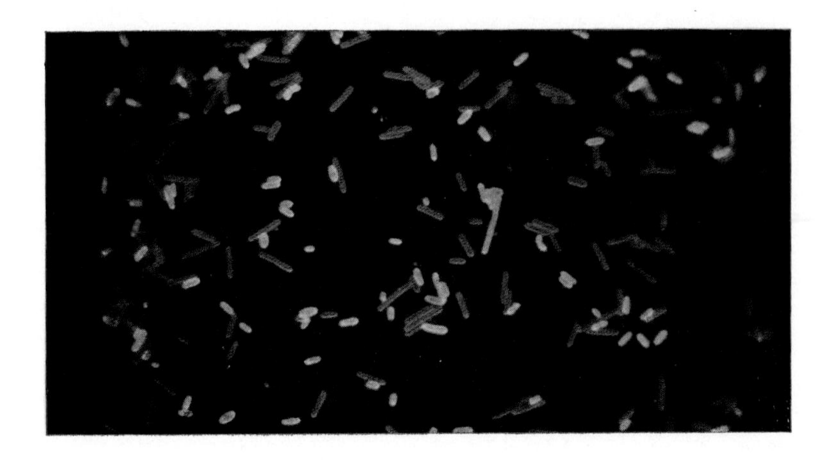

FIG. 10. A smear of a mixed *Cl. septicum* and *Cl. chauvoei* culture stained with a mixture of fluorescein isothiocyanate labelled *Cl. septicum* antiserum and lissamine rhodamine B.200 *Cl. chauvoei* labelled antiserum.

Cl. septicum (Fig. 7*d*), where some smooth colonies can be seen amongst the more usual spreading rhizoidal colonies.

If *Cl. septicum* is grown on 1·8% agar, like *Cl. tetani*, it spreads over the surface. After 24 h the spreading edge can be clearly seen (Fig. 8*a*), but after 48 h the growth has completely covered the surface of the agar plate (Fig. 8*b*). If, however, the same strain is grown on 3% agar discrete rhizoidal colonies can be seen after 24 h, which increase in size during the next 24 h, but remain discrete (Fig 8*c* and *d*).

Clostridium tertium/Cl. histolyticum

These two organisms are micro-aerophilic. Morphologically *Cl. tertium* has characteristic large terminal oval spores which swell the sporangium (Fig. 9*a*). When grown anaerobically it produces smooth round colonies with a crenated edge (Fig. 9*c*). The colonies which develop under aerobic conditions are much smaller (Fig. 9*e*).

Clostridium histolyticum is widely distributed and is strongly proteolytic. It also produces oval spores which swell the sporangium (Fig. 9*b*). The colonies are opaque and greyish-white with entire margins (Fig. 9*d*). Colonies grown aerobically are again much smaller (Fig. 9*f*) than those grown anaerobically.

Fluorescent staining

In a series of papers (Batty & Walker 1963*a*, *b*, 1964; Batty *et al.* 1964; Walker & Batty 1964, 1966), the fluorescent-labelled antibody staining technique and its application as a diagnostic aid are discussed. The main conclusions are summarized as follows:

Clostridium septicum/chauvoei

Twenty-two strains of *Cl. septicum* collected from all over the world could be divided into two groups on the basis of 'O' agglutination and fluorescent staining, whilst 26 strains of *Cl. chauvoei*, again collected from all over the world, fell into a single group both by the 'O' agglutination or by fluorescent staining. Neither species showed any cross reaction in 'O' agglutination or by fluorescent staining. Strains of both species could be identified in a single smear by labelling the *Cl. septicum* antiserum with fluorescein isothiocyanate (green fluorescence) and the *Cl. chauvoei* antiserum with lissamine rhodamine B.200 (orange fluorescence) (Fig. 10).

Clostridium novyii

The 25 strains of *Cl. novyii* representing all types fluoresced with an antiserum prepared against *Cl. novyii* type B. Attempts to produce type specific antisera by absorption tests have so far been unsuccessful.

These three fluorescent antisera have been used over a considerable period in epizootiological studies and have yielded most valuable information on the distribution of these organisms in nature.

Between 1963 and 1972 these antisera have been used to examine smears sent in from all parts of the UK to identify organisms present in the specimens. The results have been correlated with those of clinical symptoms and other investigations to give an indication of the value of fluorescent labelled antibodies in the diagnosis of clostridial infections. It can be seen (Table 2) that over this period the percentage of *Cl. septicum* strains identified in specimens remained fairly consistent between 10 and 15%. Those of *Cl. chauvoei* remained consistent in the first half of the period but increased towards the latter part of the study and the percentage of *Cl. oedematiens* identified, although varying, was consistently higher than the other two organisms. The relatively high percentage of negative smears present throughout the period, particularly during the first six years, supports the specificity of the method and absence of non-specific staining.

TABLE 2. Number of specimens and percentages of *Clostridium septicum*, *Cl. chauvoei* and *Cl. oedematiens* identified in smears during the years 1963–1972

Year	No. of specimens	Percentage of *Cl. septicum*	Percentage of *Cl. chauvoei*	Percentage of *Cl. oedematiens*	Negative
1963–1964	140	12	6	39	43
1964–1965	684	10	5	43	42
1965–1966	429	10	9	39	42
1966–1967	186	14	16	53	12
1967–1968	37	13	13	46	28
1968–1969	98	10	10	60	20
1969–1970	83	11	13	70	6
1970–1971	65	14	15	64	7
1971–1972	85	15	24	48	13

Of the *Cl. septicum* positive smears, the majority came from sheep, and slightly fewer from cattle. The number of positive slides for porcines was relatively low (Table 3). Of the *Cl. chauvoei* positive smears, the majority came from cattle with fewer from ovines and a relatively small number of smears from porcines (Table 4). Of the *Cl. oedematiens* posi-

TABLE 3. Percentage distribution of *Clostridium septicum* positive smears among animal species

Year	Ovine	Bovine	Porcine
1964–1965	59	34	7
1965–1966	47	47	6
1966–1967	33	50	17
1967–1968	60	40	—
1968–1969	80	20	—

—: not found.

TABLE 4. Percentage distribution of *Clostridium chauvoei* positive smears among animal species

Year	Ovine	Bovine	Porcine	Others
1964–1965	21	63	—	16
1965–1966	6	92	2	—
1966–1967	7	65	28	—
1967–1968	100	—	—	—
1968–1969	10	90	—	—

—: not found.

TABLE 5. Percentage distribution of *Clostridium oedematiens* positive smears among animal species

Year	Ovine	Bovine	Porcine	Others
1964–1965	36	49	12	3
1965–1966	23	60	17	—
1966–1967	27	48	15	10
1967–1968	25	50	25	—
1968–1969	44	48	8	—

—: not found.

tive smears (Table 5) the majority came from bovines with fewer from ovines, but the percentage from porcines was much higher than that observed with *Cl. septicum* and *Cl. chauvoei*.

Clostridium tetani

The 26 strains of *Cl. tetani* fell into a single group on the basis of 'O' agglutination and fluorescent staining. It was suggested that this anti-serum might prove useful in the examination of material from cases of tetanus and from contaminated dressings or other materials implicated in a case.

Clostridium botulinum

Fluorescent labelled antisera can also be used as a rapid method for differentiating between various types of *Cl. botulinum*. Three groups were distinguished, one group comprising types A, B and F, another group types C and D, and a single group consisting of type E only. These results have been confirmed by other workers (Boothroyd & Georgala 1964; Lee & Hawirko 1972; Lynt *et al.* 1972).

Discussion

It can be seen, provided that appropriate media are used, that many species of *Clostridium*, including extremely fastidious strains, can be grown in surface culture. The use of stiff agar results in discrete colonies in species which tend to form a spreading film on normal agar. All other methods for preventing spread are inhibitory to a greater or lesser degree. Both the use of suitable media and the inhibition of spreading are of great importance in the purification of cultures. Colonial and microscopic appearance are useful guides in identification.

Fluorescent staining techniques also make possible a precise identification of certain species of *Clostridium*. Stained slides made from the suspected lesion or other appropriate sites in animals found dying or recently dead are likely to give a picture of the relative numbers of the various clostridia present more nearly approximating to the situation at the time of death than can be obtained by culturing material taken *post mortem* and conveyed to the laboratory, often under conditions which are selective for certain organisms. Quite apart from the saving in time, labour and materials resulting from the use of the fluorescent staining technique, the method has, where applicable, very real advantages over all other methods of identification. This is clearly seen in the survey of specimens during the years 1963–1972 when accurate identification of three species of clostridia in bovines, ovines and porcines was achieved. In the case of *Cl. chauvoei* the smears were mainly from cases of classical blackleg in cattle where the organism has been incriminated using cultural methods for some years. In the case of smears of *Cl. septicum* these were mainly from cases of gangrene or from 'rotten lamb disease' where the presence of the organism is consistent with the symptoms produced. In the case of *Cl. oedematiens*, particularly in view of the difficulty of cultivating this organism, the results of fluorescent staining represent a significant advance. Not only has *Cl. oedematiens* been identified in smears of liver associated with fluke areas, i.e. traditional black disease, but also in bovines and porcines where intensive feeding appears to lead to similar liver damage to that caused by flukes allowing dormant spores of the organism to germinate and multiply. These factors in *Cl. oedematiens* disease were previously unappreciated (Batty *et al.* 1964).

References

Batty, I. & Walker, P. D. 1963*a* The differentiation of *Clostridium septicum* and *Clostridium chauvoei* by the use of fluorescent labelled antibodies. *Journal of Pathology and Bacteriology* **85,** 517–527.

Batty, I. & Walker, P. D. 1963*b* Fluorescent labelled clostridial antisera as specific stains. *Bulletin. Office international des épizooties* **59,** 1499–1513.

Batty, I. & Walker, P. D. 1964 The identification of *Clostridium novyi* (*Clostridium oedematiens*) and *Clostridium tetani* by the use of fluorescent labelled antibodies. *Journal of Pathology and Bacteriology* **88,** 327–328.

Batty, I., Buntain, D. & Walker, P. D. 1964 *Clostridium oedematiens*: A cause of sudden death in sheep, cattle and pigs. *Veterinary Record* **76,** 1115–1117.

Boothroyd, M. & Georgala, D. L. 1964 Immunofluorescent identification of *Clostridium botulinum. Nature, London* **202,** 515–516.

Brewer, J. H. & Allgeier, D. L. 1966 Safe self-contained carbon dioxide-hydrogen anaerobic system. *Applied Microbiology* **14,** 985–988.

Brooks, M. E. & Epps, H. B. G. 1959 Taxonomic studies of the genus *Clostridium*: *Clostridium bifermentans* and *C. sordellii. Journal of General Microbiology* **21,** 144–155.

Buchanan, R. E. & Gibbons, N. E. (eds) 1974 *Bergey's Manual of Determinative Bacteriology*, 8th edn. Baltimore: Williams & Wilkins.

Bulatova, T. I. & Kabanova, Y. A. 1960 On the problem of pathogens of identification of the botulism with the aid of luminescent sera. *Zhurnal mikrobiologii* **31,** 18–22.

Dolman, C. E. 1957 Recent observations of type E botulism. *Canadian Journal of Public Health* **48,** 187–198.

Ellner, P. D. & Green, S. S. 1963 Serological relationship between *Clostridium bifermentans* and *Clostridium sordellii* based upon soluble antigens. *Journal of Bacteriology* **86,** 605.

Fredette, V. 1956 Methods for the isolation and identification of the anaerobic bacteria of medical importance. *Vermont Journal of Medical Technology* **1,** 8.

Geck, P. & Szanto, R. 1961 Examination of *Clostridium perfringens* with the fluorescent tracer technique. *Acta microbiologica Academiae scientiaram hungaricae* **8,** 423–425.

Heller, C. L. 1954 A simple method for producing anaerobiosis. *Journal of Applied Bacteriology* **17,** 202.

Hobbs, G., Williams, K. & Willis, A. T. 1971 Basic methods for the isolation of clostridia. In *Isolation of Anaerobes*. Society for Applied Bacteriology Technical Series No. 5, eds Shapton, D. A. & Board, R. G. London & New York: Academic Press.

Kalitina, T. A. 1960 The detection of *Clostridium botulinum* by means of luminescent antibodies. Communication I: the production of specific luminescence in *Clostridium botulinum* by treatment with luminescent immune serum. *Bulletin of Experimental Biology and Medicine, U.S.S.R.* **49,** 278–280.

LEE, W. K. & HAWIRKO, R. Z. 1972 Spore antigen common to type 6 strains of *Clostridium botulinum*: purification and partial characterization. In *Spores V*, eds Halvorson, H. O., Hanson, R. & Campbell, I. American Society for Microbiology.

LYNT, R. K. JR., SOLOMON, H. M. & KAUTTER, D. A. 1972 Specificity of somatic antisera for detection of *Clostridiumi botulinum* by immunofluorescence. In *Spores V*, eds Halvorson, H. O., Hanson, R. & Campbell, I. American Society for Microbiology.

MCINTOSH, J. & FILDES, P. 1916 A new apparatus for the isolation and cultivation of anaerobic micro-organisms. *Lancet*, i, 768–770.

MOORE, W. B. 1968 Solidified media suitable for the cultivation of *Clostridium novyi* type B. *Journal of General Microbiology* **53,** 415–423.

SHANK, J. L. 1963 Applications of the plastic film technique in the isolation and study of anaerobic bacteria. *Journal of Bacteriology* **86,** 95–100.

SMITH, L. D. S. 1955 *Introduction to Pathogenic Anaerobes*. Chicago: University of Chicago Press.

TATAKI, H. & HUET, M. 1953 Valeur du test de l'urease pour la différentiation de *Cl. sordellii* et *Cl. bifermentans. Annales de l'Institut Pasteur* **84,** 890–894.

WALKER, P. D. & BATTY, I. 1964 Fluorescent studies in the genus *Clostridium*. II. A rapid method for differentiating *Clostridium botulinum* types A, B and F, types C and D, and type E. *Journal of Applied Bacteriology* **27,** 137–139.

WALKER, P. D. & BATTY, I. 1966 The serology of *Cl. botulinum* with reference to fluorescent staining. In *Botulism* 1966, eds Ingram, M. & Roberts, T. A. London: Chapman & Hall Ltd.

WILLIS, A. T. 1977 *Anaerobic Bacteriology, Clinical & Laboratory Practice*. London: Butterworths.

Identification of the Lactic Acid Bacteria

M. Elisabeth Sharpe

National Institute for Research in Dairying, Shinfield, Reading RG2 9AT, UK

Lactic acid bacteria, which include both cocci and rods, are now classified in the families Streptococcaceae and Lactobacillaceae, respectively (Buchanan & Gibbons 1974). They are commonly found in foods, including dairy products, in the mouth and intestinal tracts of man and other animals and in plant material; they are particularly common in fermenting products. Some pathogenic species are found among the streptococci, including *Streptococcus pneumoniae* which will not be considered further here. Anaerobic streptococci, discussed by Barnes *et al.* (1977) and Jones (1978) are also excluded here as their taxonomic position is presently uncertain.

These organisms are Gram positive, non-sporing cocci, coccobacilli or rods, dividing in one plane only, with the exception of the pediococci which divide in two planes; catalase is absent (some strains have a 'pseudo catalase' detectable when cultures are grown on low sugar-containing media; see section '*Pediococcus*'); a fermentable carbohydrate is required for growth and glucose is converted mainly to lactic acid or to lactic acid, CO_2, ethanol and/or acetic acid.

Lactic acid bacteria are subdivided into the following genera:

Streptococcus. Homofermentative cocci in pairs or chains.

Leuconostoc. Heterofermentative cocci which are sometimes oval, in pairs or chains.

Pediococcus and *Aerococcus*. Homofermentative cocci dividing in two planes, often forming tetrads or clusters (Gunther 1959).

Lactobacillus. Homofermentative or heterofermentative rods (a few of the latter can be coccobacilli).

Characteristics Used to Recognize Genera of Lactic Acid Bacteria

(1) Microscopic appearance using Gram strain.

(2) Catalase test.

(3) Fermentative use of carbohydrate. The Hugh & Leifson (1953) test as described by Skerman (1959) is used. Acid is produced from the carbohydrate in the medium under both aerobic and anaerobic conditions, whilst acid production only under aerobic conditions excludes lactic acid bacteria. For most strains glucose is a suitable substrate but some organisms, particularly when freshly isolated, require fructose or a disaccharide.

(4) Homofermentative or heterofermentative. Homofermentative species growing in a good nutrient medium ferment glucose to form almost entirely lactic acid, whilst heterofermentative species form lactic acid, CO_2, ethanol and/or acetic acid. In practice these two groups are distinguished by testing for the production of gas from glucose, using the method of Gibson & Abd-el-Malek (1945). Cultures are inoculated into a well-buffered nutrient semisolid agar medium containing a high concentration of glucose and an agar seal is poured on the surface. Gas production may be indicated by only small bubbles or pockets, but some strains will force the seal up the tube. It is essential to use a heavy inoculum of a vigorously growing culture. The milk-based medium of Gibson & Abd-el-Malek (1945) is often used for this test; some homofermentative lactic acid bacteria can form gas bubbles from the citrate present e.g. *Streptococcus lactis* subsp. *diacetylactis*, occasionally *Lactobacillus casei*.

Carbohydrate fermentation

In all groups of lactic acid bacteria carbohydrate fermentation patterns are useful to distinguish between species. These fermentation patterns can be determined by (1) the conventional tube method, (2) a miniplate method (Jayne-Williams 1975) which is more quickly set up and performed, gives more rapid results and uses less equipment and media, and (3) the L-50 API Lactobacillus Identification System (API Laboratory Products, Rayleigh, Essex, UK) which is purchased ready for use and also gives rapid results. The basal media recommended for (1) and (2) are described by Jayne-Williams (1975).

Determination of isomer of lactate formed from glucose

This is not often necessary but if required, total lactic acid can be determined chemically (Barker & Sommerson 1941), L(+) lactic acid enzymically (Garvie 1967*a*) and D(−) acid being taken as the difference. Owing to the size of analytical error difficulty arises when one of the isomers constitutes less than 10% of the total lactate formed.

Streptococcus

The criteria of Sherman (1937) divided the streptococci into four groups, enterococcus, lactic, viridans and pyogenic. More recent work, particularly on the viridans group has cut across some of these major groups and recognized new species. Streptococci are now broadly grouped as faecal, lactic, oral and pyogenic (Table 1); species in the old viridans group are included in the oral, faecal or pyogenic groups, whilst some remain unclassified (Colman & Williams 1972; Buchanan & Gibbons 1974; Wilson & Miles 1975; Hardie & Bowden 1976*a*). Recent aspects of the taxonomy of the streptococci are discussed by Jones (1978). Cell structure, biochemical aspects, enzymic activities and applied aspects are reviewed by Hahn *et al.* (1970).

In the past, division of streptococci into serological groups, based on the presence of specific antigens tended to be emphasized more than species differentiation. This method is still of great value, particularly with pyogenic and faecal strains. However, the discovery of many new grouping antigens which could not be associated with particular physiological characteristics of the organisms, possession of more than one such antigen, and the lack of grouping antigens in some newly defined species has led to a more cautious use of this method for identification. A combination of physiological, serological and genetic tests is the best basis for defining species. However, it is possible to characterize many species of streptococci by a number of simple physiological tests.

Habitat

Streptococci are widely distributed as human and animal pathogens, in the mouth and intestine of man and animals, in raw milk and dairy products and on plant material (see Table 1). These aspects are discussed in detail by a number of workers in a *Streptococcus* Symposium (Skinner & Quesnel 1978). Although the current names of three of the broad groups, faecal, oral and lactic, suggest the main habitat of the organisms, there are many exceptions to this and it should never be assumed that the habitat denotes the only species to be isolated, e.g. the oral group species *S. salivarius* may occur in the pig intestine (Barrow *et al.* 1978); the faecal group species *S. faecium* subsp. *casseliflavus* occurs on plant material (Mundt & Graham 1968); dental plaque from different animals may contain enterococci or *S. bovis* as well as oral group species (Dent *et al.* 1978). Strains belonging to groups other than pyogenic may manifest themselves as pathogens (Parker 1978) and such isolates from clinical sources have been identified by simple physiological tests (Parker & Ball 1976).

TABLE 1. Main habitats of species of *Streptococcus*

Species	Habitat
faecalis } *faecium* }	Intestine, man and animals
avium	Intestine, birds
faecium subsp. *casseliflavus*	Plant material
bovis	Intestine, animals, man
equinus	Intestine, horse
thermophilus	Heat treated milk
lactis	Milk and dairy products, plant material
lactis subsp. *diacetylactis* } *cremoris* } *raffinolactis* }	Milk and dairy products
mutans	Human mouth: dental plaque and caries lesions
sanguis	Human mouth: dental plaque
mitior	Human mouth: dental plaque
milleri	Human mouth: dental plaque. Brain abscesses, other lesions
salivarius	Human mouth: saliva and tongue. Intestine of animals
	All pathogenic
pyogenes	human
anginosus	human
zooepidemicus	animals
equi	horse
equisimilis	human, animals
agalactiae	mastitic udder, human vagina and intestine
dysgalactiae } *uberis* } *acidominimus* }	mastitic udder

Media and methods. Those generally used are described by Cowan (1974) and Jayne-Williams (1976). Many tests are usefully discussed by Hardie & Bowden (1976*a*). Isolation media for streptococci from different habitats are described in detail in the *Streptococcus* Symposium (Skinner & Quesnel 1978).

Serological tests. Where suitable antisera are available many streptococci can be allocated serologically (Table 2) into Lancefield (1933) groups using precipitin tests described by Cruickshank *et al.* (1975). For identification of streptococci see Table 2.

Faecal streptococci

Taxonomy of this group, and detailed media and methods are described by Smith & Shattock (1962), Deibel (1964), Sharpe *et al.* (1966) and Jones *et al.* (1972). Facklam (1972) gives details of strains of human origin, and suggests the use of a bile-aesculin medium (Facklam & Moody 1970) for presumptive identification of group D streptococci.

Streptococcus faecalis, *S. faecium*, *S. avium* and *S. faecium* subsp. *casseliflavus* belong to the enterococcus group of Sherman (1937). Strains of *S. faecium* subsp. *casseliflavus* are almost always motile and are yellow pigmented (Mundt & Graham 1968). Other strains of *S. faecium* are occasionally motile. *Streptococcus bovis* and *S. equinus*, although possessing the group D antigen and having other characteristics in common with the enterococci (see Table 1), form a distinct subgroup with many differences from the enterococci. *Streptococcus bovis* has been subdivided into more than one biotype but the differences between them are not clearcut. There is some evidence that non-mannitol-fermenting strains produce capsular slime whilst mannitol-fermenting strains do not, but this requires further confirmation. Not all strains of *S. bovis* can be identified serologically.

Lactic streptococci

Media for physiological tests are described by Swartling (1951), and Garvie (1953) and isolation media by Sharpe (1978).

For practical purposes the thermophilic *S. thermophilus* has been included with the mesophilic lactic streptococci, although its physiological characteristics are widely different. It is not suggested that there is a close taxonomic relationship. Strains of *S. thermophilus* do not always ferment glucose and should be isolated on a lactose based medium (Sharpe 1978).

The ability to form acetoin or diacetyl from citrate is important to the dairy industry and for this reason *S. lactis* subsp. *diacetylactis* is distinguished as a variant of *S. lactis*, which it otherwise closely resembles. Not all strains of *S. lactis* grow at 40°C, and 39·5°C is used as the preferred growth temperature distinguishing it from *S. cremoris*, using litmus milk medium.

So far *S. raffinolactis* (Garvie 1953, 1978) has been reported only from raw milk, whilst other mesophilic starters are also found in starter cultures and dairy products. This species can be differentiated from *S. cremoris* by its sugar fermentation pattern, further substantiated by more fundamental studies (Garvie 1978).

Metabolic aspects of the lactic streptococci are discussed by Law & Sharpe (1978).

TABLE 2. Differentiating characteristics among the genus *Streptococcus*

Group	Species	Growth at pH 9·6	Growth in 6·5 % NaCl	Survive 60°C for 30 min	Growth at 10°C	Growth at 45°C	0·1 % methylene blue	NH_3 from arginine	Aesculin hydrolysis	Growth on 40 % bile	Type of haemolysis	Serological group	Reduction of tetrazolium	0·4 % K tellurite	Ferments: Lactose	Ferments: Mannitol	Ferments: Melezitose	Ferments: Melibiose	Ferments: Raffinose	Ferments: Starch	Ferments: Glycerol
Faecal	*faecalis*[a]					+	+	+	+	+	λ, α, β	D	+	+	+	+	+	—	—	—	+A
	faecium					+	+	+	+	+	α	I	—	—	+	d	—	+	d	—	—
	avium	+	+	+	+	+	—	—	+	+	α	D & Q	—	—	+	+	+	d	—	—	+AO
	faecium subsp. *casseliflavus*					d	+	d	+	+	α	D	w	w	+	+	—	+	+	—	dA
	bovis											D	d	—	+	d	—	+	+	+	—
	equinus	—	—	—	—	+	—	—	+	+	α, λ	D	—	—	—	—	—	—	+	d	—
Lactic	*thermophilus*	—	—	—	—	+	—	—	—	—	λ										
	lactis							+	+	—	λ	N									
	lactis subsp. *diacetylactis*	—	—	—	+	—	+	+	+	—	λ	N									
	cremoris							—	+	—	λ	N									
	raffinolactis							—	+	—	λ	N									

Group	Species	Growth at 50°C	Growth at 39·5°C	Growth at pH 9·2	Growth in 4 % NaCl	Citrate utilized	Acetoin/diacetyl produced	Ferments: Maltose	Ferments: Raffinose	Ferments: Sucrose	Levan from sucrose	Dextran from sucrose	Acetoin produced	H_2O_2 produced	Ferments: Mannitol	Ferments: Sorbitol	Ferments: Inulin	Ferments: Salicin	Ferments: Trehalose	Ferments: Glycerol	Hippurate hydrolysis
Faecal	*faecalis*[a]	—																			
	faecium	+																			
	avium	—																			
	faecium subsp. *casseliflavus*	d																			
	bovis	—																			
	equinus	—																			
Lactic	*thermophilus*	+	+	—	—	—	—	—	—	+											
	lactis	—	+	+	+	—	—	+	—	d											
	lactis subsp. *diacetylactis*	—	+	+	+	+	+	+	—	—											
	cremoris	—	—	—	—	—	—	—	—	—											
	raffinolactis	—	—	—	—	—	—	+	+	+											

Oral	*mutans*	— — d d	d			— + + — + +
	sanguis	d + + d	d, β	H		— + — + — —
	mitior	— — — — — d — — —	d			— d d + — —
	milleri	— + + d	d			— — + — — —
	salivarius	+ — + d	d	K		+ — d — — —
Pyogenic	*pyogenes*	+ — —	β	A	—	— — — + — — —
	anginosus[b]	+ — —	α, β, λ	F or G	d	— — — + + — —
	zooepidemicus	— + + —	β	C	—	— — — + — — —
	equi	+ — —	β	C	—	— — — + — — —
	equisimilis	— — — — — — — + d —	β	C	—	— — — d + + O —
	agalactiae	— + — —	β	B	— —	— — — + d d +
	dysgalactiae	+ + — —	α	C	— —	— d — d + — d
	uberis	d + + —	α, λ		— —	+ + + + + — +
	acidominimus[c]	— — —	α		— —	— — — — w — —

A: ferments glycerol anaerobically; O, ferments glycerol aerobically; d: variable reaction; w: weak reaction.

[a] Two subspecies, *S. faecalis* subsp. *liquefaciens* which liquefies gelatin, and subsp. *zymogenes*, which is β-haemolytic are recognized.

[b] This requires CO_2 for growth on blood agar.

[c] Not always pathogenic.

Oral streptococci

The physiological classification of these streptococci is described in detail by Colman (1968) and discussed by Hardie & Bowden (1976*a*); media for isolation are described by Hardie & Marsh (1978*a*). These streptococci are now separated into five species which can be identified by a set of simple physiological tests (Table 1); each species is phenotypically homogeneous but genotypically heterogeneous, as shown by DNA studies (Coykendall 1977). Although grouping antisera can be used to identify some of the species, interspecies cross-reactions make this difficult (Hardie & Bowden 1976*b*). Aesculin hydrolysis in *S. mutans* is variable in the species as a whole. However, most of the isolates in the UK are aesculin positive and belong to serotype *c*. Other serotypes such as *d* and *g* tend to be aesculin negative. Sorbitol is typically fermented by *S. mutans*, but there are negative strains, also usually serotypes *d* and *g* (J. M. Hardie pers. comm.).

Individual species are quantitatively distributed at different sites in the mouth, related to their ability to adhere to different surfaces (Hardie & Marsh 1978*b*).

Pyogenic streptococci

Many of these streptococci can be typed by both precipitin and agglutination reactions, and serological grouping (Lancefield 1933) is the reliable method for identification, especially for the human pathogens of groups A and C. Cowan (1974) gives useful references and tables for identification by physiological tests.

For streptococci associated with the bovine udder, growth on aesculin-blood agar and sugar fermentation patterns separate *S. agalactiae*, *S. dysgalactiae*, *S. uberis* and *S. acidominimus*, as shown by McDonald & McDonald (1976). Serological methods are useful for *S. agalactiae* (group B) and *S. dysgalactiae* (group C). All these species may occur in raw milk.

Until recently group B streptococci were regarded only as a pathogen affecting the bovine udder. It is now known that some strains may colonize the human intestine and female genito-urinary tract, and thus act as a common pathogen in the etiology of human neonatal infections (Badri *et al.* 1977; Ross 1978). These group B streptococci are different serotypes from the animal strains. Many of the human strains produce an orange pigment.

Unclassified streptococci

The somewhat ill-defined, both serologically and physiologically, pyogenic species *S. lentus* and *S. suis* are discussed by Jones (1978).

They are both associated with clinical conditions in pigs.

Not all strains in a given species have the reactions indicated in Table 1, and occasionally fermentation of key sugars or other reactions may vary. In addition there are at present many streptococci being isolated from animal intestinal sources which do not fit any of the recognized groups and cannot yet be assigned to any species.

Pediococcus and *Aerococcus*

Pediococcus

These organisms occur in fermenting plant material, milk and dairy products and in spoiled beer.

The nomenclature of this genus has been clarified by a ruling of the Judicial Commission of the International Committee on Systematic Bacteriology (Anon. 1976) that the species name *cereviseae* was not validly published. The generic name has been conserved over *Pediococcus* (Claussen), with *Pediococcus damnosus* as the type species. This ruling post-dates the publication of Bergey's Manual (Buchanan & Gibbons 1974) so that some species names given there are now incorrect.

Previously the same name *cereviseae* was used to describe different species by Western and Japanese workers. Now that this name has been removed it is possible to combine the two systems. Table 3 shows the previous and the present nomenclature (Gunther & White 1961; Coster & White 1964; Sharpe *et al.* 1966; Buchanan & Gibbons 1974; Garvie 1974). Details of the history of this nomenclature and of characteristics of some of these species involved are described by Garvie (1974).

TABLE 3. Changes in nomenclature of *Pediococcus* sp.

Gunther & White (1961) used by Sharpe *et al.* (1966)	Kitahara (Buchanan & Gibbons 1974)	Garvie (1974) IJSB[a] ruling (1976) and present scheme
damnosus	*cereviseae*	*damnosus*
cereviseae	*pentosaceus*	*pentosaceus*
parvulus	included with *pentosaceus*	*parvulus*
	acidilactici	As in Bergey's Manual (1974)[b]
halophilus	*halophilus*	As in Bergey's Manual (1974)
	urinae-equi	As in Bergey's Manual (1974)
Gunther & White Group III	not described	Gunther & White Group III

[a] International Journal of Systematic Bacteriology.
[b] Buchanan & Gibbons (1974).

TABLE 4. Characteristics differentiating species of the genus *Pediococcus*, and describing the genus *Aerococcus*

Species of *Pediococcus*	Growth at 37°C	Growth at 45°C	Growth at pH 4·4	Growth in 6·5% NaCl	Growth in 10% NaCl	Lactic acid isomer	Diacetyl from glucose	Arginine	Arabinose	Dextrin	Maltose	Raffinose	Sorbitol	Sucrose	Catalase production	Growth in Rogosa AcA	Folinic acid required
damnosus	—	—	+	—	—	DL	+	—	—	—	—	—	—	—	—	+	—
pentosaceus	+	+	+	+	—	DL	+	+	+	—	+	—	—	—	d	+	+
acidilactici	+	+	+	+	—	DL	+	+	+	—	—	w	—	w	—	+	—
parvulus	+	—	+	—	—	DL	—	—	—	—	+	—	—	—	—	+	.
halophilus[a]	+	—	—	+	+	L(+)	d	+	+	—	+	—	+	+	—	—	+
urinae-equi	+	—	.	+	+	L(+)	—	—	d	+	+	d	—	+	—	.	+
Gunther & White Group III	+	—	—	.	—	L(+)	—	—	—	+	d	—	—	—	—	+	.
Aerococcus viridans	+	—	—	+	+	L(+)	—	—	—	—	+	+	—	+	—	—	—

All grow at 12°C.

[a] grows only poorly in the absence of 6–8% NaCl.

Little further work has been done on the taxonomy of the pediococci.

For identification of species the media and methods of Gunther & White (1961) and Coster & White (1964) are used except that MRS broth (de Man *et al.* 1960) is recommended instead of tomato juice broth for general cultivation and as a basal medium. The API Lactobacillus System (API Ltd., Rayleigh, Essex) has been used successfully for characterization of pediococci (Dolezil & Kirsop 1977). Species are differentiated as shown in Table 4. Serological methods have been used (Dolezil & Kirsop 1976) to identify beer spoilage types.

Pediococcus damnosus, the type species, causes spoilage of beer, generally by diacetyl production and is readily distinguished from other pediococci. It is highly tolerant to antiseptic hop constituents. The most widely occurring species in fermented foods and dairy products is *P. pentosaceus* which has many characteristics in common with *P. acidilactici* (Garvie 1974); there is no clear evidence that they should be separate species although there is a difference of 6% in G + C ratio (Buchanan & Gibbons 1974) and many strains of *P. pentosaceus* have an absolute requirement for folinic acid. *Pediococcus halophilus* and *P. urinae-equi* are halophilic organisms not easy to distinguish from each other. The still unnamed Gunther and White Group III underlines the lack of taxonomic studies on the genus since 1964.

Absence of cytochrome-containing respiratory systems

Many strains of *P. pentosaceus* when grown on a low sugar-containing medium produce a 'pseudocatalase' which decomposes H_2O_2 but differs from catalase in being insensitive to cyanide and azide and in not containing a haem prosthetic group (Delwiche 1961; Whittenbury 1964; Johnston & Delwiche 1965). The detection of cytochrome-containing systems by a benzidine test for iron porphyrins (described by Sharpe *et al.* 1966) should *not* be used, as benzidine is highly carcinogenic.

Aerococcus

The systematic position of aerococci is still not resolved (Buchanan & Gibbons 1974) but as these organisms are fermentative, unable to grow without carbohydrate and form lactic acid (Deibel & Niven 1960), they have been included here. The characteristics of *Aerococcus viridans* (synonym *P. homari*) are given in Table 3.

Leuconostoc

Leuconostocs are found on herbage, vegetables and fruit, in fermenting

TABLE 5. Differentiating characteristics of species of the genus *Leuconostoc*

Species of *Leuconostoc*	Growth at 37°C	Growth in 10% EtOH	Growth at pH 4·2	Dextran from sucrose	Aesculin hydrolysis	Citrate utilization	Arabinose	Fructose	Lactose	Maltose	Sucrose	Trehalose	G + C content (%)	Habitat
lactis	+	—	—	—	—	+	—	+	+	+	+	—	42–44	milk and dairy products
paramesenteroides	d	—	—	—	d	d	d	+	w	+	+	+	38–39	herbage, fermented vegetables, milk and dairy products
mesenteroides[a]	d	—	—	+	d	d	+	+	w	+	+	+	38–41	sugar factories, fruit, vegetables, milk, dairy products
dextranicum[a]	+	—	—	+	d	d	—	+	+	+	+	+	38–39	fruit, vegetables, milk, dairy products
cremoris[a]	—	—	—	—	—	+	—	—	+	d	—	—	38–41	starters, milk and dairy products
oenos	d	+	+	—	+	d	d	+	—	—	—	+	38–39	wine

All grow at 10°C, none grows at 45°C.
None produces NH_3 from arginine, all form D(—) lactic acid.
+: > 90% strains positive; d: 10–90 strains positive; —: > 90% strains negative; w: weak or delayed action.
[a] Genotypically a single species.

vegetable matter, in sugar solutions, particularly in the sugar industry, in milk and dairy products and one species occurs specifically in wine. The ability of some species to form profuse dextran slime from sucrose may result in extensive spoilage of products.

The media and methods used for identification are those described by Garvie (1960, 1967*b*, *c*). The scheme of identification (Table 5) is based on this and later work (Buchanan & Gibbons 1974; Garvie 1976), and gives differentiating characteristics of the species. Some of these species may be difficult to differentiate as phenotypic characteristics tend to merge. Separation between the two dextran forming species *Leuc. mesenteroides* and *Leuc. dextranicum* is unimportant from a practical point of view.

Genetic DNA homology studies (Garvie 1976) and immunological relationships of enzymes (Hontebeyrie & Gasser 1975) indicate that *Leuc. mesenteroides* and *Leuc. dextranicum* and also *Leuc. cremoris* are closely related and may be a single genotypic species. Thus *Leuc. cremoris* although very different phenotypically from *Leuc. mesenteroides* is genotypically related. However' its different physiological characteristics and restricted habitat (milk and dairy products) indicate the usefulness of phenotypic differentiation (Table 5).

Leuconostoc lactis, *Leuc. paramesenteroides* and *Leuc. oenos* are all genotypically as well as phenotypically separate species (Garvie 1976; Hontebeyrie & Gasser 1975).

Leuconostoc oenos is important in the malolactate fermentation of wine (reviewed Kunkee 1975) and is so far found specifically in this habitat. Owing to the low pH of its normal habitat and the dependence of some strains on a form of pantothenate found in tomato juice and other plant material (Garvie & Mabbitt 1967; Amachi 1975) special media are required for isolation and culture (Garvie 1967*b*) with an initial pH of 4·8 or less and the inclusion of tomato juice. Strains of *Leuc. oenos* are also differentiated by their ability to grow in the presence of 10% ethanol.

Dextran production, characteristic of many strains of *Leuc. mesenteroides* and *Leuc. dextranicum* is observed by production of slime on a 5% sucrose-containing agar (Garvie 1960), optimal temperature, 20°–25°C.

Leuconostoc and heterofermentative lactobacilli

Leuconostocs and heterofermentative lactobacilli may frequently occur in the same habitat and have similar nutritional requirements. As the leuconostocs, with the exception of *Leuc. cremoris* and *Leuc. oenos* will all grow on the selective acetate medium generally used for isolating lactobacilli (Rogosa *et al.* 1951), lactobacilli and leuconostocs are likely to be isolated together. They may not be readily distinguished as some species of lactobacilli can occur as coccobacilli. Leuconostocs can be differentiated from most betabacteria by fermenting trehalose (except *Leuc. cremoris* and *Leuc. oenos*), by not producing NH_3 from arginine and by forming D(−) lactic acid from glucose, whilst most heterofermentative lactobacilli do not ferment trehalose or hydrolyse arginine and all form DL-lactic acid.

However, *Lactobacillus viridescens* and *L. confusus* share many properties with *Leuc. dextranicum* and *Leuc. mesenteroides*, respectively, including slime production and sugar fermentation patterns, and *L. viridescens* does not hydrolyse arginine. These two species of lactobacilli are considered to be closely related to *Leuc. dextranicum* and *Leuc. mesenteroides*, respectively (Sharpe *et al.* 1972; Garvie 1976), although the leuconostocs form only D(−) lactic acid and have a single lactate dehydrogenase whilst the lactobacilli form DL-lactic acid and have both a D(−) and L(+) LDH. However, *L. viridescens* may form only a small amount (*ca.* 10% of the total) of L(+) lactate.

Lactobacillus

The genus was divided by Orla Jensen (1919, 1943) into three main subgroups, thermobacteria, streptobacteria and betabacteria, on the basis of their growth temperatures and end-products of fermentation. The genus is still divided in this way and the division is now confirmed by further tests (Table 6). Whilst Rogosa (Buchanan & Gibbons 1974) does not mention these subgroups by name, the main subdivisions of lactobacilli in his chapter are based on the concept of the three main subgroups (Rogosa 1970). Other recent reviews on taxonomy also recognize these subdivisions (Sharpe 1974; London 1976; Wilkinson & Jones 1977). Thus, lactobacilli are divided into homofermentative, producing almost entirely lactic acid from glucose, and heterofermentative (betabacteria), producing lactic acid, CO_2, acetic acid and/or ethanol; the division is confirmed by testing for a thiamine requirement for growth and presence or absence in the cells of fructose diphosphate aldolase (Table 6). Homofermentative lactobacilli are further divided into thermobacteria and streptobacteria by growth temperatures and simple biochemical tests (Rogosa 1970). Rogosa (1974) calls the thermobacteria and streptobacteria groups IA or IB, respectively, as we do here, whilst the betabacteria are divided into groups II and III, the latter being much less active biochemically. As this division of betabacteria seems to imply a wider distinction than can be assessed on present data, they have been divided here into IIA and IIB (Table 6).

TABLE 6. Subdivision of genus *Lactobacillus*

Test no.	I Homofermentative		II Heterofermentative	
	Glucose fermented almost entirely (over 85%) to lactic acid (LA)		Glucose fermented to LA (50%) + CO² + acetic acid + ethanol	
1[a]	—		+	
2	—		+	
3	+		—	
	Thermobacterium IA	*Streptobacterium* IB	*Betabacterium* II	
4[a]	+	d	dependent on species usually	
5[a]	—	+		
6[a]	—	+	+	
7[a]	—	+	+	
			All form DL LA from glucose	
			IIA	IIB acidophilic, ethanol tolerant, inactive to most carbohydrates
	acidophilus	*casei*	*fermentum*	*hilgardii*
	helveticus	*plantarum*	*cellobiosus*	*trichodes*
	bulgaricus	*xylosus*	*brevis*	*fructivorans*
	lactis	*curvatus*	*buchneri*	*desidiosus*
	delbrueckii	*coryneformis*	*viridescens*	*heterohiochi*
	leichmannii	*homohiochi*	*confusus*	
	salivarius	*yamanashiensis*		
	jensenii			
	ruminis, *vitulinus* } anaerobic			

d: variable reaction.
1: CO_2 from glucose; 2: thiamine required for growth; 3: fructose diphosphate aldolase present; 4: growth at 45°C; 5: growth at 15°C; 6: Ribose fermented; 7: CO_2 from gluconate.
[a] Simple test used for initial identification of isolates.

Each subgroup contains a number of species (Table 6) differentiated by means of simple biochemical and physiological tests, these species being confirmed by other more complex determinations, such as vitamin requirements (Rogosa *et al.* 1961); the chemical nature of cell wall and cell membrane antigenic determinants (serological grouping) (Sharpe 1970; Knox & Wicken 1973); amino acid sequences of the peptidoglycan (Kandler 1970); DNA composition (Gasser & Mandel 1968); electrophoretic mobility of lactic dehydrogenases (Gasser 1970); and DNA homology (Simonds *et al.* 1971; Dellaglio *et al.* 1973, 1975). Other reviews on the taxonomy of lactobacilli include those of Rogosa & Sharpe 1959; Sharpe 1962, 1974; Rogosa 1970; London 1976).

In addition to the widely accepted species mentioned previously (Sharpe *et al.* 1966), identification schemes for 13 further species now listed in Bergey's Manual (Buchanan & Gibbons 1974) are included here. Strains which cannot be identified at species level can usually at least be assigned to the appropriate subgroup.

Habitat

Table 7 gives the main habitats of many species of lactobacilli. Species such as *L. plantarum*, *L. brevis* and *L. casei* are widespread in their occurrence. Others have very restricted habitats, e.g. *L. homohiochi* and *L. heterohiochi* have so far been isolated only from spoiled saké, *L. acidophilus* is usually found mainly in the intestinal tract.

Physiological tests

The MRS medium of deMan *et al.* (1960) is used for general cultivation of strains, determination of growth temperatures and in a modified form as the basal medium for biochemical tests. The methods referred to are discussed by Sharpe (1962) and many media recipes are given by Harrigan & McCance (1976). Most lactobacilli grow better at an increased CO_2 tension and should be isolated in such an atmosphere. The slow-growing species of Group IIA may require more specialized media containing tomato juice or natural substrate as an ingredient, and a low pH (reviewed by Sharpe, in press).

Thermobacteria

Seven species are recognized by these tests, the species designation being confirmed by further data (Table 8). The species *L. jensenii*, phenotypically similar to *L. leichmannii*, can be distinguished from the latter only by different electrophoretic mobilities of the lactic dehydrogenases and by its different G + C content (Gasser *et al.* 1970). In Bergey's Manual (Buchanan & Gibbons 1974) it is suggested that *L. lactis* and *L. bulgaricus* may be variants of a single species.

Two species of anaerobic lactobacilli (Sharpe *et al.* 1973; Sharpe & Dellaglio 1977) not included in Table 8 are further differentiated from other thermobacteria by the presence of *meso* DAP in the peptidoglycan, type of lactic acid formed and G + C content. Some strains of these organisms which have an intestinal habitat, may appear as Gram positive club forms, thus resembling the anaerobic bifidobacteria. To distinguish anaerobic homofermentative lactobacilli from bifidobacteria, it is necess-

TABLE 7. Distribution of *Lactobacillus* spp. in some of their main habitats

Milk and milk products	Meat	Mouth or intestine of man and animals	Fermented vegetable products, silage	Pickles, mayonnaise	Fermented grain mashes	Beer, wines, cider, saké
helveticus	*viridescens*	*salivarius*	*plantarum*	*plantarum*	*delbrueckii*	*brevis*
bulgaricus	*plantarum*	*casei*	*brevis*	*brevis*	*leichmannii*	*buchneri*
lactis	*brevis*	*plantarum*	*buchneri*	*buchneri*	*fermentum*	*hilgardii*
casei	*buchneri*	*brevis*	*fermentum*	*fructivorans*	*brevis*	*trichodes*
plantarum	*casei*	*cellobiosus*	*confusus*			*yamamashiensis*
brevis		*acidophilus*	*coryniformis*			*homohiochi*
xylosus		*fermentum*				*heterohiochi*
curvatus		*lactis*				
		jensenii				
		curvatus				

Table 8. Differentiating characteristics in species of thermobacteria[a]

Species of *Lactobacillus*	Presence of granules	% acid in milk	Lactic acid isomer	NH_3 from arginine	Aesculin hydrolysis	Amygdalin	Cellobiose	Galactose	Lactose	Maltose	Mannitol	Melibiose	Salicin	Sorbitol	Sucrose	Trehalose	Serological group[c]	Riboflavin[b, c]	Pyridoxal[b, c]	Folic acid[b, c]	Thymidine[b, c]	G + C content (%)[c]
helveticus	—	+2·7	DL	—	—	—	—	+	+	+	—	—	—	—	—	d	A	+	+	—	—	40
bulgaricus	+	1·7	D(—)	—	—	—	—	+	+	—	—	—	—	—	—	—	E	—	—	—	—	50–51
lactis	+	+1·75	D(—)	—	d	—	—	+	+	+	—	—	+	—	+	+	E	+	—	—	—	50–51
acidophilus	—	0·3–1·7	DL	—	+	+	+	+	+	+	—	d	+	—	+	+	.	+	—	+	—	36–37
leichmannii	+	0	D(—)	d	+	+	+	—	+	+	—	—	+	—	+	+	.	—	—	+	—	50–51
delbrueckii	—	0	D(—)	d	—	—	—	w	—	d	—	—	—	—	+	—	.	+	—	—	+	50
salivarius	—	0·9	L(+) & DL	—	d	—	—	+	+	+	+	+	d	+	+	+	G	+	—	+	—	34–35

None produces CO_2 from glucose or gluconate. None ferments ribose, arabinose, xylose, melezitose, or requires thiamine for growth.
d: variable reaction; w: weak slow or negative reaction; .: no data available.
[a] Group IA in Bergey's Manual (Buchana & Gibbons 1974).
[b] Vitamin requirements.
[c] Confirmatory, more specialized tests, not necessary for routine identification.

ary to confirm that they are anaerobic and that they do not produce CO_2 from glucose, and to determine the major fermentation end-products by g.l.c. Bifidobacteria do not produce CO_2 from glucose and produce mainly acetic acid (2 acetate:1 lactate) whilst these anaerobic homofermentative lactobacilli produce $> 85\%$ of lactate.

Streptobacteria

In addition to the classical species *L. casei* and its subspecies and *L. plantarum*, five other species, four of which are listed in Bergey's Manual and a further one *L. yamanashiensis* (Carr *et al.* 1977) previously described as *L. mali* are tabulated here (Table 9). When isolating from dairy and other food and plant sources, strains of streptobacteria with less active fermentation patterns than *L. casei* and *L. plantarum* are frequently found and it may be possible to assign them to one of these new species. When they cannot be identified they can be described as unidentified streptobacteria. *Lactobacillus curvatus*, *L. coryniformis*, and 2 subspecies of *L. casei* not tabulated here, all originally described by Abo-Elnaga & Kandler (1965) may be the same as atypical strains isolated by Keddie (1959) from silage and herbage (Rogosa 1974).

Betabacteria

The heterofermentative lactobacilli are now divided into two groups (Table 6).

Group IIA contains the well-recognized fermentatively active species *L. fermentum*, *L. cellobiosus*, *L. brevis*, *L. buchneri*, *L. viridescens* and also *L. confusus* (Table 10). *Lactobacillus fermentum* and *L. cellobiosus* are now considered to be closely related to each other (Gasser 1970; Sharpe 1974); *L. brevis* and *L. buchneri* are only distinguished by melezitose fermentation and the vitamin requirements of some strains. *Lactobacillus viridescens* and *L. confusus* have some characteristics in common with the leuconostocs (Sharpe *et al.* 1972). *Lactobacillus confusus* was originally named *L. coprophilus* subsp. *confusus* but the original strains of *L. coprophilus* are no longer extant and no new strains have been isolated; therefore *L. confusus*, a widely distributed and well-recognized organism, has been raised to species level.

Group IIB. This comprises species which are inert to most carbohydrates, acidophilic and tolerate organic acid and ethanol (15%) (Table 11). These species have been less widely studied and appear to occur in more restricted habitats, *L. hilgardii* being concerned in the malolactic fermentation of wine, *L. heterohiochi* only as a spoilage organism in saké.

TABLE 9. Differentiating characteristics of species of streptobacteria[a]

Species of *Lactobacillus*	Growth at 45°C	% acid in milk	Lactic acid isomer	Aesculin hydrolysis	Fermentation of: Amygdalin	Arabinose	Cellobiose	Galactose	Lactose	Mannitol	Melezitose	Melibiose	Raffinose	Rhamnose	Ribose	Xylose	Serological group[c]	*Meso* DAP in cell wall[b, c]	Pyridoxal[b, c]	Folic[b, c]	G + C content (%)[c]
casei subsp. *casei*	±	1·2–1·5	L(+)[d]	+	+	—	+	+	+	+	+	—	—	—	+	—	B, C	—	+	+	45
casei subsp. *rhamnosus*	+	1·2–1·5	L(+)[d]	+	+	—	+	+	+	+	+	—	—	+	+	—	C	—	+	+	45
casei subsp. *alactosus*	—	0	L(+)[d]	+	+	—	+	+	—	+	+	—	—	—	+	—	usually B	—	+	+	45
plantarum	±	0·3–1·2	DL	+	+	d	+	+	+	+	d	+	+	—	+	±	D	+	—	—	46
xylosus	.	.	L(+)	—	+	—	+	+	—	+	—	—	—	—	+	+	.	—	.	.	40
curvatus	—	.	DL	+	—	—	+	+	w	—	—	—	—	—	+	—	.	—	.	.	44
coryneformis	—	.	(DL or D(—)	—	—	—	—	+	—	+	—	d	d	d	—	—	.	—	.	.	45
homohiochi	—	.		.	—	—	—	—	—	+	—	—	—	—	—	—	.	—	.	.	46
yamanashiensis	—	.	.	+	.	—	d	±	—	—	—	.	—	d	—	—	.	+	.	.	32–34

All grow at 15°C. All ferment fructose, maltose, mannose, produce CO_2 from gluconate. None requires thiamine for growth. None produces CO_2 from glucose; none produces NH_3 from arginine under the specific test conditions.
d: Variable reaction; w: weak, slow or negative reaction; .: no data available; +: requirement; —: no requirement; ±: usually negative.
[a] Group IB in Bergey's Manual (Buchanan & Gibbons 1974).
[b] Vitamin requirements.
[c] Confirmatory, more specialized tests, not necessary for routine identification.
[d] May form a very small amount (1–5%) of D (—) lactic acid.

TABLE 10. Differentiating characteristics of species of betabacteria IIA[a]

Species of *Lactobacillus*	Growth at 15°C	Growth at 45°C	NH_3 from arginine	Slime from sucrose	Aesculin hydrolysis	Fermentation of Amygdalin	Arabinose	Cellobiose	Mannitol	Melizitose	Melibiose	Raffinose	Xylose	Serological group[c]	Riboflavin,[b c]	Folic acid,[b c]	G + C content (%)[c]
fermentum	—	+	+	—	—	—	d	—	—	—	+	+	d	F	—	—	53
cellobiosus	±	±	+	—	+	+	+	+	—	—	+	+	d	.	—	—	53
brevis	+	—	+	—	d	—	+	—	w	—	+	w	d	E	—	+	43–46
buchneri	+	—	+	—	d	—	+	—	w	+	+	w	d	E	d	d	45
viridescens	+	—	—	d	—	—	—	—	—	—	—	—	—	.	+	+	42
confusus	+	+	d	+	+	+	±	+	—	—	—	—	+	.	d	+	44

All produce CO_2 from glucose and gluconate, form DL lactic acid, and require thiamine for growth. All ferment ribose, maltose and fructose. None ferments rhamnose or sorbitol.

d: variable reaction; w: weak, slow or negative reaction; .: no data available.

[a] Group II of Bergey's Manual (Buchanan & Gibbons 1974).

[b] Vitamin requirements.

[c] Confirmatory, more specialized tests, not necessary for routine identification.

TABLE 11. Differentiating characteristics of species of betabacteria IIB[a]

Species of *Lactobacillus*	Growth at 15°C	Growth at 25°C	Growth in 15% EtOH	Fermentation of Arabinose	Fructose	Glucose	Maltose	Ribose	Xylose	dissimulated Malate	Citrate	Optimum pH for growth	G + C content (%)
hilgardii[b]	—	+	+	—	+	+	+	+	+	+	+	4·5–5·0	40
trichodes	w	+	+[c]	—	+	+	d	.	—	+	—	4·5–5·0	39–40
fructivorans[b]	—	+	+	—	+	w	d	—	—	d	—	4·5–5·0	40
desidiosus	+	+	+	+	w	w	—	.	.	+	d	5·0–7·0	.
heterohiochi	.	+	+	—	+	+	—	+	—	.	.	4·5–5·0	.

All produce CO_2 from glucose and (of those examined) CO_2 from gluconate. All form DL lactic acid, none grows at 45°C.
Not fermented: amygdalin, cellobiose, lactose, mannitol, mannose, melezitose, melibiose, raffinose, rhamnose, salicin, sorbitol, trehalose; aesculin not hydrolysed.
d: variable reaction; w: weak, slow or negative reaction; .: no data available.

[a] Group III in Bergey's Manual (Buchanan & Gibbons 1974).
[b] Growth stimulated by CO_2 for initial isolation.
[c] Ethanol highly stimulatory for growth, many strains produce only trace growth without it.

Interesting data of Dakin & Radwell (1971) suggests that *L. trichodes*, the wine spoilage organism, and *L. fructivorans* which causes spoilage in acetic acid preserves, are the same organism. In much work with Group IIB organisms there has been a tendency to study characteristics of isolates from one particular habitat only and not to make comparisons with Group IIB isolates from other sources. Dakin & Radwell's work shows that further comparative studies would be of value. Additional data on a larger number of strains is also necessary.

The author wishes to thank Dr. E. I. Garvie for much useful information, helpful advice and discussion.

References

ABO-ELNAGA, I. G. & KANDLER, O. 1965 Zur Taxonomie der Gattung *Lactobacillus* Beijerinck. I. Das Subgenus *Streptobacterium* Orla Jensen. *Zentrablatt für Bakteriologie, Parasitenkunde, Infektionskrankheiten und Hygiene, Abt. II* **119,** 1–36.

AMACHI, T. 1975 Chemical structure of a growth factor (TJF) and its physiological significance for malo-lactic bacteria. In *Lactic Acid Bacteria in Beverages and Food*, eds Carr, J. G., Cutting, C. V. & Whiting, G. C. London, New York & San Francisco: Academic Press.

ANON. 1976 International Committee on Systematic Bacteriology Opinion 52. Conservation of the generic name *Pediococcus* Claussen with the type species *Pediococcus damnosus* Claussen. *International Journal of Systematic Bacteriology* **26,** 392.

BADRI, M. S., ZAWANEH, S., CRUZ, A. C., MANTILLA, G., BAER, H., SPELLACY, W. N. & AYOUB, E. M. 1977 Rectal colonization with group B *Streptococcus*. Relation to vaginal colonization of pregnant women. *Journal of Infectious Diseases* **135,** 308–312.

BARKER, S. B. & SOMMERSON, W. H. 1941 The colorimetric determination of lactic acid in biological materials. *Journal of Biological Chemistry* **138,** 535–554.

BARNES, E. M., IMPEY, C. S., STEVENS, B. J. H. & PEEL, J. L. 1977 *Streptococcus pleomorphus* sp.nov. An anaerobic streptococcus isolated mainly from the caeca of birds. *The Journal of General Microbiology* **102,** 45–53.

BARROW, P. A., FULLER, R. & NEWPORT, M. J. 1978 Changes in the microflora and physiology of the anterior intestinal tract of pigs weaned at 2 days, with special reference to the pathogenesis of diarrhoea. *Infection & Immunity* **18,** 586–595.

BUCHANAN, R. E. & GIBBONS, N. E. (eds) 1974 *Bergey's Manual of Determinative Bacteriology*, 8th edn. Baltimore: Williams & Wilkins.

CARR, J. G., DAVIES, P. A., DELLAGLIO, F., VESCOVA, V. & WILLIAMS, R. A. D. 1977 The relationship between *Lactobacillus mali* from cider and *Lactobacillus yamanashiensis* from wine. *Journal of Applied Bacteriology* **42,** 219–228.

COLMAN, G. 1968 The application of computers to the classification of streptococci. *Journal of General Microbiology* **50,** 149–158.

COLMAN, G. & WILLIAMS, R. E. O. 1972 Taxonomy of some human viridans streptococci. In *Streptococci and Streptoccal Diseases*, eds Wannamaker, L. W. & Matsen, J. M. New York: Academic Press.

Coster, E. & White, H. R. 1964 Further studies of the genus *Pediococcus*. *Journal of General Microbiology* **37,** 15–31.

Cowan, S. T. (ed.) 1974 *Cowan & Steel's Manual for the Identification of Medical Bacteria* 2nd edn. London: Cambridge University Press.

Coykendall, A. L. 1977 Proposal to elevate the subspecies of *Streptococcus mutans* to species status, based on their molecular composition. *International Journal of Systematic Bacteriology* **27,** 26–30.

Cruickshank, R., Duguid, J. P., Marmion, B. P. & Swain, R. H. A. 1975 *Medical Microbiology* 12th edn. Edinburgh, London & New York: Churchill Livingstone.

Dakin, J. C. & Radwell, J. Y. 1971. Lactobacilli causing spoilage of acetic acid preserves. *Journal of Applied Bacteriology* **34,** 541–545.

Deibel, R. H. 1964 The group D streptococci. *Bacteriological Reviews* **28,** 330–366.

Deibel, R. H. & Niven, C. F. Jr. 1960 Comparative study of *Gaffkya homari*, *Aerococcus viridans*, tetrad-forming cocci from meat curing brines, and the genus *Pediococcus*. *Journal of Bacteriology* **79,** 175–180.

Dellaglio, F., Bottazzi, V. & Trovatelli, L. D. 1973 Deoxyribonucleic acid homology and base composition in some thermophilic lactobacilli. *Journal of General Microbiology* **74,** 289–297.

Dellaglio, F., Bottazzi, V. & Vescova, M. 1975 Deoxyribonucleic acid homology among *Lactobacillus* species of the sub genus *Streptobacterium*. Orla-Jensen. *International Journal of Systematic Bacteriology* **25,** 160–172.

Delwiche, E. A. 1961 Catalase of *Pediococcus cereviseae*. *Journal of Bacteriology* **81,** 416–418.

deMan, J. C., Rogosa, M. & Sharpe, M. E. 1960 A medium for the cultivation of lactobacilli. *Journal of Applied Bacteriology* **23,** 130–135.

Dent, V. E., Hardie, J. M. & Bowden, G. H. 1978 Streptococci isolated from dental plaque of animals. *Journal of Applied Bacteriology* **44,** 249–258.

Dolezil, L. & Kirsop, B. H. 1976 The detection and identification of *Pediococcus* and *Micrococcus* in breweries using a serological method. *Journal of the Institute of Brewing* **82,** 93–95.

Dolezil, L. & Kirsop, B. H. 1977 The use of the A.P.I. Lactobacillus system for the characterization of pediococci. *Journal of Applied Bacteriology* **42,** 213–217.

Facklam, R. R. 1972 Recognition of group D streptococcal species of human origin by biochemical and physiological tests. *Applied Microbiology* **23,** 1131–1139.

Facklam, R. R. & Moody, M. D. 1970 Presumptive identification of group D streptococci: the bile esculin test. *Applied Microbiology* **20,** 245–250.

Garvie, E. I. 1953 Some group N streptococci isolated from raw milk. *Journal of Dairy Research* **20,** 41–44.

Garvie, E. I. 1960 The genus *Leuconostoc* and its nomenclature. *Journal of Dairy Research* **27,** 283–292.

Garvie, E. I. 1967*a* The production of L(+) and D(−) lactic acid in culture of some lactic acid bacteria with a special study of *Lactobacillus acidophilus* NCDO 2. *Journal of Dairy Research* **34,** 31–38.

Garvie, E. I. 1967*b* *Leuconostoc oenos* sp. nov. *Journal of General Microbiology* **48,** 431–438.

Garvie, E. I. 1967*c* The growth factor and amino acid requirements of species of the genus *Leuconostoc* including *Leuconostoc paramesenteroides* sp.nov. and *Leuconostoc oenos*. *Journal of General Microbiology* **48,** 439–447.

Garvie, E. I. 1974 Nomenclatural problems of the pediococci. Request for an opinion. *International Journal of Systematic Bacteriology* **24,** 301–306.

Garvie, E. I. 1976 Hybridization between the deoxyribonucleic acids of some strains of heterofermentative lactic acid bacteria. *International Journal of Systematic Bacteriology* **26,** 116–122.

Garvie, E. I. 1978 *Streptococcus raffinolactis* (Orla Jensen and Hansen) a group N streptococcus found in raw milk. *International Journal of Systematic Bacteriology* (in press).

Garvie, E. I. & Mabbitt, L. A. 1967 Stimulation of the growth of *Leuconostoc oenos* by tomato juice. *Archiv für Mikrobiologie* **55,** 398–407.

Gasser, F. 1970 Electrophoretic characterization of lactic dehydrogenases in the genus *Lactobacillus*. *Journal of General Microbiology* **62,** 223–239.

Gasser, F. & Mandel, M. 1968 Deoxyribonucleic acid base composition of the genus *Lactobacillus*. *Journal of Bacteriology* **96,** 580–588.

Gasser, F., Mandel, M. & Rogosa, M. 1970 *Lactobacillus jensenii* sp.nov., a new representative of the subgenus *Thermobacterium*. *Journal of General Microbiology* **62,** 219–222.

Gibson, T. & Abd-el-Malek, Y. 1945 The formation of CO_2 by lactic acid bacteria and *Bacillus licheniformis* and a cultural method of detecting the process. *Journal of Dairy Research* **14,** 35–44.

Gunther, H. L. 1959 Mode of division of pediococci. *Nature, London* **183,** 903–904.

Gunther, H. L. & White, H. R. 1961 The cultural and physiological character of the pediococci. *Journal of General Microbiology* **26,** 185–197.

Hahn, G., Heeschen, W. & Tolle, A. 1970 Streptococcus: a study of structure, biochemistry, culture and classification. *Kieler Milchwirtshaftliche Forschungsberichte* **22,** 333–546.

Hardie, J. M. & Bowden, G. H. 1976*a* Physiological classification of oral viridans streptococci. *Journal of Dental Research* **55,** *Special issue A*, A166–A176.

Hardie, J. M. & Bowden, G. H. 1976*b* Some serological cross-reactions between *Streptococcus mutans*, *S. sanguis*, and other dental plaque streptococci. *Journal of Dental Research* **55,** *Special issue C*, C50–C58.

Hardie, J. M. & Marsh, P. D. 1978*a* Isolation media for oral streptococci. In *Streptococci*, eds Skinner, F. A. & Quesnel, L. B. London: Academic Press.

Hardie, J. M. & Marsh, P. D. 1978b Streptococci and the human oral flora. In *Streptococci*, eds Skinner, F. A. & Quesnel, L. B. London: Academic Press.

Harrigan, W. F. & McCance, M. E. 1976 *Laboratory Methods in Food and Dairy Microbiology*. London & New York: Academic Press.

Hontebeyrie, M. & Gasser, F. 1975 Comparative immunological relationships of two distinct sets of isofunctional dehydrogenases in the genus *Leuconostoc*. *International Journal of Systematic Bacteriology* **25,** 1–6.

Hugh, R. & Leifson, E. 1953 The taxonomic significance of fermentative versus oxidative metabolism of carbohydrates by various Gram-negative bacteria. *Journal of Bacteriology* **66,** 24–26.

JAYNE-WILLIAMS, D. J. 1975 Miniaturized methods for the characterization of bacterial isolates. *Journal of Applied Bacteriology* **38,** 305–309.

JAYNE-WILLIAMS, D. J. 1976 The application of miniaturized methods for the characterization of various organisms isolated from the animal gut. *Journal of Applied Bacteriology* **40,** 189–200.

JOHNSTON, M. A. & DELWICHE, E. A. 1965 Distribution and characteristics of the catalases of Lactobacillaceae. *Journal of Bacteriology* **90,** 347–351.

JONES, D. 1978 Composition and differentiation of the genus *Streptococcus.* In *Streptococci,* eds Skinner, F. A. & Quesnel, L. B. London: Academic Press.

JONES, D., SACKLIN, M. J. & SNEATH, P. H. A. 1972 A numerical taxonomic study of streptococci of serological group D. *Journal of General Microbiology* **72,** 1–12.

KANDLER, O. 1970 Amino acid sequence of the murein and taxonomy of the genera *Lactobacillus, Bifidobacterium, Leuconostoc* and *Pediococcus. International Journal of Systematic Bacteriology* **20,** 491–508.

KEDDIE, R. M. 1959 The properties and classification of lactobacilli isolated from grass and silage. *Journal of Applied Bacteriology* **22,** 403–416.

KNOX, K. W. & WICKEN, A. J. 1973 Immunological properties of teichoic acids *Bacteriological Reviews* **37,** 215–257.

KUNKEE, R. E. 1975 A second enzymatic activity for decomposition of malic acid by malo-lactic bacteria. In *Lactic Acid Bacteria in Beverages and Food,* eds Carr, J. G., Cutting, C. V. & Whiting, G. C. London, New York & San Francisco: Academic Press.

LANCEFIELD, R. C. 1933 A serological differentiation of human and other groups of haemolytic streptococci. *Journal of Experimental Medicine* **57,** 571–595.

LAW, B. A. & SHARPE, M. E. 1978 Streptococci in the dairy industry. In *Streptococci,* eds Skinner, F. A. & Quesnel, L. B. London: Academic Press.

LONDON, J. 1976 The ecology and taxonomic status of the lactobacilli. *Annual Reviews of Microbiology* **30,** 279–301.

MCDONALD, T. J. & MCDONALD, J. S. 1976 Streptococci isolated from bovine intramemmary infections. *American Journal of Veterinary Research* **37,** 377–381.

MUNDT, J. O. & GRAHAM, W. F. 1968 *Streptococcus faecium* var *casseliflavus. nov. var. Journal of Bacteriology* **95,** 2005–2009.

ORLA-JENSEN, S. 1919 *The Lactic Acid Bacteria.* Copenhagen: Andr. Fred Host & Son.

ORLA JENSEN, S. 1943 *The Lactic Acid Bacteria.* Copenhagen: Einar Munksgaard.

PARKER, M. T. 1978 The pattern of streptococcal disease in man. In *Streptococci,* eds Skinner, F. A. & Quesnel, L. B. London: Academic Press.

PARKER, M. T. & BALL, L. C. 1976 Streptococci and aerococci associated with systemic infection in man. *Journal of Medical Microbiology* **9,** 275–302.

ROGOSA, M. 1970 Characters used in the classification of lactobacilli. *International Journal of Systematic Bacteriology* **20,** 519–534.

ROGOSA, M. 1974 Lactobacillus. In Bergey's Manual of Determinative Bacteriology, 8th edn, eds Buchanan, R. E. & Gibbons, N. E. Baltimore: Williams & Wilkins.

ROGOSA, M., FRANKLIN, J. G. & PERRY, K. D. 1961 Correlation of the vitamin requirements with cultural and biochemical characters of *Lactobacillus* spp. *Journal of General Microbiology* **25,** 473–482.

ROGOSA, M., MITCHELL, J. A. & WISEMAN, R. F. 1951 A selective medium for the isolation of oral and faecal lactobacilli. *Journal of Bacteriology* **62,** 132–133.

ROGOSA, M. & SHARPE, M. E. 1959 An approach to the classification of the lactobacilli. *Journal of Applied Bacteriology* **22,** 329–340.

ROSS, P. W. 1978 The ecology of Group B streptococci. In *Streptococci*, eds Skinner, F. A. & Quesnel, L. B. London: Academic Press.

SHARPE, M. E. 1962 Taxonomy of the lactobacilli. *Dairy Science Abstracts* **24,** 109–118.

SHARPE, M. E. 1970 Cell wall and cell membrane antigens used in the classification of lactobacilli. *International Journal of Systematic Bacteriology* **20,** 509–518.

SHARPE, M. E. 1974 Recent aspects of taxonomy of the lactobacilli. *Il ruolo terapeutico e nutrizionale dei lattobacilli: Seminario Internazionale, Roma 1974* 10, 15. Fondazione Giovanni Lorenzini.

SHARPE, M. E. 1978 Isolation media for dairy streptococci. In *Streptococci*, eds Skinner, F. A. & Quesnel, L. B. London: Academic Press.

SHARPE, M. E. & DELLAGLIO, F. 1977 Deoxyribonucleic acid homology in anaerobic lactobacilli and in possible related species. *International Journal of Systematic Bacteriology* **27,** 19–21.

SHARPE, M. E., FRYER, T. F. & SMITH, D. G. 1966 Identification of the lactic acid bacteria. In Identification Methods for Microbiologists, Part A, eds Gibbs, B. M. & Skinner, F. A. London & New York: Academic Press.

SHARPE, M. E., GARVIE, E. I. & TILBURY, R. H. 1972 Some slime-forming heterofermentative species of the genus *Lactobacillus*. *Applied Microbiology* **23,** 389–399.

SHARPE, M. E., LATHAM, M. J., GARVIE, E. I., ZIRNGIBL, J. & KANDLER, O. 1973 Two new species of *Lactobacillus* isolated from the bovine rumen, *Lactobacillus ruminis* sp.nov. and *Lactobacillus vitulinus* sp.nov. *Journal of General Microbiology* **77,** 37–49.

SHERMAN, J. M. 1937 The Streptococci. *Bacteriological Reviews* **1,** 3–97.

SIMONDS, J., HANSEN, P. A. & LAKSHMANAN, S. 1971 Deoxyribonucleic acid hybridization among strains of lactobacilli. *Journal of Bacteriology* **107,** 382–384.

SKERMAN, V. B. D. 1959 A Guide to the Identification of the Genera of Bacteria. Baltimore: Williams & Wilkins.

SKINNER, F. A. & QUESNEL, L. B. (eds) 1978 *Streptococci*. Society for Applied Bacteriology Symposium Series No. 7. London: Academic Press.

SMITH, D. G. & SHATTOCK, P. M. F. 1962 The serological grouping of *Streptococcus equinus*. *Journal of General Microbiology* **29,** 731–736.

SWARTLING, P. F. 1951 Biochemical and serological properties of some citric acid fermenting streptococci from milk and dairy products. *Journal of Dairy Research* **18,** 256–267.

WHITTENBURY, R. 1964 Hydrogen peroxide formation and catalase activity in the lactic acid bacteria. *Journal of General Microbiology* **35,** 13–26.

WILKINSON, B. J. & JONES, D. 1977 A numerical Taxonomic survey of *Listeria* and related bacteria. *Journal of General Microbiology* **98,** 399–421.

WILSON, G. S. & MILES, A. A. 1975 *Principles of Bacteriology, Virology and Immunity*, 6th edn. London: Edward Arnold.

Identification Methods for *Nocardia, Actinomadura* and *Rhodococcus*

M. GOODFELLOW

Department of Microbiology, The Medical School, The University, Newcastle upon Tyne NE1 7RU, UK

AND

K. P. SCHAAL

Institute of Hygiene, University of Cologne, Cologne, FRG

The long and turbulent history of the genus *Nocardia* has been outlined by Lechevalier (1976). Recently, the application of modern taxonomic methods has underlined the heterogeneity of *Nocardia sensu* Waksman (1961) and led to radical changes in the classification of the taxon (Goodfellow & Minnikin 1977, 1978). In particular, the genera *Actinomadura*, *Oerskovia*, *Rhodococcus* and *Rothia* have been proposed for actinomycetes previously classified in the genus *Nocardia*. Acceptance of these proposals leave *Nocardia* as a relatively homogeneous taxon.

The data derived from recent investigations have not only contributed to improvements in the classification of nocardiae and related actinomycetes but have also highlighted properties of possible value in identification. Unfortunately, few of these diagnostic properties have been the subject of critical reproducibility studies (Wayne *et al.* 1976) and cannot, therefore, be recommended with complete confidence at present. However, most of the methods cited here for the identification of *Nocardia*, *Actinomadura* and *Rhodococcus* have been found effective in several laboratories.

Identification at Generic Level

Nocardia, *Actinomadura* and *Rhodococcus* can be distinguished from one another and from related taxa by a number of chemical, morphological and physiological properties (Table 1). The chemical tests provide the most reliable data and are particularly useful in distinguishing nocardiae, mycobacteria, rhodococci and corynebacteria which cannot always be separated using other taxonomic criteria. Both sophisticated and simple chemical methods have been developed for the classification and identification of actinomycetes and related bacteria (Minnikin & Goodfellow 1976; Minnikin *et al.* 1978*c*) but only the latter are of value in routine diagnostic laboratories and are considered here. However, most aerobic actinomycetes can be identified to the genus level by examining whole-organism hydrolysates for marker amino-acids, sugars and lipids.

Whole-organism hydrolysate analysis

Actinomycetes can be classified into nine broad groups based on the distribution of certain amino-acids and sugars found in major amounts in their walls (Lechevalier & Lechevalier 1970; Lechevalier 1976). In many cases the wall chemotype of strains can be obtained without the necessity of preparing pure wall fractions. Thus, wall chemotypes I to IV can be determined indirectly by the analysis of whole-organism hydrolysates for the isomer of diaminopimelic acid (DAP) and sugars present (Table 1). Several simple hydrolysis methods are available for the identification of aerobic actinomycetes (Lechevalier 1968; Berd 1973; Staneck & Roberts 1974).

Whole-organism methanolysate analysis

In recent years analyses of mycolic acids have been particularly useful in clarifying the relationships between the wall chemotype IV genera *Corynebacterium*, *Mycobacterium*, *Nocardia* and *Rhodococcus* (Minnikin & Goodfellow 1976; Minnikin *et al.* 1978*b*). It can be seen from Table 1 that the mycolic acids from nocardiae and rhodococci are intermediate in size between those of corynebacteria and mycobacteria (Minnikin *et al.* 1978*b*). This difference in overall mycolic acid size is the basis of several diagnostic methods developed to aid the separation of the mycolic acid containing taxa (Mordarska *et al.* 1972; Hecht & Causey 1976).

Thin-layer chromatography (t.l.c.) is the method of choice for the preliminary qualitative analysis of mycolic acid composition. A simple procedure (Minnikin *et al.* 1975), involving t.l.c. analysis of whole-organism methanolysates, allows the rapid analysis of mycolic acid composition. Methanolysates of most mycobacteria produce a multispot mycolate pattern in contrast to single spots produced by mycolates of corynebacteria, nocardiae and rhodococci. Nocardiae and corynebacteria can, however, be distinguished as they produce mycolic esters of different mobility (Goodfellow *et al.* 1976).

Another useful diagnostic method is based upon the characteristic breakdown of mycolic esters on pyrolysis to give straight-chain esters and long-chain aldehydes (Lechevalier *et al.* 1971; Lechevalier *et al.* 1973). On pyrolysis gas chromatography the mycolic esters of mycobacteria yield C_{22} to C_{26} fatty acid esters whereas corynebacteria, nocardiae and rhodococci have mycolates which produce C_{12} to C_{18} esters.

Analysis of isoprenoid quinones

Menaquinones are the characteristic isoprenoid quinone type found in aerobic actinomycetes, and the distribution of their various structural forms is proving valuable in classification (Minnikin *et al.* 1978*a*). Although menaquinone analyses have not yet been used for identification preliminary data suggest that they will be of value in helping to establish the identity of difficult isolates (Collins *et al.* 1977; Collins *et al.* 1979). In particular, menaquinone analyses distinguish *Nocardia*, which contain tetrahydromenaquinones with eight isoprene units as main component, abbreviated as MK-8(H_4), from *Corynebacterium*, *Mycobacterium* and *Rhodococcus*, and *Actinomadura dassonvillei* from *A. madurae* and *A. pelletieri* (Table 1).

Micromorphology

The growth and stability of the mycelium of fragmenting actinomycetes can be markedly affected by the consistency and composition of the growth medium, the incubation temperature and similar factors (Williams *et al.* 1976). The morphology of undisturbed growth can be recorded from glucose yeast extract (Gordon & Mihm 1962), Bennett's (Jones 1949) and Diagnostic Sensitivity Test (Oxoid, CM261) agars after 3, 7 or 14 days incubation at 30°C using an ordinary light microscope. Blocks from such cultures can be used for scanning electron microscopy and Gram-stained smears examined for the presence of stable mycelia, fragmenting hyphae, and rod and coccoid elements. However, cultures grown on slides and coverslips provide a more sensitive way of detecting fragmentation and distinguishing between substrate and aerial mycelium (see Williams & Cross 1971). These preparations can also be examined by scanning electron microscopy (Williams & Davies 1967).

Actinomycetes may be acid-fast, partially acid-fast or non-acid-fast. The degree of acid-fastness can be determined using the modified Kinyoun stain (Georg 1974).

TABLE 1. Differentiation of *Nocardia*, *Actinomadura*, *Rhodococcus* and related taxa using chemical, morphological and biochemical characters

Taxon	*Corynebacterium*	*Mycobacterium*	*Nocardia*
Chemical characters			
Whole-organism hydrolysate analysis:			
meso-diaminopimelic acid	+	+	+
LL-diaminopimelic acid	—	—	—
Arabinose	+	+	+
Galactose	+	+	+
Madurose	—	—	—
Wall chemotype	IV	IV	IV
Whole-organism sugar pattern[b]	A	A	A
Mycolic acids[c]	20–38	60–90	46–60
(no. of carbons)	(12–18)	(22–26)	(12–18)
Major isoprenoid quinones[d]	MK-*8* (H_2) MK-*9* (H_2)	MK-*9* (H_2)	MK-*8* (H_4)
Morphological characters			
Substrate mycelium formed	—	∓	+
Extensive fragmentation of substrate mycelium	—	+	+
Aerial hyphae formed	—	∓	±
Spore surface	—	—	smooth
Acid-fastness	negative	acid-fast	partially acid-fast
Physiological tests			
Arylsulphatase production	—	+	—
Casein degradation	—	—	—[f]

+: positive; —: negative; ±: usually positive; ∓: rarely positive.

[a] *Nocardiopsis* has been proposed for *Actinomadura dassonvillei* strains (Meyer 1976).

[b] A: arabinose plus galactose present; B: madurose present; C and NC: no characteristic sugar (Lechevalier 1976).

[c] Parentheses show number of carbons in fatty acids released on pyrolysis.

[d] MK-*n* (H_x), menaquinone with n isoprene units and x additional hydrogens.

[e] Unpublished data (Goodfellow *et al.*).

[f] *Nocardia brasiliensis* positive.

Rhodococcus	*Saccharopolyspora*	*Streptomyces*	*Actinomadura*	*Actinomadura dassonvillei*[a]
+	+	−	+	+
−	−	+	−	−
+	+	−	−	−
+	+	−	−	−
−	−	−	+	−
IV	IV	I	III	III
A	A	NC	B	C
34–66				
(12–18)	−	−	−	−
MK-*8* (H_4)	MK-*9* (H_4)[e]	MK-*9* (H_4, H_6, H_8)	MK-*9* (H_2, H_6, H_8)	MK-*10* (H_2, H_4, H_6)
MK-*8* (H_4)				
±	+	+	+	+
+	−	−	−	+
−	+	+	±	+
−	hairy	hairy, smooth spiny, warty	smooth, spiny warty	smooth
partially acid-fast	negative	negative	negative	negative
−	−	−	−	−
−	+	+	+	+

Physiological tests

Casein decomposition can be detected using the medium and method of Gordon (1967) and arylsulphatase production after 3 and 10 days following Vestal (1969).

Identification at Specific Level

Nocardia

Goodfellow & Minnikin (1977) defined *Nocardia* as 'aerobic, Gram positive, acid-fast to partially acid-fast actinomycetes that produce a primary mycelium that fragments into rod- and coccoid-like elements. Aerial hyphae are usually formed, strains contain mycolic acids and have a wall chemotype IV, and the G + C content of their DNA ranges from 64–69%.' This definition includes the established taxa shown in Table 2 but excludes 'Nocardia' strains which have a wall chemotype IV but lack mycolic acids (Goodfellow & Minnikin 1977). *Nocardia aerocolonigenes*, *N. autotrophica* and *N. orientalis* feature prominently amongst the latter (Gordon *et al.* 1978).

Nocardia brasiliensis and *N. otitidis-caviarum* (*N. caviae*) are good taxospecies but the other well-established taxon *N. asteroides* is markedly heterogeneous and has been divided into varying numbers of subgroups (Lechevalier 1976). Since few strains are common to more than one investigation it is difficult to know whether or not, and to what extent, the *N. asteroides* subgroups overlap and systematic studies are required. The revision of the taxon will probably have medical implications, for Schaal and his colleagues (Pulverer & Schaal 1978; Schaal & Reutersberg 1978) found two distinct subgroups, *N. asteroides* A and B, which were equivalent in rank to *N. brasiliensis* and *N. otitidis-caviarum* but contained strains with different pathogenic properties. Infections due to *N. asteroides* B are more frequent and have a higher degree of malignancy than those caused by *N. asteroides* A. The characteristics of *Nocardia* and related taxa are shown in Table 2. *Nocardia asteroides* subgroup A can be distinguished from *N. asteroides* subgroup B by its inability to grow on *iso*-amyl alcohol, 2,3 butylene glycol, 1,2 propylene glycol or rhamnose, and its ability to use gluconate as sole carbon source (Schaal 1977).

Actinomadura

The genus *Actinomadura* (Lechevalier & Lechevalier 1970) was proposed to accommodate *Nocardia* species with walls containing *meso*-DAP but lacking arabinose and galactose. Initially three species, *N. dassonvillei*, *N. madurae* and *N. pelletieri*, were included in the genus but recently many new species have been described mainly on the basis of morphology and wall chemotype (Lacey *et al.* 1978). The new genus *Nocardiopsis* (Meyer 1976) has been proposed for strains previously classified as *A. dassonvillei*.

In a recent numerical phenetic survey (Goodfellow *et al.* 1979*a*) strains of *A. pelletieri* formed a heterogeneous cluster, and *A. dassonvillei* and *A. madurae* were considered good taxospecies while the single representatives of *A. helvata*, *A. pusilla*, *A. roseoviolacea*, *A. spadix* and *A. verrucosospora* seemed to form new centres of variation. These species can be separated from one another and from the related taxon *Streptomyces somaliensis* by a number of morphological and physiological properties (Table 3).

The genus *Actinomadura* is clearly heterogeneous for *A. dassonvillei* and can be distinguished from *A. madurae* and *A. pelletieri* by data from chemical, serological and morphological studies (Lacey *et al.* 1978). There seem to be two possible ways of treating the genus as constituted by Lechevalier & Lechevalier (1970). The first is to retain the original concept and not to reclassify *A. dassonvillei* until further evidence is available. The second is to accept the transfer of *A. dassonvillei* to *Nocardiopsis* when there would be as much justification from numerical taxonomic analyses for erecting a new genus for *A. pelletieri* (Goodfellow *et al.* 1979*a*). Given the taxonomic history of *Actinomadura* strains and the lack of systematic studies between established and newly described species it is preferable to settle for the more cautious alternative at present.

Rhodococcus

The genus *Rhodococcus* (Goodfellow & Alderson 1977) was resurrected to accommodate a heterogeneous group of bacteria previously classified as *Gordona*, '*Mycobacterium*' *rhodochrous* and the '*rhodochrous*' complex. Goodfellow & Minnikin (1977) defined *Rhodococcus* as 'aerobic, non-sporing actinomycetes that are pleomorphic but often form a primary mycelium that soon fragments into rod and coccoid elements. A secondary mycelium is not produced but strains contain mycolic acids and have a wall chemotype IV, and the G + C content of their DNA ranges from 59–69%'.

Ten species of *Rhodococcus* are recognized and all of those studied to date form genetically homogeneous taxa in DNA:DNA reassociation analyses (Goodfellow *et al.* 1978*b*). Rhodococci can be distinguished by a number of biochemical, physiological and growth characters (Table 4)

TABLE 2. Characteristics of *Nocardia* spp. and related actinomycetes

Taxon	Mycolic acids present								Mycolic acids absent		
	N. amarae	*N. asteroides* subgroup A	subgroup B	*N. brasiliensis*	*N. carnea*	*N. otitidis-caviarum*	*N. transvalensis*	*N. vaccinii*	*N. aerocolonigenes*	*N. autotrophica*	*N. orientalis*
Morphological and staining characters:											
Acid-fastness	—	v	v	v	v	v	v	+	—	—	—
Substrate mycelium colour[a]	Cr	Y, O	Y, O, R	C, O, R	Cr, Pe	W, C	Cr, Pu	Cr, O, R	Cr, O	Y, B	Cr, Pe
Decomposition of:											
Adenine	—	—	—	—	—	—	v	—	—	+	—
Casein	—	—	—	+	—	—	—	—	+	—	+
Elastin	—	—	—	+	—	—	—	—	—	—	—
Hypoxanthine	—	—	—	+	—	+	+	—	+	+	+
Testosterone	—	+	+	+	+	—	—	—	+	ND	ND
Tyrosine	—	—	—	+	—	—	+	—	+	+	+
Xanthine	—	—	—	—	—	+	v	—	—	v	v

Resistance to:											
Lysozyme	—	+	+	+	+	+	+	+	+	—	—
Rifampin	v	+	+	+	+	+	+	v	—	—	v
Urease production	+	+	+	+	—	+	+	+	+	+	+
Acid from:[b]											
Adonitol	—	ND	ND	—	—	—	+	—	v	+	+
Cellobiose	—	ND	ND	—	—	—	—	—	+	v	+
meso-Erythritol	—	ND	ND	—	—	—	+	—	—	+	+
Lactose	—	ND	ND	—	—	—	—	—	+	—	+
Maltose	+	ND	ND	—	—	—	—	v	+	+	v
Melezitose	—	ND	ND	—	—	—	—	—	—	+	v
α-Methyl-D-glucoside	—	ND	ND	—	—	—	—	—	—	v	+

ND: not determined.
[a] B: brown; C: colourless; Cr: cream; O: orange; Pe: peach; Pu: purple; R: red; W: white; Y: yellow.
[b] *Nocardia asteroides* does not produce acid from these sugars (Gordon *et al.* 1978).

TABLE 3. Characteristics of *Actinomadura* spp. and related actinomycetes

Taxon	*A. dassonvillei*	*A. helvata*	*A. madurae*	*A. pelletieri*	*A. pusilla*	*A. roseoviolacea*	*A. spadix*	*A. verrucosospora*	*Streptomyces somaliensis*
Cultural and morphological characters									
Substrate mycelium colour[a]	Y, O, B RB, Gy	Y, B	W, Y, R	R	Y, B	Y, O, R, B, Bk	Gy, B	Y, O Pk, B	Cr, B
Aerial mycelium colour[a]	W, Gy, BlGy, Bl, BlG	W	W	—	W, Pk	Gy, Y, Pk	Gy	W, Gy, Y, Pk, Bl	W, Y, BlBk
Soluble pigment[a]	Y, B, Gy, Pu	None	Y, B, Pu	R, B	B	Pu, R, B	R, Gy, B	Y	Bl, Y
Spore chain morphology	straight-zigzag, long	hooks, pseudo-sporangia, short	hooks, spirals, short	hooks, spirals, short (rare)	tight spirals, pseudo-sporangia, short	hooks, spirals, pseudo-sporangia, short	pseudo-sporangia	hooks, short	flexuous

Spore surface	smooth	smooth	smooth	smooth	smooth	smooth	smooth	warty	smooth
Decomposition of:									
Adenine	+	—	—	—	—	—	—	—	—
Aesculin	+	—	+	—	+	—	—	+	—
Elastin	+	+	+	+	—	—	—	+	—
Guanine	+	—	—	—	—	—	—	—	—
Hypoxanthine	+	—	+	+	+	+	—	+	—
RNA	+	+	+	—	+	+	+	+	+
Testosterone	—	+	—	—	—	—	—	+	v
Tyrosine	+	—	+	+	+	+	—	+	+
Xanthine	+	—	—	—	—	—	—	—	—
Sensitivity to:									
Pencillin (10 i.u.)	—	+	—	+	—	—	—	—	+

[a] Bk: black; Bl: blue; G: green; Gy: grey; Pk: pink.

but are in need of further systematic study to highlight additional characters for identification.

Tests for distinguishing Nocardia, Actinomadura *and* Rhodococcus *species*

Incubation conditions

Tests should be read after 3, 7 and 14 days incubation at 30°C unless otherwise stated.

Degradation tests

The decomposition of adenine, tyrosine (0·5%, w/v), elastin (0·3%, w/v), guanine (0·1%, w/v), hypoxanthine, xanthine (0·4%, w/v) and testosterone (0·1%, w/v) are detected in the basal medium of Gordon *et al.* (1974). Plates should be heavily inoculated and observed for the disappearance of the insoluble compounds for up to 21 days. Aesculin decomposition is detected after Sneath (1966), and ribonucleases in a basal medium (tryptone, 20 g; sodium chloride, 5 g; agar, 15 g; distilled water, 1 l; pH 7·3) supplemented with RNA (0·3%, w/v) after 7 days using the method of Jeffries *et al.* (1957).

Acid production from carbohydrates

The medium of Gordon *et al.* (1974) is used. Cultures on slants of the carbohydrate agars are observed for acid colour of the indicator after 7 and 28 days incubation at 28°C.

Biochemical tests

Urease production is detected after 28 days incubation at 28°C using the medium of Gordon *et al.* (1974).

Sole carbon sources

The media and methods described by Goodfellow (1971) are used.

Tolerance to chemical inhibitors and temperature

Growth at 10°C, and in the presence of crystal violet and sodium azide, are examined in glucose yeast extract agar (GYEA) (Goodfellow 1971). Sensitivity to penicillin discs (10 i.u., Oxoid) is detected after 2 to 7 days incubation on GYEA plates flooded with broth suspensions (0·1 ml). Resistance to lysozyme is observed after 4 weeks incubation at 28°C using the media and methods of Gordon & Barnett (1977). The rifampin test described by these authors is read after 2 weeks incubation at 28°C and should be used for taxonomic purposes only.

TABLE 4. Characteristics of *Rhodococcus* spp.

Taxon	*R. bronchialis*	*R. coprophilus*	*R. corallinus*	*R. equi*	*R. erythropolis*	*R. rhodnii*	*R. rhodochrous*	*R. ruber*	*R. rubropertinctus*	*R. terrae*
Decomposition of:										
Adenine	—	—	—	+	+	—	v	v	—	—
Tyrosine	—	—	—	—	v	+	+	+	—	—
Enzymic activity:										
α-esterase	—	ND	—	ND	ND	ND	—	—	ND	+
β-esterase	+	ND	+	ND	ND	ND	+	—	ND	+
Sole carbon sources (1 %, w/v):										
Glycerol	+	—	+	+	+	v	+	+	+	+
Inositol	+	—	—	—	v	—	—	—	—	—
Maltose	—	—	—	—	+	—	+	v	+	—
Rhamnose	—	—	—	—	—	—	—	—	—	+
Sorbitol	—	—	+	—	+	+	+	+	v	+
Trehalose	+	—	+	—	+	v	+	+	+	+
m-Hydroxybenzoic acid (0·1 %, w/v)	—	+	—	—	—	—	+	+	—	—
Growth at 10°C	—	v	—	+	+	—	+	v	+	—
Predominant menaquinone	MK-*9* (H_2)	MK-*8* (H_2)[a]	MK-*9* (H_2)	MK-*8* (H_2)	MK-*8* (H_2)	MK-*8* (H_2)[a]	MK-*8* (H_2)	MK-*8* (H_2)	ND	MK-*9* (H_2)
Growth in the presence of:										
Crystal violet (0·0001 %, w/v)	+	+	+	+	—	v	v	+	v	+
Sodium azide (0·02 %, w/v)	+	v	+	v	v	v	—	+	v	+

ND: not determined.
[a] Unpublished data.

One of us (K.P.S.) gratefully acknowledges receipt of a Travelling Scholarship awarded by the Robert Koch Stiftung.

References

BERD, D. 1973 Laboratory identification of clinically important aerobic actinomycetes. *Applied Microbiology* **25,** 665–681.

COLLINS, M. D., GOODFELLOW, M. & MINNIKIN, D. E. 1979 Isoprenoid quinones in the classification of coryneform and related bacteria. *Journal of General Microbiology* **110,** 127–136.

COLLINS, M. D., PIROUZ, T., GOODFELLOW, M. & MINNIKIN, D. E. 1977 Distribution of menaquinones in actinomycetes and corynebacteria. *Journal of General Microbiology* **100,** 221-230.

GEORG, L. K. 1974 *Nocardia* species as opportunists and current methods for their identification. In *Opportunistic Pathogens*, eds Prier, J. E. , & Friedman, H. Baltimore: University Press.

GOODFELLOW, M. 1971 Numerical taxonomy of some nocardioform bacteria. *Journal of General Microbiology* **69,** 33–80.

GOODFELLOW, M. & ALDERSON, G. 1977 The actinomycete-genus *Rhodococcus*: a home for the '*rhodochrous*' complex. *Journal of General Microbiology* **100,** 99–122.

GOODFELLOW, M. & MINNIKIN, D. E. 1977 Nocardioform bacteria. *Annual Review of Microbiology* **31,** 159–180.

GOODFELLOW, M. & MINNIKIN, D. E. 1978 Numerical and chemical methods in the classification of *Nocardia* and related taxa. In *Nocardia and Streptomyces*, eds Mordarski, M., Kuryłowicz, W., Jeljaszewicz, J. Stuttgart & New York: Gustav Fischer Verlag.

GOODFELLOW, M., ALDERSON, G. & LACEY, J. 1979 Numerical taxonomy of *Actinomadura* and related actinomycetes. *Journal of General Microbiology* **112,** 95–111.

GOODFELLOW, M., COLLINS, M. D. & MINNIKIN, D. E. 1976 Thin-layer chromatographic analysis of mycolic acid and other long-chain components in whole-organism methanolysates of coryneform and related taxa. *Journal of General Microbiology* **96,** 351–358.

GOODFELLOW, M., MORDARSKI, M., SZYBA, K. & PULVERER, G. 1978*b* Relationships among rhodococci based upon deoxyribonucleic acid reassociation. In *Genetics of the Actinomycetales*, eds Freerksen, E., Tárnok, I. & Thumin, J. H. Stuttgart & New York: Gustav Fischer Verlag.

GORDON, R. E. 1967 The taxonomy of soil bacteria. In *The Ecology of Soil Bacteria*, eds Gray, T. R. G. & Parkinson, D. Liverpool: Liverpool University Press.

GORDON, R. E. & BARNETT, D. A. 1977 Resistance to rifampin and lysozyme of strains of some species of *Mycobacterium* and *Nocardia* as a taxonomic tool. *International Journal of Systematic Bacteriology* **27,** 176–178.

GORDON R. E. & MIHM, J. H. 1962 Identification of *Nocardia caviae* (Erikson) nov. comb. *Annals of the New York Academy of Sciences* **98,** 628–636.

GORDON, R. E., MISHRA, S. K. & BARNETT, D. A. 1978 Some bits and pieces of the genus *Nocardia*: *N. carnea, N. vaccinii, N. transvalensis, N. orientalis* and *N. aerocolonigenes*. *Journal of General Microbiology* **109,** 69–78.

GORDON, R. E., BARNETT, D. A., HANDERHAN, J. E. & PANG, C. H.-N.

1974 *Nocardia coeliaca, Nocardia autotrophica,* and the nocardin strain. *International Journal of Systematic Bacteriology* **24,** 54–63.

HECHT, S. T. & CAUSEY, W. A. 1976 Rapid method for the detection and identification of mycolic acids in aerobic actinomycetes and related bacteria. *Journal of Clinical Microbiology* **4,** 284–287.

JEFFRIES, C. D., HOLTMAN, D. F. & GUSE, D. G. 1957 Rapid method for determining the activity of micro-organisms on nucleic acids. *Journal of Bacteriology* **73,** 590–591.

JONES, K. L. 1949 Fresh isolates of actinomycetes in which the presence of sporogenous aerial mycelia is a fluctuating characteristic. *Journal of Bacteriology* **57,** 141–145.

LACEY, J., GOODFELLOW, M. & ALDERSON, G. 1978 The genus *Actinomadura* Lechevalier and Lechevalier. In *Nocardia and Streptomyces,* eds Mordarski, M., Kuryłowicz, W. & Jeljaszewicz, J. Stuttgart & New York: Gustav Fischer Verlag.

LECHEVALIER, H. A. & LECHEVALIER, M. P. 1970 A critical evaluation of the genera of aerobic actinomycetes. In *The Actinomycetales,* ed. Prauser, H. Jena: Gustav Fischer Verlag.

LECHEVALIER, M. P. 1968 Identification of aerobic actinomycetes of clinical importance. *Journal of Laboratory and Clinical Medicine* **71,** 934–944.

LECHEVALIER, M. P. 1976 The taxonomy of the genus *Nocardia*: some light at the end of the tunnel? In *The Biology of the Nocardiae,* eds Goodfellow, M., Brownell, G. H. & Serrano, J. A. London & New York: Academic Press.

LECHEVALIER, M. P. & LECHEVALIER, H. 1970 Chemical composition as a criterion in the classification of aerobic actinomycetes. *International Journal of Systematic Bacteriology* **20,** 435–444.

LECHEVALIER, M. P., HORAN, A. C. & LECHEVALIER, H. 1971 Lipid composition in the classification of nocardiae and mycobacteria. *Journal of Bacteriology* **105,** 313–318.

LECHEVALIER, M. P., LECHEVALIER, H. & HORAN, A. C. 1973 Chemical characteristics and classification of nocardiae. *Canadian Journal of Microbiology* **19,** 965–972.

MEYER, J. 1976 *Nocardiopsis,* a new genus of the order Actinomycetales. International *Journal of Systematic Bacteriology* **26,** 487–493.

MINNIKIN, D. E. & GOODFELLOW, M. 1976 Lipid composition in the classification and identification of nocardiae and related taxa. In *The Biology of the Nocardiae,* eds Goodfellow, M., Brownell, G. H. & Serrano, J. A. London & New York: Academic Press.

MINNIKIN, D. E., ALSHAMAONY, L. & GOODFELLOW, M. 1975 Differentiation of *Mycobacterium, Nocardia,* and related taxa by thin-layer chromatographic analysis of whole-organism methanolysates. *Journal of General Microbiology* **88,** 200–204.

MINNIKIN, D. E., COLLINS, M. D. & GOODFELLOW, M. 1978*a* Menaquinone patterns in the classification of nocardioform and related bacteria. In *Nocardia* and *Streptomyces,* eds Mordarski, M., Kuryłowicz, W. & Jeljaszewicz, J. Stuttgart & New York: Gustav Fischer Verlag.

MINNIKIN, D. E., GOODFELLOW, M. & ALSHAMAONY, L. 1978*b* Mycolic acids in the classification of nocardioform bacteria. In *Nocardia* and *Streptomyces,* eds Mordarski, M., Kuryłowicz, W. & Jeljaszewicz, J. Stuttgart & New York: Gustav Fischer Verlag.

MINNIKIN, D. E., GOODFELLOW, M. & COLLINS, M. D. 1978c Lipid composition in the classification and identification of coryneform and related taxa. In *Coryneform Bacteria*, eds Bousfield, I. J. & Callely, A. G. London & New York: Academic Press.

MORDARSKA, H., MORDARSKI, M. & GOODFELLOW, M. 1972 Chemotaxonomic characters and classification of some nocardioform bacteria. *Journal of General Microbiology* **71,** 77–86.

PULVERER, G. & SCHAAL, K. P. 1978 Pathogenicity and medical importance of aerobic and anaerobic actinomycetes. In *Nocardia* and *Streptomyces*, eds Mordarski, M., Kuryłowicz, W. & Jeljaszewicz, J. Stuttgart & New York: Gustav Fischer Verlag.

SCHAAL, K. P. 1977 *Nocardia*, *Actinomadura* and *Streptomyces*. In *CRC Handbook Series in Clinical Laboratory Sciences, Section E, Clinical Microbiology*, Vol 1, ed. von Graevenitz, A. Cleveland: CRC Press.

SCHAAL, K. P. & REUTERSBERG, H. 1978 Numerical taxonomy of *Nocardia asteroides*. In *Nocardia* and *Streptomyces*, eds Mordarski, M., Kuryłowicz, W. & Jeljaszewicz, J. Stuttgart & New York: Gustav Fischer Verlag.

SNEATH, P. H. A. 1966 Identification methods applied to *Chromobacterium*. In *Identification Methods for Microbiologists, Part A*, eds Gibbs, B. M. & Skinner, F. A. London & New York: Academic Press.

STANECK, J. L. & ROBERTS, G. D. 1974 Simplified approach to identification of aerobic actinomycetes by thin-layer chromatography. *Applied Microbiology* **28,** 226–231.

VESTAL, A. L. 1969 *Procedures for the Isolation and Identification of Mycobacteria*. U.S. Department of Health, Education and Welfare, Public Health Publication. No. 1995.

WAKSMAN, S. A. 1961 *The Actinomycetes*, Vol. II. Baltimore: Williams & Wilkins.

WAYNE, L. G. & 18 colleagues. 1976 Highly reproducible techniques for use in systematic bacteriology in the genus *Mycobacterium*: tests for niacin and catalase and for resistance to isoniazid, thiophene-2-carboxylic acid hydrazide, hydroxylamine and p-nitro-benzoate. *International Journal of Systematic Bacteriology* **26,** 311–318.

WILLIAMS, S. T. & CROSS, T. 1971 Isolation, purification, cultivation and preservation of actinomycetes. In *Methods in Microbiology*, vol. 4, ed. Booth, C. London & New York: Academic Press.

WILLIAMS, S. T. & DAVIES, F. L. 1967 Use of a scanning electron microscope for the examination of actinomycetes. *Journal of General Microbiology* **48,** 171–177.

WILLIAMS, S. T., SHARPLES, G. P., SERRANO, J. A., SERRANO, A. A. & LACEY, J. 1976 The micromorphology and fine structure of nocardioform organisms. In *The Biology of the Nocardiae*, eds Goodfellow, M., Brownell, G. H. & Serrano, J. A. London & New York: Academic Press.

The Estimation of Base Compositions, Base Pairing and Genome Sizes of Bacterial Deoxyribonucleic Acids

R. J. Owen and L. R. Hill

National Collection of Type Cultures, Central Public Health Laboratory, Colindale Avenue, London NW9 5HT, UK

Characteristics of deoxyribonucleic acids, such as overall base content (% G + C) and nucleotide sequence similarity (base pairing), are widely used in microbial taxonomy. As the experimental techniques are relatively specialized and, for some, require high capital outlay in equipment, they are not readily applicable to the identification of fresh isolates in a routine diagnostic laboratory. However, the techniques have been developed to an extent where they should now be within the scope of a reference laboratory which deals with the identification of more difficult strains (Brenner 1976). In this paper, some of the techniques used to determine DNA characteristics are outlined and experimental details are given of those methods that we have used for taxonomic studies and for identification of isolates in our laboratory.

DNA Extraction and Purification

Cultivation of bacteria

The bacteria are grown in liquid medium (200 ml medium in a 1 l flat-bottomed flask) or on solid medium (150 ml in a Roux bottle). The growth medium and incubation conditions should be selected to give efficient rates of growth and a high yield of cells. The advantages of a liquid medium are: ease of aeration (an orbital incubator is very effective), radioactive labels (when needed) can be added simply as growth medium supplements in the early part of the logarithmic phase of growth and it is generally easier to harvest cells, which is important when handling pathogenic organisms. The yield of bacterial cells will depend on the bacterium but as a guide quantities in the order of 3–5 g wet weight of packed cells are generally needed. The bacterial cells are collected by centrifugation (10 000 rev/min for 20 min), washed and re-suspended in 0·15 mol l^{-1} NaCl and 0·1 mol l^{-1} EDTA buffer at pH 8·0, and a density of 2–3 g wet weight of cells/25 ml of buffer.

Lysis of cells

Gram negative and some Gram positive bacteria are readily lysed by treating the cell suspension with 2% SLS (sodium lauryl sulphate) solution for 15 min at 60°C. Lysis is observed by a clearing and increased viscosity of the suspension. A more effective method to lyse cells of Gram positive bacteria involves the combined action of lysozyme (10 mg/25 ml bacterial cells for 1 h at 37°C) followed by treatment with 2% SLS. Some bacteria (for example, coryneforms and lactobacilli) are resistant to the above methods of lysis and call for other techniques. The bacteria can be rendered more susceptible to lysis by cultivating in a medium that contains penicillin (Silvestri & Hill 1965), or glycine (Yamada & Komagata 1970), by lyophilization of harvested cells (De Ley 1971), or by resuspending cells in a modified buffer system (Crombach 1972; Garvie 1976). Alternatively, more drastic methods may have to be adopted to break up the cells: cycles of freezing and thawing, enzyme digestion, sonication, mechanical disruption by passing through a pressure cell or shaking with glass beads. Where possible these physical disruption methods should be avoided because they are likely to shear the DNA molecules too, thus causing problems in purification as it is difficult to obtain fibrous precipitates if the DNA has a low molecular weight.

Purification of DNA

Without radioactive label

The most commonly used method to prepare high molecular weight DNA from bacterial cells is the procedure of Marmur (1961). The main steps to purify DNA are illustrated in Fig. 1 although there are numerous modifications to this basic procedure such as those used by Brenner *et al.* (1969) and De Ley *et al.* (1970). To ensure that DNA is of sufficiently high purity for DNA-DNA binding experiments, the Marmur procedure can be supplemented with the following purification stages: additional treatment with pancreatic ribonuclease (50 μg ml^{-1} for 1 h at 37°C), incubation with pronase (self-digested: 50 μg ml^{-1} for 2 h at 37°C), and deproteinization with aqueous phenol. For phenol treatment, the DNA is dissolved in 0·1 mol l^{-1} NaCl + 0·1 mol l^{-1} EDTA + 0·05 mol l^{-1} Tris buffer. An equal volume of phenol saturated with

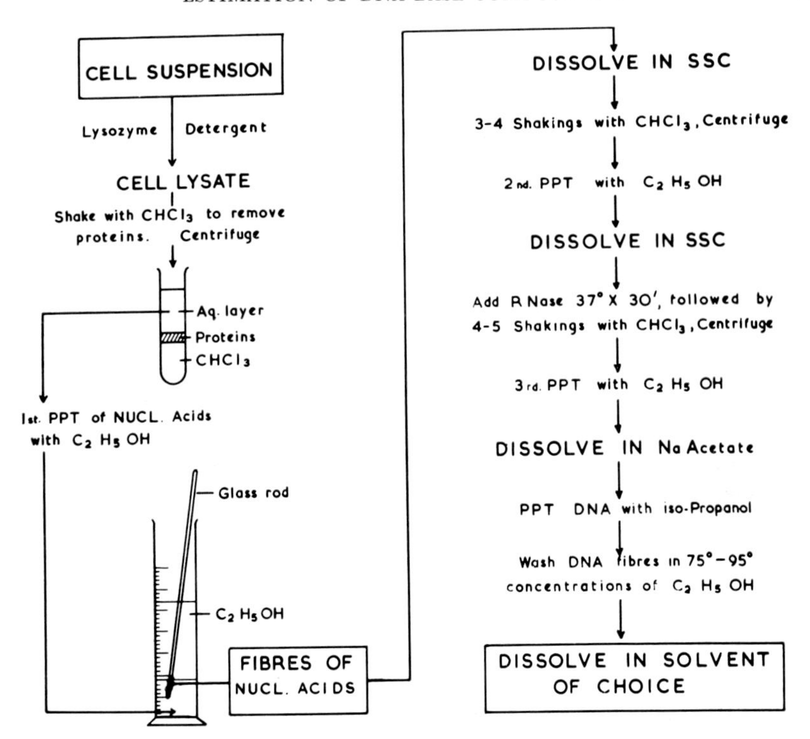

FIG. 1. DNA preparation. SSC: 0·15 mol l^{-1} NaCl + 0·015 mol l^{-1} Na_3 citrate; PPT: Precipitation. Notes: (1) 2–3 g wet weight of logarithmic phase cells required. Yield of pure DNA very variable; 1–4 mg is a good yield. (2) 6–8 h from cell lysate to pure DNA fibres. (3) Reference: Marmur J. (1961).

0·05 mol l^{-1} Tris is then mixed with the DNA solution and shaken for 5 min. The phenol should include 8-hydroxyquinoline to eliminate the accumulation of harmful peroxides. Extreme care should be taken when handling phenol, and in ensuring that all the phenol is eventually removed from the DNA phase by repeated chloroform treatments and precipitations with 95% ethanol. The DNA is finally precipitated from standard saline citrate (SSC) buffer with *iso*-propanol in the presence of 3 mol l^{-1} sodium acetate + 0·001 mol l^{-1} EDTA, or with 2-ethoxyethanol. To remove traces of protein and RNA impurities, De Ley & De Smedt (1975) recommended a step involving CsCl density ultracentrifugation of the purified DNA in a fixed angle rotor. Hydroxyapatite (HA) chromatography is also applicable to DNA purification (Britten *et al.* 1969).

With radioactive label

Similar procedures are used for the purification of labelled DNA (De Ley 1970) except that rigorous safety precautions should be observed in handling materials, in disposing of waste, and in washing glassware. The two isotopes commonly used to label DNA specifically are ^{14}C and ^{3}H in the forms [2- ^{14}C]-thymidine and [methyl-^{3}H]-thymidine. Incorporation of thymidine may be poor due to the induction of thymidine-destroying enzymes, in which case it may be necessary to use [^{32}P]-orthophosphate or nucleic acid precursors such as [2-^{14}C]-uracil or [8-^{14}C]-adenine. The isotopes are added as growth medium supplements in the early logarithmic phase of growth. We have found the method outlined below convenient for preparing small quantities of labelled DNA with high specific activities. The bacteria are grown in a medium containing: M9 salts solution (prepared at tenfold concentration: Na_2HPO_4, 60 g; KH_2PO_4, 30 g; NaCl, 5 g; NH_4Cl, 10 g; distilled water, 1 l), 10 ml; 0·1 mol l^{-1} $MgSO_4$, 1 ml; 0·01 mol l^{-1} $CaCl_2$, 1 ml; Tryptone (Difco), 1 g, 40% (w/v) maltose, 2·5 ml; thiamine (sterilized separately), 200 μg; distilled water, 85 ml. The strains are grown overnight at 37°C in 5 ml medium, subcultured into 10 ml of fresh medium, and then incubated at 37°C with aeration. Deoxyadenosine (2·5 mg) and [methyl-^{3}H] thymidine (50 μCi) of specific activity 215 mCi/mg are added early in the exponential phase of growth, and again after 30 min and 1 h. The DNA is purified by CsCl density gradient centrifugation at 90 000 g (18°C, 60 h) in a fixed angle rotor. At the end of the run, each tube is pierced and fractions are collected. The fractions that contain DNA, identified by the ^{3}H-label, are pooled and dialysed for 1 h at 4°C against distilled water, then overnight at 4°C against 0·14 mol l^{-1} sodium phosphate buffer. Each 10 ml of lysate yielded *ca.* 30 μg of highly pure, labelled DNA with a specific activity of *ca.* $4{\cdot}0 \times 10^4$ counts min^{-1} μg^{-1}.

Storage of DNA

Solutions of high mol. wt DNA (1 mg ml^{-1} or more) in SSC or 0·1 × SSC buffer may be stored for several months at 4°C with a drop of chloroform as an antifungal agent. The DNA should be re-precipitated and dissolved again in 0·1 × SSC before use if stored for periods exceeding 1 month. However if the DNA fails to precipitate, it should be discarded as degradation may have occurred during storage. The DNA can be stored for longer periods of up to 1 year after quick freezing at − 70°C and storage at −20°C (Crombach 1973). The DNA can also be stored at 4°C as precipitated fibres on glass rods in 95% alcohol although the efficacy of this method of storage over long periods is unknown.

Base Composition Estimation

Base composition of DNA is conveniently expressed as the molar percentage of guanine and cytosine to total bases (% G + C). The various methods that have been developed to estimate % G + C are listed in Table 1, and the ranges of common bacterial genera are listed in Table 2. The methods most commonly used in taxonomic studies are measurement of the denaturation temperature and measurement of buoyant density in CsCl density gradient. These physico-chemical characteristics of DNA are related empirically to % G + C by formulae based on the direct chemical determination of base composition. The latter method is not suitable for routine use and has been largely superseded by the physico-chemical methods.

Melting temperature determination

When DNA in aqueous solution is heated, the complementary strands separate so that the double-stranded helix (native configuration) changes to single-stranded DNA (denatured configuration). This is accompanied by an increase in the absorbance at 260 nm of the nucleotide bases (the so-called hyperchromic effect). The thermal denaturation (or melting) temperature (T*m*), which is defined as the temperature that corresponds to 50% of the hyperchromicity, can be determined by monitoring the increase in absorbance as a function of temperature. In the thermal denaturation technique (Marmur & Doty 1962), the DNA is dissolved in a saline-citrate buffer at concentrations in the range 20 to 50 μg ml^{-1}. Essential equipment for the T*m* determination comprises: 1 cm light-path silica cells, which are stoppered to prevent evaporation; a cell holder the temperature of which can be thermostatically controlled; and a spectrophotometer that is preferably equipped with a chart recorder. In our laboratory, a thermister thermometer is used to measure the temperature directly in the DNA solution during an experiment. This thermometer measures temperatures in the 70–100°C range with an accuracy of $\pm$ 0·1°C, and is calibrated against a standard reference thermometer. The DNA solutions are heated continuously by means of a temperature programme controller adjusted to give a rate of increase that does not exceed 0·25°C min^{-1}. The T*m* is then calculated from the resulting melting curve (see Fig. 2). The data may alternatively be expressed as normal probability plots or as differential melting curves.

TABLE 1. Methods used to estimate bacterial DNA base compositions

Description of method	Characteristic determined
Basic methods	
(1) Hydrolysis and chromatographic analysis (reviewed by Bendich 1957)	Molar concentration of bases
(2) Thermal denaturation[a] (Marmur & Doty 1962)	Melting temperature (T*m*)
(3) Analytical ultra-centrifugation in a neutral caesium chloride density gradient (Schildkraut *et al*. 1962)	Buoyant density (ρ)
Other methods	
(4) Gas-liquid chromatography (Mitruka 1975)	Molar concentration of bases
(5) Analysis of native DNA spectrum[a] (Hirshman & Felsenfeld 1966; Ulitzur 1972)	Direct measurement of u.v. absorbance
(6) Analysis of denatured DNA spectrum (Hirshman & Felsenfeld 1966)	Direct measurement of u.v. absorbance
(7) Reactivity to bromination (Wang & Hashagen 1964)	Direct measurement of u.v. absorbance
(8) Acid denaturation and spectral analysis (Frédericq *et al*. 1961)	Direct measurement of u.v. absorbance
(9) Depurination in dilute acid and spectral analysis (Huang & Rosenberg 1966)	Direct measurement of u.v. absorbance
(10) Doublet frequency analysis (Russell *et al*. 1976)	Molar concentration of bases

[a] Method can be used in conjunction with simplified methods of DNA isolation and purification for routine % G + C estimation (from thermal denaturation, Owen & Lapage 1976; from ultraviolet spectroscopy, Meyer & Schleifer 1975).

TABLE 2. Base composition ranges of common bacterial genera[a]

(a) Gram negative genera	G + C (mol %)		G + C (mol %)
Fusobacterium	26–34	*Erwinia*	50–58
Flavobacterium	30–42	*Klebsiella*	52–56
Pasteurella	36–43	*Enterobacter*	52–59
Haemophilus	38–42	*Hafnia*	52–57
Proteus	38–42	*Serratia*	53–59
Vibrio	40–50	*Azomonas*	53–59
Actinobacillus	40–42	*Beijerinckia*	54–61
Bacteroides	40–45	*Acetobacter*	55–64
Branhamella	40–50	*Brucella*	56–58
Moraxella	40–46	*Alcaligenes*	57–70
Acinetobacter	40–47	*Aeromonas*	57–63
Veillonella	40–44	*Pseudomonas*	58–70
Yersinia	45–47	*Rhizobium*	59–66
Zymomonas	47–48	*Agrobacterium*	59–63
Neisseria	47–52	*Gluconobacter*	60–64
Nitrosomonas	47–51	*Nitrobacter*	60–62
Escherichia	50–51	*Azotobacter*	63–66
Nitrosococcus	50–51	*Xanthomonas*	63–69
Thiobacillus	50–68	*Chromobacterium*	63–72
Salmonella	50–53		
(b) Gram positive genera			
Clostridium	23–43	*Planococcus*	48–52
Sarcina	28–31	*Corynebacterium*	52–68
Staphylococcus	30–40	*Bifidobacterium*	57–65
Bacillus	32–62	*Propionibacterium*	59–66
Streptococcus	33–42	*Arthrobacter*	60–72
Lactobacillus	34–54	*Nocardia*	60–72
Pediococcus	34–44	*Mycobacterium*	62–70
Aerococcus	37–41	*Micrococcus*	66–75
Listeria	38–56	*Streptomyces*	69–73
(c) Some other miscellaneous genera			
Mycoplasma	23–40		
Spiroplasma	25–26		
Acholeplasma	30–33		
Rickettsia	30–33		
Flexibacter	31–43		
Cytophaga	33–42		
Leptospira	36–39		
Spirillum	38–65		
Myxococcus	63–71		

[a] Data from *Bergey's Manual of Determinative Bacteriology*, 8th edn (Buchanan & Gibbons 1974). Values for individual species are also listed in this edition; also see Normore (1973)

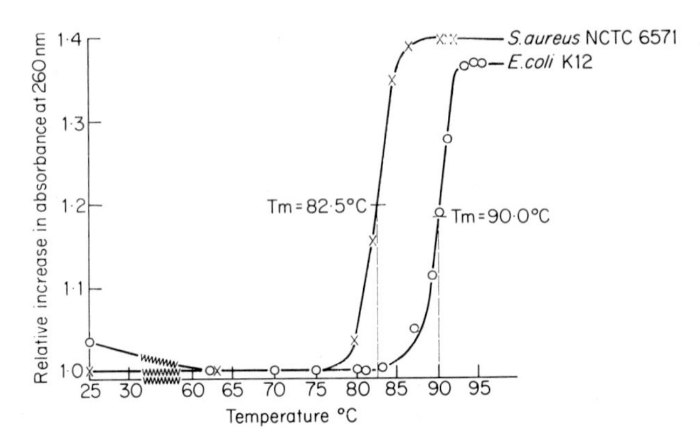

FIG. 2. Two DNA heat-denaturation curves. Temperature of DNA samples in the spectrophotometer is raised slowly, absorbance recorded either periodically or, with an X-Y recorder, continuously: as DNA denatures (melts), the absorbance increases. T*m* (melting temperature): mid-point of increase of absorbance. Solvent: SSC. u.v. light wavelength: 260 nm.

Calculation of % G + C

Under the experimental conditions described above the denaturation temperature of DNA is mainly dependent on two factors: the base composition of the DNA, and the sodium ion concentration of the buffer. At constant ionic strength, there is a linear relationship between T*m* and % G + C (Marmur & Doty 1962). The gradient of this relationships decrease slightly as the ionic strength is lowered because there is also a linear relationship between T*m* and the logarithm of the sodium molarity (Schildkraut & Lifson 1965; Owen *et al.* 1969). The two saline-citrate buffers commonly used in T*m* determinations for taxonomic purposes are standard saline citrate (SSC; 0·15 mol l^{-1} NaCl + 0·015 mol l^{-1} tri-sodium citrate, pH 7·0) and a one-tenth dilution of this buffer (0·1 × SSC). Base compositions are calculated from the melting temperatures according to the following equations:

In SSC, % G + C = 2·44 T*m* — 169·3 (De Ley 1970)
In 0·1 × SSC, % G + C = 2·08 T*m* — 106·4 (Owen & Lapage 1976)

It is advisable to include in a series of determinations a reference DNA for which the % G + C has been chemically determined. The DNA of *Escherichia coli* strain B (50·9% G + C) or *E. coli* strain K12 (51·2% G + C) are suitable for this purpose. The % G + C of the unknown

DNA can be expressed relative to the reference DNA by the following equations:

In SSC, % G + C = 50·9 + 2·44 (T*m* unknown — T*m* *E. coli* strain B)
In 0·1 × SSC, % G + C = 50·9 + 2·08 (T*m* unknown — T*m* *E. coli* strain B)

Buoyant density determination

When DNA molecules are centrifuged at high speeds in a CsCl solution, they will eventually band at an equilibrium position in the linear CsCl density gradient that forms. The band is positioned where the density of the DNA molecules is equal to the density of the CsCl in that part of the gradient. The density of the DNA at the mid-point of this band is called its buoyant density (ρ), and is linearly related to the G + C content. In this technique, solid CsCl is dissolved in a suitable buffer such as 0·05 mol l^{-1} Tris pH7·3 at a density of *ca.* 1·700 g cm^{-3}; this is achieved by dissolving 6·3g of CsCl in a volume of buffer adjusted to a *final weight of* 10 g. To 1 ml of this solution, 2–3 μg of the DNA to be investigated and a similar amount of marker DNA are added and the density is re-adjusted to 1·700 g cm^{-3}. This is simply checked by using a refractometer; a density of 1·700 is equivalent to a refractive index of 1·3994. The solution is transferred to a 10 mm single sector cell, placed in an analytical rotor and then accelerated to 45 000 rev/min in an analytical ultra-centrifuge. Equilibrium in the gradient normally takes 20 h at 25°C. The buoyant density of the unknown DNA is calculated using the difference between its position and that of the marker DNA; the buoyant density of *E. coli* DNA is taken to be 1·710 g cm^{-3}. The base composition of the unknown can then be calculated from an equation relating buoyant density to % G + C derived by Schildkraut *et al.* (1962):

$$\rho = 1{\cdot}660 + 0{\cdot}098\ (G + C)$$

This technique has several advantages: (a) small quantities of partly purified nucleic acids can be used satisfactorily, (b) extra-chromosomal DNA that differs in base composition from the main population of molecules is readily observed as a separate band or as a 'shoulder' on the main band, (c) in a modern analytical ultra-centrifuge, usually up to five DNA samples can be run simultaneously.

DNA-DNA Base Pairing

Determination of the extent that single-strand DNA molecules from two different bacterial strains will pair to form double-strand interstrain duplexes *in vitro* is a widely used taxonomic tool. The technique of DNA-DNA pairing (also referred to as reassociation, hybridization and binding) enables estimates to be made of the relative number of nucleotide sequences held in common by two micro-organisms. This in turn is generally assumed to be a measure of their overall genetic relatedness. DNA-DNA pairing was introduced as a taxonomic approach by Bolton & McCarthy (1962), and since then considerable advances have been made technically and in understanding the chemical kinetics involved (Wetmur 1976). Table 3 summarizes the main methods developed since 1962 to determine DNA-DNA pairing. Two basic types of methods are used: (a) DNA-DNA duplexes are formed and retained in agar gels or on membrane-filters (unlabelled DNA is initially fixed to a support with labelled DNA in solution), and (b) duplexes are formed in free solution. The methods using agar gels and membrane-filters have been the most widely used, but the free solution techniques are preferred by several groups of workers. Table 3 also indicates some of the advantages and disadvantages of the various systems. We shall limit our description here to the free solution systems. Accounts of the experimental details, and an evaluation of the techniques involving immobilized DNA are given elsewhere (De Ley & Tijtgat 1970; De Ley 1971).

Hydroxyapatite batch procedure

The principle of this method is that small amounts of the labelled DNA are reassociated in solution with a large excess of unlabelled DNA; this keeps to a minimum the extent that the labelled DNA reassociates with itself. The double-stranded DNA is then separated from the single strands by hydroxyapatite chromatography.

Fragmentation of DNA

The double-stranded DNA is fragmented by sonication (MSE ultrasonic disintegrator, 100 W model). The DNA concentration is adjusted to 400 $\mu g\ ml^{-1}$ in 2–5 ml volumes of 0·1 × SSC buffer. Each solution is sonicated for three periods of 1 min at 4°C at a frequency of 20–25 KHz and an amplitude of 2–5 μm. This procedure yields fragments with an average mol. wt of 2×10^5 to 3×10^5 daltons (Owen & Snell 1976). Disruption of DNA in a French pressure cell is also a reproducible method (Gillis *et al.* 1970).

TABLE 3. Methods to detect and assay DNA-DNA duplexes

(A) Historical interest: detection of duplexes by ultra-centrifugation (Schildkraut *et al.* 1961)
(B) Unlabelled DNA immobilized[a]
  (1) Agar-gel method[b] (Bolton & McCarthy 1962; Hoyer *et al.* 1964)
  (2) Membrane-filter method[b] (Gillespie & Spiegelman 1965). There are several modifications of this technique: filters are pre-treated with an albumen-polymer mixture (Denhardt 1966); filters washed with 3×10^3 mol l^{-1}-tris pH 9·4 after reassociation (Warnaar & Cohen 1966); DN Are-associated in the presence of dimethyl-sulphoxide (Legault-Démare *et al.* 1967)
(C) Free solution reassociation[c]
  (1) Hydroxyapatite batch procedure[b] (Brenner *et al.* 1969)
  (2) S1 endonuclease assay[b] (Crosa *et al.* 1973)
  (3) Renaturation rate method[d] (DeLey *et al.* 1970; Bradley 1973)

[a] Competition experiments possible. Methods have disadvantage that leaking of DNA from supports may occur at elevated temperatures. Modifications introduced to reduce this effect (see B.2.).
[b] With these methods the DNA-DNA duplexes formed can be analysed by thermal analysis and estimates of base mismatching made.
[c] Reassociation kinetics are essentially second-order during initial reaction in free solution; kinetics are more complicated when DNA is immobilized, hence specific binding levels are lower.
[d] Radioactive labelled DNA is not needed.

DNA reassociation

About 150 to 200 μg of unlabelled, sheared DNA in 0·06 mol l^{-1} sodium phosphate are mixed with 0·05 μg of sheared, labelled DNA. The ratio of the concentrations of unlabelled to labelled DNA should be *ca.* 3000:1. The DNA mixture is denatured at 95–100°C for 15 min and the buffer concentration adjusted to 0·28 mol l^{-1} with 0·5 mol l^{-1} sodium phosphate. The reassociation reaction is allowed to proceed to 100 Cot for the unlabelled DNA in a 20 h period of incubation, where Cot is the acronym for the product of the initial DNA concentration (Co) and the incubation time (t). The optimum renaturation temperature (TOR) is calculated from the equation of Gillis *et al.* (1970):

$$\text{TOR} = 0{\cdot}51\ (\%\ \text{G} + \text{C}) + 47{\cdot}0$$

The TOR is equivalent to approximately Tm — 25°C, and for less stringent reassociation conditions Tm — 35°C may be used. The effect of temperature on the specificity of binding has been investigated by several authors (Johnson & Ordal 1968; De Ley *et al.* 1973).

Duplex fractionation and binding assays

In the procedure of Brenner *et al.* (1969), the unbound single-stranded DNA fragments are fractionated from the reassociated DNA duplexes. The DNA mixtures are adjusted to 0·14 mol l^{-1} sodium phosphate buffer and each mixture is added to 10 ml suspension of hydroxapatite (Bio-Gel HT; Bio-Rad Laboratories), mixed well and equilibrated at 65°C for 10 min. The volume is increased to 20 ml (with 0·14 mol l^{-1} PO_4 buffer) and the mixture again thoroughly mixed. The samples are centrifuged at 60–65°C and the aqueous supernatant fractions collected. The samples are then washed twice with 20 ml portions of 0·4 mol l^{-1} sodium phosphate buffer at 65°C to elute the double-strand DNA from the HA. The supernatant fractions are collected by centrifugation and pooled. Yeast RNA (200 μg) is added as carrier to each of the pooled supernatant fractions, and the DNA precipitated with 5% (w/v) trichloracetic acid (TCA) at 4°C. The nucleic acid precipitates are collected by suction filtration on Whatman GF/B discs (2·5 cm diam.) and washed with 2·5% (w/v) TCA followed by 95% (v/v) ethanol. The discs are dried for 30 min in a hot air oven and each suspended in 5 ml of scintillation fluid containing: 2,5-diphenyloxazole, 4 g; 1,4 bis-[2-(4 methyl-5-phenyloxazolyl)]-benzene, 0·5 g; toluene, 1 l. They are then counted repeatedly in a scintillation spectrometer. The observed homologous binding for each labelled DNA is arbitrarily designated 100% and the reassociation values of the heterologous DNA reactions are determined relative to the binding values of homologous DNA.

Thermal stability of duplexes

The DNA duplexes bound to the HA, after removal of the single-strand DNA fragments in 0·14 mol l^{-1} sodium phosphate buffer, are analysed by thermal denaturation. Samples are washed with 20 ml volumes of 0·14 mol l^{-1} sodium phosphate buffer at temperature increments of 5°C over the range 65–95°C in a water bath, being equilibrated at each temperature for 10 min with mixing at 5 min intervals. The supernatant portions are collected by centrifugation at 65°C to prevent reassociation of the denatured DNA fragments and possible non-specific binding to the HA. The samples are finally washed with 20 ml 0·4 mol l^{-1} sodium phosphate buffer and boiled to remove any DNA duplexes that have not denatured. The DNA in each sample is precipitated with 5% (w/v) TCA and the radioactivity measured as before. The duplex melting temperature ($Tm(e)$) is calculated as the temperature at which 50% of the DNA fragments are dissociated and eluted from the HA. The factor used to calculate base sequence or evolutionary divergence from DNA-DNA duplex melting profiles is a decrease in T*m*(*e*) of 1·6°C for 1·0% unpaired or mismatched bases (McCarthy & Farquhar 1972).

Renaturation rate method

The principle of this method is that the rate of reassociation of a heterologous reaction will be slower than that of the homologous reaction. The reassociation of DNA fragments is determined spectrophotometrically in an instrument that should be equipped with a programme controller and automatic cell changer. The decrease in extinction at 260 nm due to the reassociation of the denatured DNA fragments is plotted on a linear recorder with the scale set to a maximum expansion of 0·10 extinction units. Two samples (A and B) of sheared DNA at a concentration of *ca.* 60 $\mu g\ ml^{-1}$ in 1·8 ml of 0·1 × SSC are denatured at 95°C for 10 min in stoppered glass tubes. A third sample containing an equimolar mixture of DNA samples A and B is treated similarly. The sodium ion concentration in each sample is adjusted to 0·51 $mol\ l^{-1}$ NaCl with 0·2 ml of 5 $mol\ l^{-1}$ NaCl. The DNA solutions are then quickly transferred to preheated 0·5 cm light path cuvettes in the spectrophotometer cell compartment at the optimum temperature of renaturation. For 0·51 $mol\ l^{-1}$ sodium buffer, the TOR is calculated from the equation:

$$\text{TOR} = 0{\cdot}47\ (\%\ \text{G} + \text{C}) + 51{\cdot}9$$

The renaturation of the DNA is monitored by the decrease in extinction during the initial 40 min of the reaction. In a mixture of two DNAs at approximately the same molar nucleotide pair concentration, the degree of binding (%D) between the DNAs can be calculated according to De Ley *et al.* (1970) from their formula:

$$\%\text{D} = \frac{4v'_{\text{M}} - (v'_{\text{A}} + v'_{\text{B}})}{2\sqrt{v'_{\text{A}} + v'_{\text{B}}}} \times 100$$

where v'_{A}, v'_{B} and v'_{M} are the apparent renaturation rates (expressed as decrease in absorbancy at 260 nm min^{-1}) of samples *A*, *B* and the mixture (*M*), respectively.

Tables 4 and 5 give examples of levels of base pairing, generally at non-stringent temperatures of renaturation, within and between selected baterial genera.

Estimation of Genome Size

The genome size of a prokaryotic organism is the amount of chromo-

TABLE 4. Levels of DNA-DNA binding within type species of selected bacterial genera

Type species	DNA-DNA binding (%)	No. strains ≥ 69% binding / No. strains examined	Reference[a]
Aerococcus viridans	69–100	15/15	1
Agrobacterium tumefaciens	51–100	about 20	2
Bacillus subtilis	18–90	10/25	3
Bifidobacterium bifidum	100	2/2	4
Bacteroides fragilis ss. *fragilis*	65–100	44/52	5
Clostridium butyricum	15–100	10/20	6
Escherichia coli	85–100	—	7
Erwinia amylovora	85–100	7/7	8
Edwardsiella tarda	84–96	20/20	9
Listeria monocytogenes	40–100	9/11	10
Leuconostoc mesenteroides	34–99	16/19	11
Mycobacterium tuberculosis	78–100	3/3	12
Micrococcus luteus	48–91	9/10	13
Myxococcus fulvus	71	2/2	14
Propionibacterium freudenreichii	85–92	10/10	15
Pseudomonas aeruginosa	90–99	4/4	16
Vibrio cholerae	82–100	18/18	17
Klebsiella pneumoniae	91–100	3/3	18
Zymomonas mobilis	76–100	35/35	19

[a] 1. Schultes & Evans 1971; 2. De Ley *et al.* 1973; 3. Seki *et al.* 1975; 4. Scardovi *et al.* 1971; 5. Johnson 1973; 6. Cummins & Johnson 1971; 7. Brenner 1973; 8. Brenner *et al.* 1974*a*; 9. Brenner *et al.* 1974*b*; 10. Stuart & Welshimer 1973; 11. Houtebeyrie & Gasser 1977; 12. Gross & Wayne 1970; 13. Ogasawara-Fujita & Sakaguchi 1976; 14. Johnson & Ordal 1968; 15. Johnson & Cummins 1972; 16. Palleroni *et al.* 1972; 17. Staley & Colwell 1973*a*; 18. Colwell *et al.* 1974; 19. Swings & De Ley 1975.

somal DNA it contains expressed in molecular weight units (daltons) or in numbers of nucleotide pairs. Direct estimates of the size of bacterial DNA can be obtained by various techniques that include electron-microscopy, auto-radiography, visco-elastic retardation time measurements, and chemical analysis. The results with these methods can be difficult to interpret because of uncertainties about the degree of fragmentation of the DNA, and about its state of replication. The spectrophotometric study of second-order reassociation kinetics has provided an alternative, and relatively simple way to estimate bacterial genome size (Bak *et al.* 1970; De Ley *et al.* 1970; Bradley 1972), because the rate of reassociation of DNA containing non-repetitive sequences is inversely proportional to its molecular complexity (Britten & Kohne 1966).

TABLE 5. Levels of DNA-DNA binding between some species of medically important bacteria

Species	DNA-DNA binding (%)	Reference[a]
Brucella abortus and *B. melitensis*	about 100	1
Clostridium botulinum and *Cl. perfringens*	18–24	2
Escherichia coli and *Shigella dysenteriae*	89	3
E. coli and *Salmonella typhimurium*	11	3
Haemophilus influenzae and *H. parainfluenzae*	44	4
Mycobacterium tuberculosis and *M. bovis*	86	5
M. tuberculosis and *M. kansasii*	38	5
M. tuberculosis and *M. smegmatis*	9–14	5
Neisseria gonorrhoeae and *N. meningitidis*	80	6
Pseudomonas aeruginosa and *Ps. fluorescens*	14–28	7
Ps. aeruginosa and *Ps. stutzeri*	5–10	7
Ps. pseudomallei and *Ps. mallei*	76–86	8
Ps. pseudomallei and *Ps. cepacia* (*multivorans*)	6–18	8
Ps. pseudomallei and *Ps. stutzeri*	2	8
Salmonella typhimurium and *S. choleraesuis*	91–94	9
S. typhimurium and *S. typhi*	88	9
S. typhimurium and *Shigella flexneri*	39	9
Vibrio cholerae and *V. parahaemolyticus*	20–33	10
Yersinia pestis and *Francisella tularensis*	0	11
Y. pestis and *Pasteurella multocida*	14	11

[a] 1. Hoyer & McCullough 1968; 2. Lee & Riemann 1970; 3. Brenner 1973; 4. Boling 1972; 5. Bradley 1973; 6. Kingsbury 1967; 7. Palleroni *et al.* 1972; 8. Rogul *et al.* 1970; 9. Crosa *et al.* 1973; 10. Staley & Colwell 1973*b*; 11. Ritter & Gerloff 1966.

In the spectrophotometric technique described by Gillis *et al.* (1970) the calculations are based on the determination of the apparent rate constant (k'), which is calculated from the equation:

$$k' = v'/c^2$$

where v is the initial renaturation rate expressed as the decrease in extinction at 260 nm min^{-1}, and c is the chemically determined DNA concentration expressed in millimolar nucleotide pairs. We usually determine v' under the conditions described for DNA-DNA pairing (see above), and the genome size is calculated relative to a reference DNA such as that of *E. coli* strain B (mol. wt $2{\cdot}71 \times 10^9$: Gillis & De Ley 1975) from the equation:

$$\text{Genome size (unknown)} = \frac{k' \text{ (\textit{E. coli} strain B)}}{k' \text{ (unknown)}} \times 2{\cdot}71 \times 10^9$$

TABLE 6. Genome size ranges of bacterial genera

	Molecular weight[a] (daltons × 10^{-9}	Number of nucleotide pairs (× 10^{-6})
Mycoplasma	0·20–0·50	0·31–0·75
Acholeplasma	0·90–1·0	1·36–1·51
Haemophilus	1·01–1·50	1·52–2·26
Neisseria	1·04–1·73	1·57–2·61
Staphylococcus	1·12–1·43	1·69–2·16
Streptococcus	1·20–1·47	1·81–2·22
Moraxella	1·20–1·70	1·81–2·56
Zymomonas	1·34–1·72	2·02–2·56
Proteus	2·02–2·83	3·05–4·27
Bacillus	2·18–2·78	3·29–4·19
Klebsiella	2·36–3·70	3·56–5·58
Escherichia	2·49–2·84	3·76–4·28
Mycobacterium	2·50–4·50	3·77–6·79
Flavobacterium	2·50–3·52	3·77–5·31
Micrococcus	2·68–2·82	4·04–4·25
Salmonella	2·77–3·15	4·18–4·75
Pseudomonas	2·99–6·96	4·51–10·50
Erwinia	3·05–3·28	4·60–4·95

[a] Data from: McCarthy 1967; Bak & Black 1968; Bak *et al.* 1970; De Ley 1970; Black *et al.* 1972; Bradley 1973; Swings & De Ley 1975; Gillis & De Ley 1975; Owen & Snell 1976.

The method of Bak *et al.* (1970) is based on similar principles, except that the second-order rate constant (k_2) is determined, and corrections are made for variations in single-strand fragment length. In the method of Bradley (1972) calculations are based on the spectrophotometric determination of $Cot_{\frac{1}{2}}$, which is the stage in the reaction when half of the initial denatured DNA has reassociated. Genome size is directly proportional to the value of $Cot_{\frac{1}{2}}$ (Britten & Kohne 1966), and is expressed relative to a suitable reference DNA. Table 6 lists the genome size ranges of various bacterial genera.

References

BAK, A. L. & BLACK, F. T. 1968 DNA base composition of human T-strain mycoplasmas. *Nature, London* **219,** 1044–1045.

BAK, A. L., CHRISTIANSEN, C. & STENDERUP, A. 1970 Bacterial genome sizes determined by DNA renaturation studies. *Journal of General Microbiology* **64,** 377–380.

BENDICH, A. 1957 Methods for characterization of nucleic acids by base composition. In *Methods in Enzymology*, Vol. III, eds Colowick, S. P. & Kaplan, N. O. New York: Academic Press.

Black, F. T., Christiansen, C. & Askaa, G. 1972 Genome size and base composition of deoxyribonucleic acid from eight human T-mycoplasmas. *International Journal of Systematic Bacteriology* **22,** 241–242.

Boling, M. E. 1972 Homology between the deoxyribonucleic acids of *Haemophilus influenzae* and *Haemophilus parainfluenzae*. *Journal of Bacteriology* **112,** 745–750.

Bolton, E. T. & McCarthy, B. J. 1962 A general method for the isolation of RNA complementary to DNA. *Proceedings of the National Academy of Sciences, New York* **48,** 1390–1397.

Bradley, S. G. 1972 Reassociation of deoxyribonucleic acid from selected Mycobacteria with that from *Mycobacterium bovis* and *Mycobacterium farcinica*. *American Review of Respiratory Disease* **106,** 122–124.

Bradley, S. G. 1973 Relationships among mycobacteria and nocardiae based upon deoxyribonucleic acid reassociation. *Journal of Bacteriology* **113,** 645–651.

Brenner, D. J. 1973 Deoxyribonucleic acid reassociation in the taxonomy of enteric bacteria. *International Journal of Systematic Bacteriology* **23,** 298–307.

Brenner, D. J. 1976 Clinical and taxonomic applications of DNA hybridization. *Public Health Laboratory* **34,** 48–55.

Brenner, D. J., Fanning, G. R., Rake, A. V. & Johnson, K. E. 1969 Batch procedure for thermal elution of DNA from hydroxypatite. *Analytical Biochemistry* **28,** 447–459.

Brenner, D. J., Fanning, G. R. & Steigerwalt, A. G. 1974*a* Deoxyribonucleic acid relatedness among Erwiniae and the *Enterobacteriaceae*: the gall, wilt and dry-necrosis organisms (genus *Erwinia* Winslow *et al., sensu strictu*). *International Journal of Systematic Bacteriology* **24,** 197–204.

Brenner, D. J., Fanning, G. R. & Steigerwalt, A. G. 1974*b* Polynucleotide sequence relatedness in *Edwardsiella tarda*. *International Journal of Systematic Bacteriology* **24,** 186–190.

Britten, R. J. & Kohne, D. E. 1966 Nucleotide sequence repetition in DNA. *Carnegie Institution, Washington, Yearbook* **65,** 78–106.

Britten, R. J., Pavich, M. & Smith, J. 1969 A new method for DNA purification. *Carnegie Institution, Washington, Yearbook* **68,** 400–402.

Buchanan, R. E. & Gibbons, N. E. (eds) 1974 *Bergey's Manual of Determinative Bacteriology*, 8th edn Baltimore: Williams & Wilkins.

Colwell, R. R., Johnson, R., Wan, L., Lovelace, T. E. & Brenner, D. J. 1974 Numerical taxonomy and deoxyribonucleic acid reassociation in the taxonomy of some Gram-negative fermentative bacteria. *International Journal of Systematic Bacteriology* **24,** 422–433.

Crombach, W. H. J. 1972 DNA base composition of soil arthrobacters and other coryneforms from cheese and sea fish. *Antonie van Leeuwenhoek* **38,** 105–120.

Crombach, W. H. J. 1973 Deep-freezing of bacterial DNA for thermal denaturation and hybridization experiments. *Antonie van Leeuwenhoek* **39,** 249–255.

Crosa, J. M., Brenner, D. J., Ewing, W. H. & Falkow, S. 1973 Molecular relationships among the *Salmonellae*. *Journal of Bacteriology* **115,** 307–315.

Cummins, C. S. & Johnson, J. L. 1971 Taxonomy of the clostridia: wall composition and DNA homologies in *Clostridium butyricum* and other butyric-acid producing clostridia. *Journal of General Microbiology* **67,** 33–46.

De Ley, J. 1970 Re-examination of the association between melting point, buoyant density, and chemical base composition of deoxyribonucleic acid. *Journal of Bacteriology* **101,** 738–754.

De Ley, J. 1971 Hybridization of DNA. In *Methods in Microbiology*, Vol. 5A, eds Norris, J. R. & Ribbons, D. W. London & New York: Academic Press.

De Ley, J. & De Smedt, J. 1975 Improvements of the membrane filter method for DNA: rRNA hybridization. *Antonie van Leeuwenhoek Journal of Microbiology and Serology* **41,** 287–307.

De Ley, J. & Tijtgat, R. 1970 Evaluation of membrane filter methods for DNA-DNA hybridization. *Antonie van Leeuwenhoek Journal of Microbiology and Serology* **36,** 461–474.

De Ley, J., Cattois, H. & Reynaerts, A. 1970 The quantitative measurement of DNA hybridization from renaturation rates. *European Journal of Biochemistry* **12,** 133–142.

De Ley, J., Tijtgat, R., De Smedt, J. & Michiels, M. 1973 Thermal stability of DNA: DNA hybrids within the genus *Agrobacterium*. *Journal of General Microbiology* **78,** 241–252.

Denhardt, D. T. 1966 A membrane-filter technique for the detection of complementary DNA. *Biochemical and Biophysical Research Communications* **23,** 641–646.

Fredericq, E., Oth, A. & Fontaine, F. 1961 The ultraviolet spectrum of deoxyribonucleic acids and their constituents. *Journal of Molecular Biology* **3,** 11–17.

Garvie, E. I. 1976 Hybridization between the deoxyribonucleic acids of some strains of heterofermentative lactic acid bacteria. *International Journal of Systematic Bacteriology* **26,** 116–122.

Gillespie, D. & Spiegelman, S. 1965 A quantitative assay for DNA-DNA hybrids with DNA immobilized on a membrane. *Journal of Molecular Biology* **12,** 829–842.

Gillis, M. & De Ley, J. 1975 Determination of the molecular complexity of double-stranded phage genome DNA from initial renaturation rates. The effect of DNA base composition. *Journal of Molecular Biology* **98,** 447–464.

Gillis, M., De Ley, J. & De Cleene, M. 1970 The determination of molecular weight of bacterial genome DNA from renaturation rates. *European Journal of Biochemistry* **12,** 143–153.

Gross, W. M. & Wayne, L. G. 1970 Nucleic acid homology in the genus *Mycobacterium*. *Journal of Bacteriology* **104,** 630–634.

Hirschman, S. Z. & Felsenfeld, G. 1966 Determination of DNA composition and concentration by spectral analysis. *Journal of Molecular Biology* **16,** 347–358.

Houtebeyrie, M. & Gasser, F. 1977 Deoxyribonucleic acid homologies in the genus *Leuconostoc*. *International Journal of Systematic Bacteriology* **27,** 9–14.

Hoyer, B. H. & McCullough, N. B. 1968 Polynucleotide homologies of *Brucella* deoxyribonucleic acids. *Journal of Bacteriology* **95,** 444–448.

Hoyer, B. H., McCarthy, B. J. & Bolton, E. T. 1964 A molecular approach in the systematics of higher organisms. *Science* **144,** 959–967.

Huang, P. C. & Rosenberg, E. 1966 Determination of DNA base composition via depurination. *Analytical Biochemistry* **16,** 107–113.

Johnson, J. L. 1973 Use of nucleic-acid homologies in the taxonomy of anaerobic bacteria. *International Journal of Systematic Bacteriology* **23,** 308–315.

JOHNSON, J. L. & CUMMINS, C. S. 1972 Cell wall composition and deoxyribonucleic acid similarities among the anaerobic coryneforms, classical propionibacteria and strains of *Arachnia propionica*. *Journal of Bacteriology* **109,** 1047–1066.

JOHNSON, J. L. & ORDAL, E. J. 1968 Deoxyribonucleic acid homology in bacterial taxonomy: effect of incubation temperature on reaction specificity. *Journal of Bacteriology* **95,** 893–900.

KINGSBURY, D. T. 1967 Deoxyribonucleic acid homologies among species of the genus *Neisseria*. *Journal of Bacteriology* **94,** 870–874.

LEE, W. H. & RIEMANN, H. 1970 The genetic relatedness of proteolytic *Clostridium botulinum* strains. *Journal of General Microbiology* **64,** 85–90.

LEGAULT-DÉMARE, J., DESSEAUX, B., HEYMAN, T., SÉROR, S. & RESS, G. F. 1967 Studies on hybrid molecules of nucleic acids. I. DNA-DNA hybrids on nitrocellulose filters. *Biochemical and Biophysical Research Communications* **28,** 550–557.

MARMUR, J. 1961 A procedure for the isolation of deoxyribonucleic acid from micro-organisms. *Journal of Molecular Biology* **3,** 208–218.

MARMUR, J. & DOTY, P. 1962 Determination of the base composition of deoxyribonucleic acid from its thermal denaturation temperature. *Journal of Molecular Biology* **5,** 109–118.

McCARTHY, B. J. 1967 Arrangement of base sequences in deoxyribonucleic acid. *Bacteriological Reviews* **31,** 215–229.

McCARTHY, B. J. & FARQUHAR, M. N. 1972 The rate of change of DNA in evolution. In *Evolution of Genetic Systems. Brookhaven Symposia in Biology* 23, ed. Smith, H. H., London: Gordon and Breach.

MEYER, S. A. & SCHLEIFER, K. H. 1975 Rapid procedure for the approximate determination of the deoxyribonucleic acid base composition of micrococci, staphylococci and other bacteria. *International Journal of Systematic Bacteriology* **25,** 383–385.

MITRUKA, B. M. 1975 In *Gas Chromatographic Applications in Microbiology and Medicine*, pp. 182–184, ed. Mitruka, B. M., New York: John Wiley & Sons.

NORMORE, W. M. 1973 Guanine-plus-cytosine (G + C) composition of the DNA of bacteria, fungi, algae and protozoa. In *CRC Handbook of Microbiology*, Vol. II *Microbial Composition*, eds Laskin, A. I. & Lechevalier, H. A. Cleveland: CRC Press.

OGASAWARA-FUJITA, N. & SAKAGUCHI, K. 1976 Classification of micrococci on the basis of deoxyribonucleic acid homology. *Journal of General Microbiology* **94,** 97–106.

OWEN, R. J. & LAPAGE, S. P. 1976 The thermal denaturation of partly purified bacterial deoxyribonucleic acid and its taxonomic applications. *Journal of Applied Bacteriology* **41,** 335–340.

OWEN, R. J. & SNELL, J. J. S. 1976 Deoxyribonucleic acid reassociation in the classification of flavobacteria. *Journal of General Microbiology* **93,** 89–102.

OWEN, R. J., HILL, L. R. & LAPAGE, S. P. 1969 Determination of DNA base compositions from melting profiles in dilute buffers. *Biopolymers* **7,** 503–516.

PALLERONI, N. J., BALLARD, R. W., RALSTON, E. & DOUDOROFF, M. 1972 Deoxyribonucleic acid homologies among some *Pseudomonas* species. *Journal of Bacteriology* **110,** 1–11.

RITTER, D. B. & GERLOFF, R. K. 1966 Deoxyribonucleic acid hybridization among some species of the genus *Pasteurella*. *Journal of Bacteriology* **92,** 1838–1839.

ROGUL, M., BRENDLE, J. J., HAAPALA, D. K. & ALEXANDER, A. D. 1970 Nucleic acid similarities among *Pseudomonas pseudomallei, Pseudomonas multivorans* and *Actinobacillus mallei. Journal of Bacteriology* **101,** 827–835.

RUSSEL, G. J., WALKER, P. M. B., ELTON, R. A. & SUBAK-SHARPE, J. M. 1976 Doublet frequency analysis of fractionated vertebrate nuclear DNA. *Journal of Molecular Biology* **108,** 1–23.

SCARDOVI, V., TROVATELLI, L. D., ZANI, G., CROCIANI, F. & MATTEUZZI, D. 1971 Deoxyribonucleic acid homology relationships among species of the genus *Bifidobacterium. International Journal of Systematic Bacteriology* **21,** 276–294.

SCHILDKRAUT, C. & LIFSON, S. 1965 Dependence of the melting temperature of DNA on salt concentration. *Biopolymers* **3,** 195–208.

SCHILDKRAUT, C. L., MARMUR, J. & DOTY, P. 1961 The formation of hybrid DNA molecules and their use in studies of DNA homologies. *Journal of Molecular Biology* **3,** 595–617.

SCHILDKRAUT, C. L., MARMUR, J. & DOTY, P. 1962 Determination of the base composition of deoxyribonucleic acid from its buoyant density in CsCl. *Journal of Molecular Biology* **4,** 430–443.

SCHULTES, L. M. & EVANS, J. B. 1971 Deoxyribonucleic acid homology of *Aerococcus viridans. International Journal of Systematic Bacteriology* **21,** 207–209.

SEKI, T., OSHIMA, T. & OSHIMA, Y. 1975 Taxonomic study of *Bacillus* by deoxyribonucleic acid-deoxyribonucleic acid hybridization and interspecific transformation. *International Journal of Systematic Bacteriology* **25,** 258–270.

SILVESTRI, L. G. & HILL, L. R. 1965 Agreement between deoxyribonucleic acid base composition and taxometric classification of Gram-positive cocci. *Journal of Bacteriology* **90,** 136–140.

STALEY, T. E. & COLWELL, R. R. 1973*a* Deoxyribonucleic acid reassociation among members of the genus *Vibrio. International Journal of Systematic Bacteriology* **23,** 316–332.

STALEY, T. E. & COLWELL, R. R. 1973*b* Polynucleotide sequence relationships among Japanese and American strains of *Vibrio parahaemolyticus. Journal of Bacteriology* **114,** 916–927.

STUART, S. E. & WELSHIMER, H. J. 1973 Intrageneric relatedness of *Listeria* Pirie. *International Journal of Systematic Bacteriology* **23,** 8–14.

SWINGS, J. & DE LEY, J. 1975 Genome deoxyribonucleic acid of the genus *Zymomonas* Kluyver and van Niel 1936: base composition, size and similarities. *International Journal of Systematic Bacteriology* **25,** 324–329.

ULITZUR, S. 1972 Rapid determination of DNA base compositions by ultraviolet spectroscopy. *Biochemica et Biophysica Acta* **272,** 1–11.

WANG, S. Y. & HASHAGEN, J. M. 1964 The determination of the base composition of deoxyribonucleic acids by bromination. *Journal of Molecular Biology* **8,** 333–340.

WARNAAR, S. O. & COHEN, J. A. 1966 A quantitative assay for DNA-DNA hybrids using membrane filters. *Biochemical and Biophysical Research Communications* **24,** 554–558.

WETMUR, J. G. 1976 Hybridization and renaturation kinetics of nucleic acids. *Annual Review of Biophysics and Bioengineering* **5,** 337–361.

YAMADA, K. & KOMAGATA, K. 1970 Taxonomic studies on coryneform bacteria. III. DNA base composition of coryneform bacteria. *Journal of General and Applied Microbiology* **16,** 215–224.

An Improved Automatic Multipoint Inoculator

P. Ridgway Watt

Beecham Pharmaceuticals Research Division, Brockham Park, Betchworth, Surrey, UK

Multiple surface inoculation of culture media in Petri dishes, with various micro-organisms, is practised in many laboratories. The procedure, when carried out manually, is tedious and time-consuming, and involves a risk of airborne contamination.

To facilitate such inoculations, several investigators have proposed mechanical devices for the simultaneous transfer of the required number of inocula to the agar surface. Thus, Beech *et al.* (1955) modified for use with bacteria and yeasts the multiple needle device proposed by Garrett (1946). A simple multipoint replicator was described by Smith (1961) for the purpose of screening cultures for lysogeny, and for studying the growth responses of mutant strains: and similar replicators, generally rigid arrays of loops, rods or tubes, have been used by many workers including Shifrine *et al.* (1954); Steers *et al.* (1959); Sneath (1962); Harris (1963); Simon & Undseth (1963); Stotzky (1965); Quadling & Colwell (1964); Hill & Stent (1965); Massey & Mattoni (1965); Gray & Goodfellow (1966); Neal *et al.* (1966); Murphy & Haque (1967); Osborn *et al.* (1967); Seman (1967); Lovelace & Colwell (1968); Hercules & Wiberg (1971).

There are, however, disadvantages in any multiple inoculation device using a fixed array of points. The rigidity of the system makes it difficult to transfer uniformly sized drops to an agar surface, and there is always a risk of damage to the surface layer in the process.

The arrangement described by Tarr (1958) overcomes these disadvantages by allowing the inoculating elements to hang freely from locating bushes in the inoculator head. Since each element is then held against the agar surface solely by its own weight, contact pressure during inoculation is necessarily light. The principle has been used by other workers, including Lidwell (1959), Ridgway Watt *et al.* (1966), Hill (1970); and, most recently, by Wiberg (1977) for genetic tests on the T4 phage—*Escherichia coli* system.

The volumes transferred by manually operated equipment are always liable to be influenced by individual variations in the speed of operation: moreover, hand movement may result in the accidental detachment of droplets from the pins or loops during transfer. A mechanically driven system, on the other hand, not only gives the best reproducibility, but can be run at the maximum practical transfer rate.

A motorized multipoint inoculator was described by Ridgway Watt *et al.* (1966), and, later, Hill (1970) mechanized the Tarr (1958) apparatus for the identification of bacteria. In each of these arrangements, the inoculator head was first driven downwards in order to pick up droplets of inocula: it was then raised, and driven horizontally to a position above the dish to be inoculated, where the vertical movements were repeated.

This system works well at moderate speeds; it cannot be run safely at high speeds, since droplets may be detached by acceleration, by mechanical vibration, or by relative air movement.

If short operating cycles of a few seconds are required, it is better to separate the horizontal and vertical movements within the equipment so that the head itself is driven only in the vertical plane. An inoculator using this principle was built by the present writer (Ridgway Watt 1975), and, with some modifications, it forms the subject of this chapter.

Construction

The machine comprises a sliding carriage, a cabinet, an inoculation head, and a transparent plastic cover (Fig. 1).

The carriage is designed to accept a Petri dish in a defined position at its front end, while a container, holding a number of separate portions of inoculating medium, fits at the rear end. A linear slide inside the cabinet allows the carriage to move backwards and forwards under the inoculation head, so that the Petri dish and the inoculum container occupy the same position in turn. Movement of the carriage along the slide is effected by means of a motor-driven crank mechanism, which provides a smooth deceleration at each end of its travel.

The inoculation head consists of a stainless steel plate, drilled with a pattern of holes. Inoculation pins, in the form of cylindrical stainless steel rods with accurately ground flat ends, hang through these holes (Fig. 2).

A ball-bearing slide is arranged to support the head assembly, so that it can be driven steadily up and down through a short distance (25 mm) by means of a second motor.

The whole system is surrounded by a transparent plastic cover, in

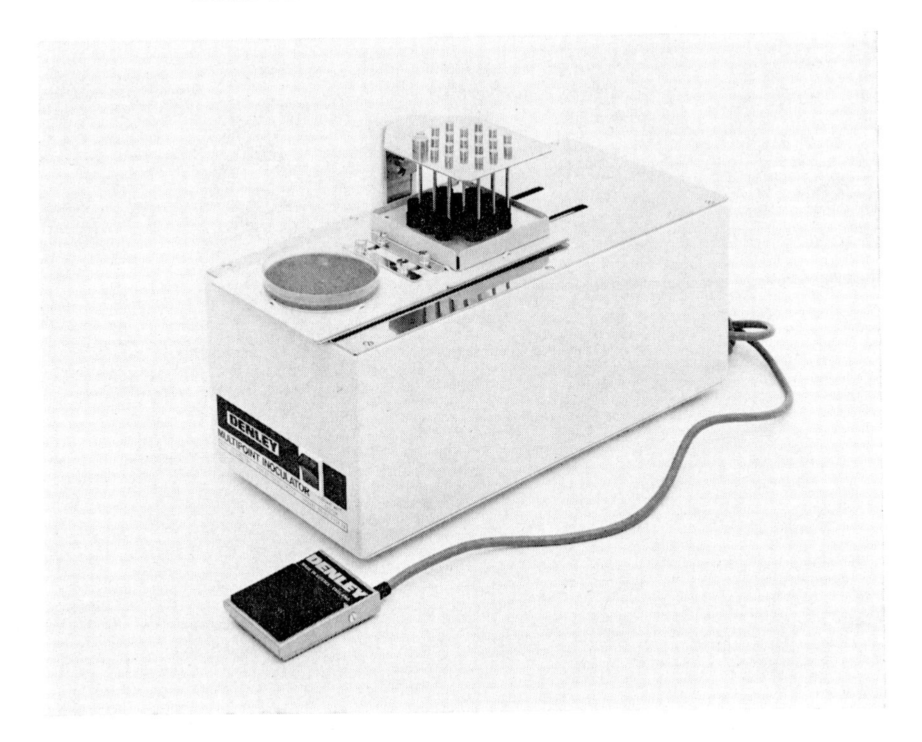

FIG. 1. General view of the apparatus showing Petri dish in forward position ready for inoculation.

order to prevent accidental contamination of the agar with airborne particles.

Operation

When a dish is to be inoculated, it is placed in the front compartment of the sliding carriage, and the lid removed under cover of the plastic canopy. The inoculation cycle is then initiated by pressure on the control switch.

The inoculator head moves down, so that the pins dip lightly into the inoculum containers. It then returns to the upper position: small uniform drops of inoculum are retained on the pins. The carriage moves backwards to its rear position, drawing the Petri dish completely inside the covered area, and the head once again travels down and up, transferring the inocula to the agar surface. Since the pins hang freely from the inoculator head, each one touches the agar surface under its own weight, regardless of the surface level of agar in the dish.

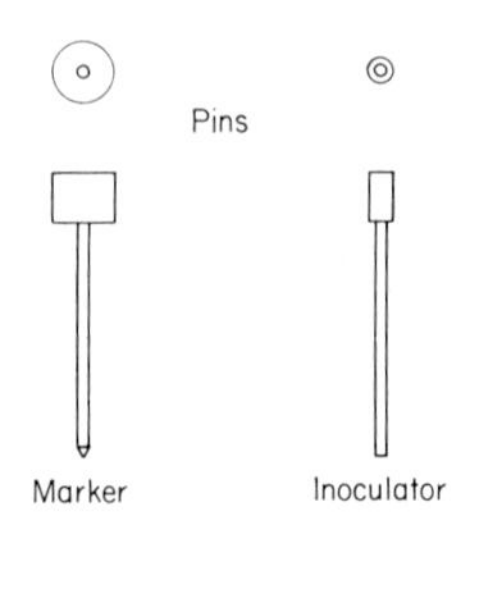

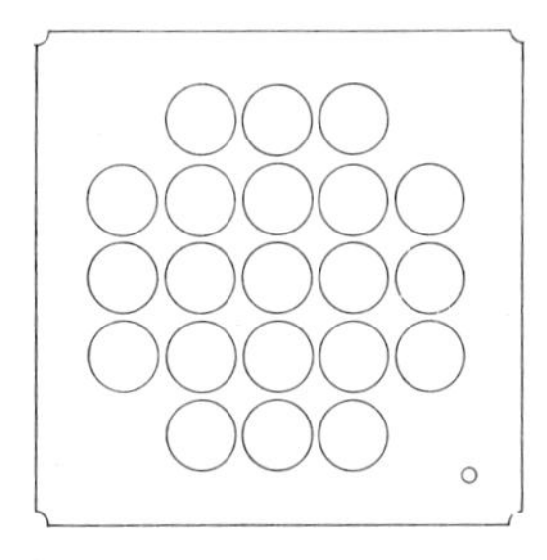

FIG. 2. Inoculation pins and 'Oxoid' cap holding plate. Holes are on 16 mm centres: pins are 2·38 mm diam.

Finally, the inoculated dish returns to its original position in the front of the apparatus, so that it may have the lid replaced, and a new dish may then be inserted.

The whole cycle occupies about four seconds, and at this operating speed there is normally no cross-contamination. The drop size is, moreover, appreciably more uniform than that quoted for hand-operated devices. The first prototype, for example, using pins of 2 mm diam., gave an average delivered volume of 0·0006 ml with a standard deviation of $\pm$ 13%. By contrast, the hand-operated apparatus of Beech *et al.* (1955) had a reported variation of $\pm$ 25%.

Storage of inocula

Three methods have been used to hold separate portions of inocula within the apparatus during a run, as follows:

'Oxoid' caps

A metal plate is punched with a pattern of holes, each large enough to accommodate an aluminium 'Oxoid' cap. The plate stands within a 100 mm square disposable Petri dish, and is held between locating projections at the back of the main slide.

'Teflon' block

It is possible to incorporate all the inoculum wells within a single block of 'Teflon', by machining with a flat-ended slot drill. The 'Teflon' withstands all normal autoclaving temperatures and is arranged as before to fit into a standard Petri dish: one corner is marked to identify the orientation of the block.

Compartmented plastic dishes

For some applications, it has been convenient to use commercially available dishes as inoculum holders. These have been either of the 'Repli-Dish' pattern, in which a 100 mm square Petri dish is divided into 25 square compartments, or variations of the 'Microtiter' plate with up to 96 moulded wells.

Whichever method is used to hold the inocula, it is a simple matter to construct a suitable inoculation head by drilling holes, on the required centres, through a blank stainless steel plate.

The head itself is removably connected to the vertical slide, using keyhole slots which slip over matching pegs on the slide (Fig. 3).

Orientation of the growth pattern

If a symmetrical set of inocula has been applied, some means of locating the resultant pattern is necessary. It is possible to use an external register mark on the inoculated dish, orienting this at the time of inoculation: alternatively, one of the inoculator pins may be made with a pointed end in order to mark the agar surface.

Applications

In its simplest form, the apparatus may be used for rapid sensitivity testing of bacteria by the agar dilution method: e.g. the screening of a new antibiotic against a large number of strains of one or more species of bacteria. (Greenwood *et al.* 1974; Greenwood & O'Grady 1975; O'Callaghan *et al.* 1976; Basker & Sutherland 1977; Comber *et al.* 1977; Greenwood *et al.* 1977; Wise *et al.* 1977). The multiple inoculation technique has also been found valuable in the study of soil microorganisms (Stotzky 1965) and rhizo-spheres (Neal *et al.* 1966), and the identification of bacteria and yeasts. (Beech *et al.* 1955; Corlett *et al.* 1965); If the cultures investigated have a tendency to spread, it may be desirable to use a compartmented dish, rather than a continuous agar surface, to receive the inocula (Sneath & Stevens 1967). Examples of this problem have been found in studies on the detection of lipase and amylase, and have been satisfactorily overcome by the use of the 'Replidish'. (Lovelace & Colwell 1968). Standard Petri dishes holding a pattern

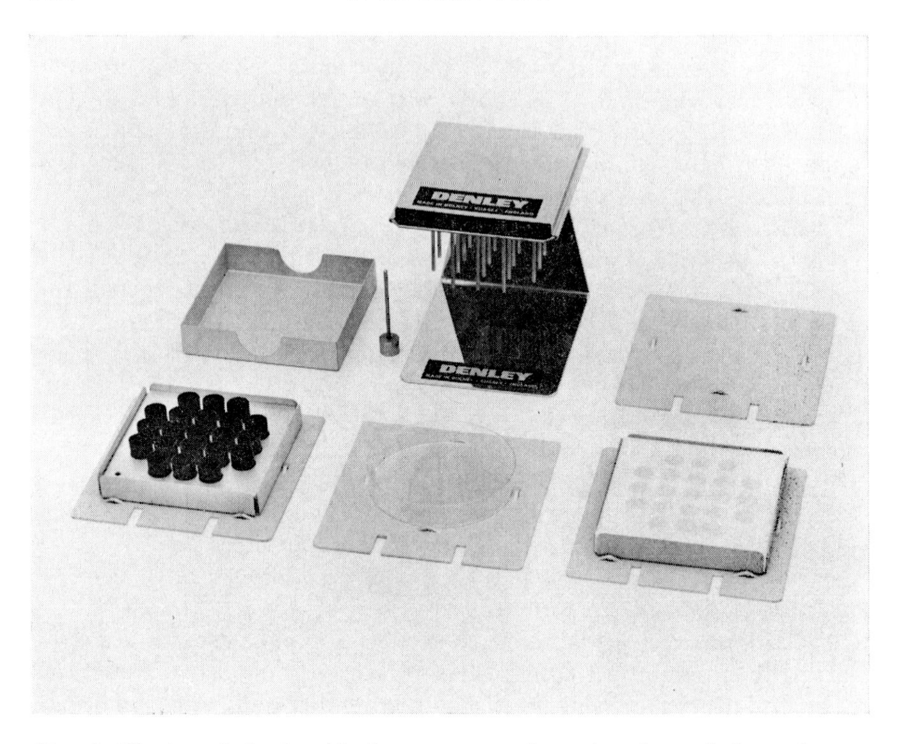

FIG. 3. The inoculation head is shown supported on a bench stand. A metal cover over the head is used to retain the pins during autoclaving. At the front are shown the 'Oxoid' cap holder for inocula, the Petri dish holder and an inoculum holder machined from 'Teflon'.

of small individual containers have also been found useful in similar applications (Goodfellow & Gray 1966). It is possible to replace the normal inoculator pins with wire loops, that may be sterilized automatically whilst in position by the passage of electric current. This adaptation extends the range of application of the equipment into, for example, bacteriophage studies.

The apparatus is now manufactured by Denley-Tech Limited, and is marketed in the United Kingdom by Denley Instruments Limited, Daux Road, Billingshurst, Sussex.

References

BASKER, M. J. & SUTHERLAND, R. 1977 Activity of amoxycillin, alone, and in combination with aminoglycoside antibiotics against streptococci associated with bacterial endocarditis. *Journal of Antimicrobial Chemotherapy* **3,** 273–282.

BEECH, F. W., CARR, J. G. & CODNER, R. C. 1955 A multipoint inoculator for plating bacteria or yeasts. *Journal of General Microbiology* **13,** 408–410.

COMBER, K. R., BASKER, M. J., OSBORNE, C. D. & SUTHERLAND, R. 1977 Synergy between ticarcillin and tobramycin against *Pseudomonas aeruginosa* and Enterobacteriacae *in vitro* and *in vivo*. *Antimicrobial Agents and Chemotherapy* **11,** 956–964.

CORLETT, D. A., LEE, J. S. & SINNHUBER, R. O. 1965 Application of replica plating and computer analysis for rapid identification of bacteria in some foods. *Applied Microbiology* **13,** 808.

GARRETT, S. D. 1946 A multiple-point inoculating needle for agar plates. *Transactions of the British Mycological Society* **29,** 171.

GOODFELLOW, M. & GRAY, T. R. G. 1966 A multipoint inoculation method for performing biochemical tests on bacteria. In *Identification Methods for Microbiologists* Part A, eds Gibbs, B. M. & Skinner, F. A. Society for Applied Bacteriology Technical Series No. 1. London & New York: Academic Press.

GRAY, T. R. G. & GOODFELLOW, M. 1966 Rapid methods of making routine physiological tests on soil bacteria. In *Proceedings of the VIIIth International Congress of Soil Science*, Commission III, Bucharest.

GREENWOOD, D., LINTON BROOKS, H., GARGAN, R. & O'GRADY, F. 1974 Activity of FL 1060, a new β-lactam antibiotic, against urinary tract pathogens. *Journal of Clinical Pathology* **27,** 192–197.

GREENWOOD, D. & O'GRADY, F. 1975 Resistance categories of Enterobacteria to β-lactam antibiotics. *Journal of Infectious Diseases* **132,** 233–240.

GREENWOOD, D., PEARSON, N. J., OLIVANT, J. & O'GRADY, F. 1977 Laboratory evaluation of pirbenicillin, a new penicillin with anti-pseudomonal activity. *Journal of Antimicrobial Chemotherapy* **3,** 195–189.

HARRIS, P. J. 1963 A replica plating culture technique. *Journal of Applied Bacteriology* **26,** 100.

HERCULES, K. & WIBERG, J. S. 1971 Specific suppression of mutations in genes 46 and 47 by *das*, a new class of mutations in bacteriophage T4D *Journal of Virology* **8,** 603–612.

HILL, J. R. 1970 Multiple inoculation technique for rapid identification of bacteria. In *Automation, Mechanization and Data Handling in Microbiology*, eds Baillie, A. & Gilbert, R. J. Society for Applied Bacteriology Technical Series No. 4. London & New York: Academic Press.

HILL, R. & STENT, G. S. 1965 Genetic factors of *E. coli* determining suppression of the amber phenotype of T4 bacteriophage. *Biochemical and Biophysical Research Communications* **18,** 757–762.

LIDWELL, C. M. 1959 Apparatus for phage-typing of *Staphylococcus aureus*. *Monthly Bulletin of the Ministry of Health Laboratory Service* **18,** 49.

LOVELACE, T. E. & COLWELL, R. R. 1968 A multipoint inoculator for Petri dishes. *Applied Microbiology* **16,** 944–945.

MASSEY, R. L. & MATTONI, R. H. T. 1965 New technique for mass assays of physiological characteristics of unicellular algae. *Applied Microbiology* **13,** 798–800.

MURPHY, R. A. & HAQUE, R. 1967 A multiple inoculating device for bacteria. *American Journal of Clinical Pathology* **47,** 554–555.

NEAL, J. L., LU, K. C., BOLLEN, W. B. & TRAPPE, J. M. 1966 Apparatus for rapid replica plating in rhizosphere studies. *Applied Microbiology* **14,** 695–696.

O'CALLAGHAN, C. H., SYKES, R. B., GRIFFITHS, A. & THORNTON, J. E. 1976 Cefuroxine, a new cephalosporine antibiotic: activity *in vitro*. *Antimicrobial Agents and Chemotherapy* **9,** 511–519.

OSBORN, M., PERSON, S., PHILLIPS, S. & FUNK, F. 1967 A determination of mutagen specificity in bacteria using nonsense mutants of bacteriophage T4. *Journal of Molecular Biology* **26,** 437–447.

QUADLING, C. & COLWELL, R. C. 1964 Apparatus for simultaneous inoculation of a set of culture tubes. *Canadian Journal of Microbiology* **10,** 87–90.

RIDGWAY WATT, P. 1975 An automatic multipoint inoculator for petri dishes. *Laboratory Practice* May 1975 pp. 350–351.

RIDGWAY WATT, P., JEFFRIES, L. & PRICE, S. A. 1966 An automatic multipoint inoculator for Petri dishes. In *Identification Methods for Microbiologists Part A*, eds Gibbs, B. M. & Skinner, F. A. Society for Applied Bacteriology Technical Series No. 1. London & New York: Academic Press.

SEMAN, J. P. 1967 Improved multipoint inoculating device for replica plating. *Applied Microbiology* **15,** 1514–1516.

SHIFRINE, M., PHAFF, H. J. & DEMAIN, A. L. 1954 Determination of carbon assimilation patterns of yeasts by replica plating. *Journal of Bacteriology* **68,** 28.

SIMON, H. J. & UNDSETH, S. 1963 Simple method for phage typing of staphylococci. *Journal of Bacteriology* **85,** 1447–1448.

SMITH, D. A. 1961 A multiple inoculation device for use with fluids. *Journal of Applied Bacteriology* **24,** 131.

SNEATH, P. H. A. 1962 The construction of taxonomic groups. In *Microbial Classification*, eds Ainsworth, G. C. & Sneath, P. H. A. 12th Symposium of the Society for General Microbiology. London: Cambridge University Press.

SNEATH, P. H. A. & STEVENS, M. 1967 A divided Petri dish for use with multipoint inoculators. *Journal of Applied Bacteriology* **30,** 495–497.

STEERS, E., FOLTZ, E. L. & GRAVES, B. S. 1959 An inocula replicating apparatus for routine testing of bacterial susceptibility to antibiotics. *Antibiotics and Chemotherapy* **9,** 307–311.

STOTZKY, G. 1965 Replica plating technique for studying microbial interactions in soil. *Canadian Journal of Microbiology* **11,** 629–636.

TARR, H. A. 1958 Mechanical aids for the phage-typing of *Staphylococcus aureus*. *Monthly Bulletin of the Ministry of Health Laboratory Service* **17,** 64.

WIBERG, J. S. 1977 Floating-loop replica-plating device. *Journal of Applied Bacteriology* **42,** 433–436.

WISE, R., ANDREWS, J. M. & BEDFORD, K. A. 1977 Pirbenicillin—a semisynthetic penicillin with antipseudomonal activity. *Journal of Antimicrobial Chemotherapy* **3,** 175–183.

Subject Index

Acetic acid bacteria, 33–45
Acetobacter, 33, 36, 38–45
Acetobacter
  aceti, 40, 41
  aceti subsp. *liquefaciens*, 40–42
  acidophilum, 43
  ascendens, 42
  estunensis, 39, 41, 42
  lovaniensis, 41, 42
  mesoxydans, 42
  paradoxus, 36, 42
  pasteurianus, 40, 41
  peroxydans, 36, 40, 42, 43
  rancens, 41, 42
  roseus, 44
  xylinum, 36, 39, 41, 42
Acetoin production, in *Staphylococcus*, 207
Acetomonas, 33, 43
Acetomonas oxidans, 43
Acholeplasma, 292
Acid-fastness, in actinomycetes, 263
Acid production, from carbohydrates, in *Micrococcus* and *Staphylococcus*, 207
Acinetobacter
  anitratus, 97
  lwoffi, 97
Actinomadura, 261, 263, 265–267
  tests for, 270, 271, 272
Actinomadura
  dassonvillei, 263, 265, 267, 270, 271
  helvata, 267, 270, 271
  madurae, 263, 267, 270, 271
  pelletieri, 263, 267, 270, 271
  pusilla, 267, 270, 271
  roseoviolacea, 267, 270, 271
  spadix, 267, 270, 271
  verrucospora, 267, 270, 271
Actinomycetes, 261–263, 266–269
  identification of, 262
Aerococcus, 233, 243
Aerococcus viridans, 242, 243, 290
Aeromonas
  biochemical and physiological characters of, 147
  identification of, 152, 153, 154, 156–161, 182
Aeromonas
  hydrophila, 154, 155, 162
  punctata, 154
  salmonicida, 154–161
  shigelloides, 154
  sobria, 155, 162
Aesculin hydrolysis
  in *Chromobacterium*, 172
  in *Streptococcus*, 240
Agglutination reaction, of *Rhizobium* strains, 60, 61
Agglutination tests, for *Brucella*, 81–83, 85, 88, 90, 91, 110, 111
Agrobacterium
  bacteriophage of, 62
  cross reaction with *Rhizobium*, 53, 55, 61, 63, 64
Agrobacterium tumefaciens, 290
Alkaligenes bronchiseptica, 97
Alteromonas, 152, 153
  differentiation from *Pseudomonas*, 1–3, 8
  motile strains of, 182, 183
Alteromonas
  citrea, 11, 12
  communis, 8, 11, 12
  espejiana, 12
  haloplanktis, 8, 11, 12
  luteoviolaceus, 11, 12, 167
  macleodii, 8, 11, 12
  marinopraesens, 8
  piscicida, 11, 12
  putrefaciens, 11, 12
  rubra, 11, 12
  undina, 12
  vaga, 8, 11, 12

Amino acid decarboxylases, testing for, 5
Anaerobes
  growth conditions for, 190, 191, 193
  identification of, 198
  properties of, 195
Antibiotics
  resistance of rhizobia to, 50
  susceptibility of Bacteroidaceae to, 192
  use of in Farrell's modified SDA medium, 77, 78
  use of multiple inoculation technique in screening of, 301
Antibiotic sensitivity tests, for *Pseudomonas*, 4
Antigens
  H and R, in vibrios, 148
  intracellular, of *Brucella* strains, 97, 98
Antisera
  against *Brucella* strains, 81–84, 109
  fluorescent labelled, for identification of clostridia, 227–230
Arginine hydrolysis, 174

Bacillus, 292
Bacillus subtilis, 290
Bacteriophage
  in *Xanthomonas*, 25
  susceptibility of *Rhizobium* strains to, 60, 62
Bacteroidaceae, 189–198
  Gram reaction of, 193
Bacteroides
  differentiation of from *Fusobacterium*, 192, 193
  succinic acid-producing species of, 198
  terminal pH of in glucose broth, 194, 195
Bacteroides
  amylophilus, 189
  biacutus, 194
  clostridiiformis, 194
  corrodens, 196, 197
  distasonis, 195
  fragilis, 195–197, 290
  melaninogenicus, 196–198
  nodosus, 196, 197
  ochraceus, 193
  oralis, 198
  ovatus, 195
  ruminicola, 189, 190, 198
  succinogenes, 189
  thetaiotamicron, 195
  vulgatus, 195
Bacteroids, of rhizobia, 52
Beer
  Lactobacillus in, 249
  spoilage of by *Pediococcus damnosus*, 243
Beneckea, 152–154
Beneckea
  alginolytica, 154
  anguillara, 154
  campbellii, 156–161
  harveyi, 154
  natriegens, 154
  neptuna, 154
  nereida, 156–161
  nigrapulchrituda, 156–161
  parahaemolytica, 154
  pelagia, 153, 156–161
  splendida, 156
Betabacterium II, 247
Bifidobacteria, differentiation from anaerobic lactobacilli, 248
Bifidobacterium bifidum, 290
Bile, growth of Bacteroidaceae in the presence of, 192, 195–198
Black rot, of crucifers, 20
Blight, of vine, bacterial causes of, 16–17
Bovine udder
  streptococci associated with, 240
Brucella, 71–117
  dissociation in, 84–86
  media for, 77–80
  oblique-light observation of, 84
  pathogenicity of for man, 86, 87
  phage typing of, 71, 72, 102–113
  precautions to be taken with, 86–90
  rodent isolates of, 76
  rough and smooth strains of, 81–85, 90, 103
Brucella
  abortus, 71–74, 76, 82, 83, 85, 86, 89, 91, 100–104, 106–108, 112–116, 291

Brucella—cont.
canis, 71, 72, 76, 101, 103, 104, 106, 108, 112, 113
melitensis, 71, 72, 74, 82, 87, 91, 100–104, 107, 108, 112, 113, 117, 291
murium, 76, 101
neotomae, 71, 75, 91, 100, 101, 103, 104, 106, 108, 112, 113
ovis, 71, 72, 75, 78, 82, 85, 91, 92, 101, 103, 104, 106, 108, 112, 113
suis, 71, 72, 74, 76, 87, 91, 100, 101, 103, 104, 106, 108, 112, 113
Buoyant density determination, of bacterial DNA, 285

Carbohydrate fermentation patterns, in lactic acid bacteria, 234
Casein hydrolysis, in *Chromobacterium*, 172
Catalase tests
for acetic acid bacteria, 38
for *Brucella*, 88, 91
for lactic acid bacteria, 234
Cattle
blackleg in, 230
Clostridium specimens from, 228–231
Cholera, vibrios causing, 143–148
Chromobacterium, 167–174
marine forms of, 167, 171, 173
Chromobacterium
fluviatile, 167, 171, 173
lividum, 167, 168, 171, 173
marinum, 167
violaceum, 167, 168, 171, 173
Cider, isolation of acetic acid bacteria from, 34
Citrobacter, diagnostic biochemical reactions of, 134
Citrus canker, study of bark from trees affected by, 18
Clostridium, 211–231
fluorescent staining of, 213, 227–230
'O' agglutination of, 227, 229
Clostridium
bifermentans, 212, 214, 215, 218
botulinum, 212, 230, 291
types of, 215, 219–221, 230
butyricum, 290
chauvoei, 212, 213, 215, 223, 227–230
histolyticum, 212, 226, 227
novyi, 212, 215
types of, 215, 222, 228
oedematiens, 212
medium, 213, 215, 228–231
parabotulinum, 215
perfringens, 214, 291
septicum, 212, 215, 223–225, 227–230
sordellii, 212, 214, 215, 218
sporogenes, 212, 214, 217
tertium, 212, 226, 227
tetani, 214, 217, 227, 229
symbiosum, 194
welchii, 212, 214, 216
Coagulase test, for identification of *Staphylococcus aureus*, 206
Cold acetone treatment, of *Brucella* cells, 98, 99
Colony form
of *Rhizobium*, 50
of *Xanthomonas*, 23
Corynebacterium, 262, 263
Coryneforms, tests for, 183
Cotton, bacterial blight of, 24
Cowpea rhizobia, 49, 52, 61, 64
Crucifers, black rot of, 20
Cysteine-lactose-electrolyte deficient medium (CLED), 163
Cytochrome absorption spectra, in *Brucella*, 101
Cytochrome-containing respiratory systems, detection of, 243
Cytophaga, 15, 184
media for, 180
Cytophaga johnsonae, 181
Cytophagaceae, 177, 178, 180, 183

Desert wood rat, pathogenicity of *Brucella neotomae* for, 75
Dihydroxyacetone, production of from glycerol, 38
Dissociation tests, for *Brucella* cultures, 84–86
DNA
duplex fractionation of, 288
fragmentation of, 286
heat-denaturation of, 284
hybridization of, 26, 27, 93, 95, 286
melting temperature determination of, 281
purification of, 278, 279

DNA—cont.
reassociation reaction of, 287
renaturation of, 289, 291
storage of, 280
with radioactive label, 280
without radioactive label, 278, 279
DNA base composition
estimation of, 92–95
ranges of for common bacterial genera, 283
DNA-DNA base pairing, 286, 290, 291
Dogs, pathogenicity of *Brucella canis* in, 76
Dye sensitivity test media
for Bacteroidaceae, 192
for *Brucella*, 79, 111
for rhizobia, 54, 55

Ears, infection of by vibrios, 144
Edwardsiella tarda, 290
Electrophoresis
for characterization of *Brucella* isolates, 95–98
of soluble proteins of *Xanthomonas*, 27, 28
Eltor vibrios, 143, 145, 146, 148, 154
Empedobacter, 185
Enterobacter, diagnostic biochemical reactions of, 134
Enterobacteriaceae, 123–138
biochemical reactions of, 125, 126, 128, 129, 131, 134–138
definition of family, 125, 126
differentiation of from Vibrionaceae and Pseudomonadaceae, 151–153
fermentation reactions of, 127
Enterococci, 235, 357
Erwinia, 292
motile strains of, 182
Erwinia
amylovora, 290
herbicola, 15–18
vitivora, 16
Erythromycin, resistance of staphylococci to, 202, 203
Escherichia, 292
biochemical reactions of, 128
Escherichia coli
differentiation from *Shigella*, 132
DNA of, 284, 285, 290, 291
multipoint inoculator for, 297
Ethanol
growth of bacteria on, 39, 41
production of acetic acid from, 34, 36, 44
Extracellular hydrolases, testing of *Pseudomonas* for, 4, 5

Faecal flora, human, identification of, 198
Fermentation tests
in Bacteroidaceae, 191, 194–196
for lactic acid bacteria, 234, 238, 239, 246, 247
Fermented grain, *Lactobacillus* in, 249
Fermented vegetable products, *Lactobacillus* in, 249
Flagellation
of *Acetobacter* and *Gluconobacter*, 40
of *Chromobacterium*, 169–171
of *Erwinia herbicola*, 16
of flavobacteria, 179, 182
of *Rhizobium*, 62, 64
of *Salmonella*, 126
Flavobacteria, 177–185
Flavobacterium, 15, 177–181, 292
gliding motility in, 177
Flavobacterium
aquatile, 177, 178, 180, 184
breve, 178
odoratum, 178
pectinovorum, 181, 185
Flexibacter, 184
Fluorescent antibody test, for *Brucella*, 85
Fluorescent staining, of clostridia, 213, 227–230
Food-poisoning, *Vibrio parahaemolyticus* as cause of, 143, 144, 148, 149
Francisella tularensis, 291
Freeze drying, of *Brucella*, 81, 83
Fusobacterium, 189, 192–195
Fusobacterium
mortiferum, 196, 197
necrophorum, 195–197
nucleatum, 196, 197
symbiosum, 194
varium, 196, 197

β-Galactosidase activity, of *Pseudomonas*, 5
Gas-liquid chromatography, for identification of *Brucella*, 101
G + C content, of bacterial genera, 284, 285
Gel diffusion reaction, of *Rhizobium* strains, 60
Genome size, estimation of, 289–292
Gliding motility, of *Flavobacterium*, 177, 180, 181, 183
Gluconobacter, 33, 36, 40, 43–45
Gluconobacter
 melanogenus, 41, 43
 oxydans, 43
Glucose, production of acid from, 39
Glycerol-dextrose agar medium, for *Brucella*, 78, 109, 113
Glycerol, dihydroxyacetone production from, 38, 39
Goats, pathogenicity of *Brucella melitensis* in, 74
Gordona, 267
Gram negative, yellow pigmented rods, 177–185
Gram staining, of *Flavobacterium*, 179
Guineapigs, inoculation of with *Brucella* organisms, 114

Haemolysis test, for micrococci and staphylococci, 209
Haemophilus, 292
Haeomophilus
 influenzae, 291
 parainfluenzae, 291
Hafnia, diagnostic biochemical reactions of, 134
Hares, infection of by *Brucella suis*, 75
Hugh & Leifson medium, 172
Hugh & Leifson test, 151, 153, 179
Hungate technique, 189–192
Hydrogen cyanide production, in *Chromobacterium*, 172
Hydrogen sulphide production
 by *Brucella* strains, 111
 by *Pseudomonas*, 5
Hydroxyapatite chromatography, 279, 286

Immunodiffusion, of *Brucella* antigens, 97–101
Inoculation, multiple, 297–302

Janthinobacterium, 167

Kanagawa test, for *Vibrio parahaemolyticus*, 148, 149
Klebsiella, 292
 diagnostic biochemical reactions of, 134, 135
Klebsiella pneumoniae, 290

Lactic acid bacteria, 233–235
 characteristics used to recognize genera of, 233–234
Lactobacilli, 233, 245–255
 anaerobic species of, 248, 255
 differentiation of Betabacteria group, 253, 254
 differentiation of Streptobacteria group, 250
 differentiation of Thermobacteria group, 250
Lactobacilli
 growth temperatures for, 246–248
 habitats of, 248, 249, 255
 heterofermentative, 245–247
 homofermentative, 246–247
 physiological tests for, 248
Lactobacillus, Betabacteria group, 246, 247
 brevis, 247, 249, 251, 253
 buchneri, 247, 249, 251, 253
 cellobiosus, 247, 249, 251, 253
 confusus, 246, 247, 249, 251, 253
 desidiosus, 247, 254
 fermentum, 247, 249, 251, 253
 fructivorans, 247, 249, 254, 255
 heterohiochi, 247, 249, 251, 254
 hilgardii, 247, 249, 251, 254
 trichodes, 247, 249, 254, 255
 viridescens, 246, 247, 249, 251, 253
Lactobacillus, Streptobacteria group, 246, 247, 251
 casei, 247, 249, 251, 252
 coryneformis, 247, 249, 251, 252
 curvatus, 247, 249, 251
 homohiochi, 247, 249, 252
 mali, 251
 plantarum, 247, 249, 251, 252
 xylosus, 247, 249, 252

Lactobacillus—cont.
yamanashiensis, 247, 249, 251, 252
Lactobacillus, Thermobacteria group, 246, 247
acidophilus, 247, 249, 250
bulgaricus, 247–250
delbrueckii, 247, 249, 250
helveticus, 247, 249, 250
jensenii, 247, 248
lactis, 247–250
leichmannii, 247–250
ruminis, 247
salivarius, 247, 249, 250
vitulinus, 247
Leaf scald, of sugar cane, 18, 19
Legumes, as hosts of rhizobia, 49, 52, 55, 56, 59, 62, 64
Leptotrichia, 189, 193, 194
Leptotrichia buccalis, 193
Leucaena, rhizobia from, 53, 55–57, 62
Leuconostoc, 233, 243–246
differentiating characteristics of, 244, 245
fermentation of trehalose by, 246
Leuconostoc
cremoris, 244, 245
dextranicum, 244–246
lactis, 244
mesenteroides, 244–246, 290
oenos, 244, 245
paramesenteroides, 244, 245
Levan production, by *Pseudomonas*, 5
Levinea
amalonatica, 134
malonatica, 134
Lipase activity, of *Pseudomonas*, 6
Liquid nitrogen, storage of *Brucella* in, 81
Listeria monocytogenes, 290
Litmus milk, as a medium for rhizobia, 55
Lotus, rhizobia from, 53, 55–57, 64
Lucibacterium, 153
Lucibacterium harveyi, 153, 154, 156–161
Luminescence, of Vibrionaceae, 162, 163
Lupinus spp., nodulation of by rhizobia, 56, 62
Lysis, methods of, 278
Lysostaphin, sensitivity of staphylococci to, 202, 203
Lysozyme, resistance of staphylococci to, 202, 203

Malt vinegar, isolation of acetic acid bacteria from, 34
Malt wort, isolation of acetic acid bacteria from, 34
Man, pathogenicity of *Brucella* for, 74–76, 86, 87
Mannitol, use of in media for rhizobia, 51, 54
Marine organisms, media for, 155, 162–165
Meat, *Lactobacillus* in, 249
Media
for acetic acid bacteria, 34–40
for Bacteroidaceae, 190, 191
for *Brucella*, 77–80, 84
for *Chromobacterium*, 171, 172, 174
for *Clostridium*, 212, 213, 230
for *Cytophaga*, 180
for Enterobacteriaceae, 123
for lactobacilli, 248
for *Leuconostoc*, 245
for marine organisms, 155, 162–165
for *Micrococcus* and *Staphylococcus*, 201–203, 205
for *Nocardia*, *Actinomadura* and *Rhodococcus* spp., 272
for pediococci, 243
for rhizobia, 50–52, 54, 55, 58
for streptococci, 237
for vibrios, 144, 145
Medicago sativa, nodulation of by rhizobia, 55, 59
Menaquinone analyses, of actinomycetes, 263
Micrococcus, 201–203, 206, 292
separation from *Staphylococcus*, 202, 203
uncommon, new species of, 206
Micrococcus
luteus, 206, 290
roseus, 206
varians, 206
Milk, *Lactobacillus* in, 249
Moraxella, 292
Moraxella non-liquefaciens, 97
Motility

Motility—cont.
gliding, 177, 180, 182, 183
of flavobacteria, 179
Mouth, *Lactobacillus* in, 249
Multipoint inoculator, 297–302
apparatus, 298–300
applications, 301, 302
storage of inocula in, 300, 301
Mycobacterium, 262–264, 292
Mycobacterium
bovis 291
kansasii, 291
rhodochrous, 267
smegmatis, 291
tuberculosis, 290, 291
Mycolic acid composition, for mycobacteria, 262, 263, 268
Mycoplasma, 292
Myxococcus fulvus, 290

Needle-puncture method, of isolation, 18
Neisseria, 292
Neisseria
gonorrhoeae, 291
meningitidis, 291
Neonatal infections, human, role of streptococci in, 240
Nitrate reduction, in micrococci and staphylococci, 208
Nitrogen fixation, by rhizobia, 58, 59
Nocardia, 261–264, 266, 268, 269
tests for, 272
Nocardia
aerocolonigenes, 266, 268, 269
amarae, 268, 269
asteroides, 266, 268, 269
autotrophica, 266, 268, 269
brasiliensis, 266, 268, 269
carnea, 268, 269
dassonvillei, 267
madurae, 267
orientalis, 266, 268, 269
otidis-caviarum, 266, 268, 269
pelletieri, 267
transvalensis, 268, 269
vaccinii, 268, 269
Nocardiopsis, 267
Nodules, root, formation of by rhizobia, 49, 52, 54, 59
Novobiocin susceptibility, in micrococci and staphylococci, 208
Nucleic acid hybridization, for the study of *Xanthomonas*, 26

Oeskovia, 261
Ooze, bacterial, from plant material, 17, 19
Oxidase production, in micrococci and staphylococci, 208
Oxidase reaction, in *Pseudomonas*, 4
Oxidation-fermentation test, for *Pseudomonas*, 4
Oxidative metabolism tests, for *Brucella* cultures, 107
Oxygen, sensitivity of Bacteroidaceae to, 189, 190

Pasteurella multocida, 291
Pediococcus, 233
changes in nomenclature of, 241
characteristics of members, 241–243
Gunther and White group III of, 241
Pediococcus
acidilactici, 241–243
cerevisae, 241
damnosus, 241–243
halophilus, 241–243
homari, 243
parvulus, 241
pentosaceus, 241–243
urinae-equi, 241, 243
Peroxide sensitivity, in *Chromobacterium*, 171
pH, terminal, of Bacteroidaceae, 194–197
Phage typing
of *Brucella*, 71, 72, 102–107
of vibrios, 145, 146
PHB granules, in *Rhizobium* cells, 50
Phosphatase test
for flavobacteria, 181
for micrococci and staphylococci, 207–208
Photobacterium, 152, 153, 155
Photobacterium
angustum, 156–161
fischeri, 154
leiognathi, 154, 156–161
mandapamensis, 154

Photobacterium—cont.
  phosphoreum, 156–161
Pickles, *Lactobacillus* in, 249
Pigment production
  in *Acetobacter*, 40, 41
  in *Chromobacterium*, 167, 168
  in *Flavobacterium*, 178, 179
  in micrococci and staphylococci, 208
  in *Pseudomonas*, 3
  in Vibrionaceae, 163
  in *Xanthomonas*, 22
Pigs
  Clostridium isolated from, 228–231
  pathogenicity of *Brucella suis* in, 74
  streptococci isolated from, 240
Plesiomonas
  biochemical and physiological characters of, 147
  identification of, 153
Plesiomonas shigelloides, 149, 154, 156–161
Polysaccharide slime production, by *Xanthomonas*, 23
Propionibacterium freudenreichii, 290
Proteus, 292
  diagnostic biochemical reactions of, 137
  G + C content of DNA of, 137
Providencia
  diagnostic biochemical reactions of, 137
  G + C content of DNA of, 137
Pseudomonadaceae, identification of, 151–153
Pseudomonas, 1–12, 15, 20, 22, 26, 44
Pseudomonas
  acidovorans, 10
  aeruginosa, 9, 168, 290, 291
  alcaligenes, 10
  cepacia, 9, 291
  cichorii, 9
  doudoroffi, 10
  fluorescens, 9, 291
  fragi, 9
  indigofera, 10
  mallei, 291
  maltophilia, 10, 27
  marina, 9
  mendocina, 10
  natriegens, 154
  nautica, 10
  paucimobilis, 10
  piscicida, 12
  pseudoalcaligenes, 10
  pseudomallei, 291
  putida, 9
  shigelloides, 154
  stutzeri, 10, 291
  syringae, 9
  testosteroni, 10
γ-Pyrones, production of by an *Acetobacter* sp., 40

Rabbits, preparation of *Brucella* antisera from, 81–85, 100
Reindeer, infection of by *Brucella suis*, 75
Rhizobium, 49–65
  fast-growing and slow-growing strains of, 51, 52
  nodulation specificity of, 56, 57
  serological differences between strains, 60
  japonicum, 52–54, 56, 57, 61
  leguminosarum, 52, 53, 56, 57, 61–63
  lupini, 53, 56, 61
  meliloti, 52–55, 57, 60–63
  phaseoli, 52, 53, 56, 57, 61–64
  trifolii, 52–54, 56, 57, 60–63
Rhizospheres, multiple inoculation technique for, 301
Rhodococcus, 261–265, 267, 273
  characteristics of species of, 273
Rice
  bacterial leaf blight of, 24
  bacterial leaf streak of, 24
Rothia, 261
Rotten lamb disease, 230
Rumen, strict anaerobes from, 189, 191

Saccharopolyspora, 265
Safety, when handling *Brucella*, 87–90
Saké, spoilage of, 249, 251
Salmonella, 292
  characteristics of, 126
  diagnostic biochemical reactions of subgenera of, 132
  isolation of, 123
  some aberrant biotypes of, 133
  choleraesuis, 126, 133

Salmonella—cont.
gallinarum, 126, 133
paratyphi, 126, 133
pullorum, 126, 133
typhi, 126, 133, 291
typhimurium, 291
Salt susceptibility, in micrococci and staphylococci, 208
Schleifer and Kloos test, 202–204
Septicaemia, isolation of *Beneckea vulnifica* from cases of, 154
Serotypes
of vibrios, 146, 148
of *Xanthomonas*, 27, 28
Serratia, diagnostic biochemical reactions of, 134, 136
Serum-dextrose agar medium, for isolation of *Brucella*, 77, 82, 91, 109, 110
Sheep
Clostridium isolated from, 228–230
isolation of *Brucella ovis* from, 78
pathogenicity of *Brucella melitensis* in, 74
pathogenicity of *Brucella ovis* in, 75, 76
Shellfish, *Vibrio parahaemolyticus* isolated from, 144, 149
Shigella
classification and nomenclature of, 130
diagnostic biochemical reactions of subgroups of, 130
differentiation from *Escherichia coli*, 132
isolation of, 123
similarity of *Plesiomonas shigelloides* to, 149
Shigella
dysenteriae, 291
flexneri, 6, 291
biotypes of, 131
Soil, survival of xanthomonads in, 20
Soil micro-organisms, multiple inoculation for, 301
Sphaerophorus, 195
Sphaerophorus necrophorus, 195
Spores, absence of, 194
Spreading growth, in flavobacteria, 180, 181
Staphylococcus, 201–209, 292
Kloos and Schleifer's 'species' of, 204
separation from *Micrococcus*, 202, 203
subgroups of, 203–206
Staphylococcus
aureus, 203–205
capitis, 204, 205
cohnii, 204, 205
epidermis, 203–205
haemolyticus, 204–205
hominis, 204–205
hyicus, 205
intermedius, 204, 205
saprophyticus, 202–205
simulans, 204
Starch hydrolysis
in *Acetobacter*, 41
in *Xanthomonas*, 25
Streptobacterium IB, 247
Streptococci, 233
faecal, 237, 238
habitats of, 235, 236
lactic, 237, 238
oral, 239, 240
physiological and biochemical tests for, 237–241
physiological groups of, 235
pyogenic, 239, 240
serological groups of, 240
serological tests for, 236
Streptococcus, 292
Streptococcus
acidominimus, 236, 239, 240
agalactiae, 236, 239, 240
anginosus, 236, 239
avium, 236–238
bovis, 235–239
cremoris, 237, 238
dysgalactiae, 236, 239, 240
equi, 236, 239
equinus, 236–238
equisimilis, 239
faecalis, 236–238
faecium, 236, 237
subsp. *casseliflavus*, 235–238
lactis, 236–239
subsp. *cremoris*, 236
subsp. *diacetylactis*, 236–238
subsp. *raffinolactis*, 236
lentus, 240

Streptococcus—cont.
milleri, 236, 239
mitior, 236, 239
mutans, 236, 239, 240
pneumoniae, 233
pyogenes, 236, 239
raffinolactis, 237, 238
salivarius, 235, 236, 239
sanguis, 236, 239
suis, 240
thermophilus, 236, 237
uberis, 236, 239, 240
Streptomyces, 265
Streptomyces somaliensis, 270, 271
Sugar cane, leaf scald of, 18, 19

Temperature, optimum renaturation, 287
Thermoagglutination test, for *Brucella*, 86
Thermobacterium IA, 247
Thermo-stable nuclease production, in *Staphylococcus aureus*, 207
Thiosulphate-citrate-bile salt (TCBS) agar, for the isolation of vibrios, 145, 146
Threonine deaminase test, in Bacteroidaceae, 192
Trema aspera, nodulation of by *Rhizobium*, 49
Trimethylamine oxide reduction, by *Pseudomonas*, 6
Turbidity from egg-yolk, in *Chromobacterium*, 174

Ultrasonic disruption, of *Brucella* cells, 99
Ultraviolet sensitivity, in flavobacteria, 181, 182
Urease test, for *Brucella*, 113

Vaccines, production of from *Brucella* strains, 113, 114
Vacuum drying, of *Brucella*, 80, 81
Vibrio, 4, 143–149, 152, 153, 156–161, 182
Vibrio
albensis, 156–161
alginolyticus, 144, 147, 153, 154, 156–161, 164
anguillarum, 154, 156–161
cholerae, 143, 145–148, 154, 156–161, 290, 291
biotype *proteus*, 153, 154
costicola, 156–161
fischeri, 154, 156–161
metschnikovii, 153, 154, 156–161
natriegens, 154, 156–161
parahaemolyticus, 143, 147, 148, 153, 154, 156–161, 291
proteus, 154
Vibrionaceae, identification of, 151–165
Vibrios
human, 143–149
biotyping of, 148
cholera, 143–148
colony characteristics of, 146
eltor and classical biotypes of 143, 145, 146, 148
non-cholera, 143, 144, 146, 148
rough colonies of, 148
Vibriostatic agent, sensitivity to, 151, 162
Vine, blight of, 16
Violacein pigment, 168, 169
Voges-Proskauer test, 164

Whole-organism hydrolysate analysis, for actinomycetes, 262
Whole-organism methanolysate analysis, 262
Wine
Lactobacillus in, 249, 251, 255
malolactate fermentation of, 245

Xanthomonas, 182
Xanthomonas
albilineans, 17–19, 21, 22, 24, 26
ampelina, 20–22
axonopodis, 20–22, 24, 26
azadirachtae, 22
campestris, 17, 20–25, 27
citri, 18
fragariae, 20–23, 26
holcicola, 25
juglandis, 22
malvacearum, 20, 24
manihotis, 21, 22
oryzae, 24, 25
pedalii, 22

Xanthomonas—cont.
phaseoli, 18
pruni, 24
ricini, 22
translucens, 24
uppalii, 22
vasculorum, 25
vesicatoria, 23, 25, 27
vitians, 25

Yeast extract-mannitol agar, for cultivation of rhizobia, 54, 55
Yersinia, diagnostic biochemical reactions of, 133
Yersinia
enterocolitica, 97, 100
pestis, 291

Zymomonas, 292
mobilis, 290